Modeling and Simulation in **SIMULINK**
for Engineers and Scientists

Mohammad Nuruzzaman

Electrical Engineering Department
King Fahd University of Petroleum & Minerals
Dhahran, Saudi Arabia

author HOUSE

1663 LIBERTY DRIVE, SUITE 200
BLOOMINGTON, INDIANA 47403
(800) 839-8640
www.authorhouse.com

First published by AuthorHouse 10/29/04

ISBN: 1-4184-9383-X (sc)

Printed in the United States of America
Bloomington, Indiana

This book is printed on acid-free paper.

To my parents

Preface

The purpose of the text is to present the systematic procedure in a tutorial fashion for many of the basic classroom modeling problems while working in SIMULINK. SIMULINK whose elaboration is SIMUlation and LINK is built in the platform of MATLAB and can be regarded as an extension to MATLAB. So to activate SIMULINK, one needs to activate MATLAB first. Most programming aspects in SIMULINK happen through dialog window or graphical user interface (GUI) instead of typing in MATLAB Command Window or entering from an M file (Source codes of MATLAB are written in an M file). Thus SIMULINK renders the programming approach by means of clicking the mouse and dialoging through the GUI. As a computational package, MATLAB is becoming more and more popular since it does not require clumsy compiling like the base language (such as C or FORTRAN), provides easy debugging convenience, offers easy to use workspace, bypasses dozens of conventional programming statements employing ample built-in functions, offers fully featured graphics environment in terms of one, two, or three dimensional plots, and includes numerous toolboxes to enhance its applicability for tackling versatile problems. Working in SIMULINK is just a digression from MATLAB.

The superb advances in computer technology, especially the easy accessibility to the personal computer based on the window system, quite factually brought the software tools for solving and visualizing the mathematical problems into user's reach and thereby increasing the interest in modeling and simulation. The process of modeling and simulation commences with a model, and that model might be a representative of the specific engineering discipline such as civil, electrical, mechanical, aeronautical ... etc, computer and telecommunication networks, integrated circuits, flight dynamics, biophysics, economics, or the system which is yet to be implemented.

Every second our so accustomed world is transforming to a newer one. No two moments and the way our world lives are ever identical. Emerging cutting-edge technologies, fast speed computers, and internet based global unification are bringing the whole universe more and more together. Now we feel, share, and express many things not the way we did several decades ago even though we live in various diversities. It is the ever changing demand and pushing the limit of human endurance that made us achieve so much. Yet, problems multiply and there are no dead end to the problems. To meet the challenges of continuous change, engineers, scientists, and researchers are working continuously to develop neoteric tools for tackling the scientific problems. Today's MS or PhD thesis is becoming the homework for the tomorrow's student. The dimension of problems is changing from one to two, three, and more. Since the base language programming, compilation, and execution consume notable time, it is very difficult to handle the complicated and ever changing problems starting from the base language programming every time. Human beings learn from mistakes, past experience, and frequently faced situations. The most dealt elements in any basic language are single scalar, in MATLAB they are vectors, whereas SIMULINK considers all problems faced in scientific world as blocks. That is how it is different from the other packages. One can imagine that SIMULINK is a library of blocks and these blocks provide the programming convenience for real life problems being represented by models.

Honestly speaking, hand on experience in SIMULINK in a simplistic way from the classroom examples is the approach of the book. The book is intended for the undergraduate as well as the graduate students of science and engineering. Toady's research and development share many concepts and implementation not from one specific discipline instead from multiple disciplines. For this reason, the text is written not showing explicit leaning to any discipline. The corporeal task of an engineer or a scientist is to design some physical system, and the design is no piece of cake rather a complex process which in essence requires some sort of creativity. Creativity does not just happen without any means. We believe that if we study afresh the knowns and extend our competence to the horizon of unknowns through the skill of knowns, the best way to nurture our creativity. SIMULINK is an appropriate tool for refreshing our memory and lexicon of knowns.

Numerous built-in and toolbox functions are accessible in the command window of MATLAB. SIMULINK replaced the writing of the MATLAB functional commands in many ways by equipping the library blocks. In particular sense SIMULINK can implement many modeling and simulation just by involving more mouse related activities like click-drag-play action rather than writing the MATLAB codes. To solve or simulate some problem in SIMULINK, necessarily one needs to contemplate what the inputs and what the outputs are. Thus SIMULINK breaks the whole programming of a particular problem into more modular form. We presuppose that the reader has some background in the disciplines related to the chapter headings and familiarity to some extent with the notations of those disciplines. Notable number of simulations ranging from the elementary to the advanced ones has been covered so that the average reader becomes benefited. Choosing the chapter heading and the article name, one can easily get through the specific simulation he is interested in.

I wish to express my acknowledgement to the King Fahd University of Petroleum and Minerals (KFUPM). I sincerely appreciate the library facilities that I received from the King Fahd University. All illustrative modeling problems covered in the text have been simulated by a Pentium IV personal computer operated on Microsoft Windows.

Implementationally subjectiveness on SIMULINK has been emphasized while modeling some problems. We hope made-easy approach of the material presentation in the text will make the reader a good user of SIMULINK library blocks and able to explore the extensive simulation facilities to tackle different types of simulation problems. I sincerely wish you a good start in SIMULINK.

Mohammad Nuruzzaman

Table of Contents

Chapter 1
Introduction to SIMULINK

Chapter 2
Modeling Mathematical Functions and Waves

Chapter 3
Modeling Ordinary Differential Equations

Chapter 4
Modeling Difference Equations

Chapter 5
Modeling Common Problems of Control Systems

Chapter 6
Modeling Some Signal Processing Problems

Chapter 7
Modeling Common Matrix Algebra Problems

Chapter 8
Modeling Common Statistics and Conversion Problems

Chapter 9
Fourier Analysis Problems

Chapter 10
Miscellaneous Modeling and Some Programming Issues

Appendix

Introduction to SIMULINK

We warmly welcome you to SIMULINK. Engineers and scientists can effectively bring the real world's dynamic system into being the computer model through SIMULINK. SIMULINK is designed to function in the platform of MATLAB, which is the easiest and quickest way to compute and visualize the scientific and technical problems. Much of the MATLAB computational style presupposes that the element to be handled is a vector or matrix. Whereas, SIMULINK quickly and accurately maps a real world problem into computer model through elementary blocks. Even though MATLAB can perform the symbolic computation, SIMULINK entitles users only to handle a system numerically. In this chapter, our principal objective is to introduce the SIMULINK basics or the general questionnaires that a beginner needs to know before he or she gets started with SIMULINK operated in Microsoft Windows System.

1.1 What is simulink?

Simulink is an additional part of MATLAB which provides an easeful way to model, simulate, and analyze of a dynamical system. Usually we define a dynamical system in terms of the mathematical model and the model is mostly characterized by some inputs and outputs. A particular input-output relationship can be assigned to some block. One can interpret that simulink is a vast collection of blocks of this kind. Although the blocks might stand for simple mathematical relationship, being concatenated they can build a much complicated system. The elaboration of simulink is <u>simu</u>lation and <u>link</u>. Initially simulink was intended particularly to handle the linear time invariant continuous systems. Progressively, the discrete time systems as well as the hybrid ones surfaced in simulink to become more pragmatic in modeling the real world's dynamic systems.

1.2 How can I get into SIMULINK?

Since simulink is an extension of MATLAB, we assume that at least you can get into MATLAB. Like DOS (prompted by C:\) and UNIX (prompted by $), MATLAB also has its own command prompt which is >>. Here in

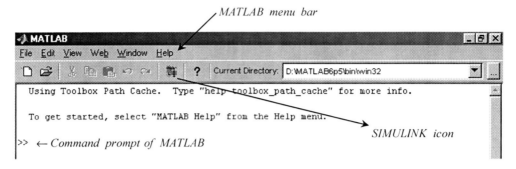

Figure 1.1(a) *Upper portion of MATLAB Command Window*

Figure 1.1(b) *Simulink library browser window*

this prompt you can type anything which is understandable to MATLAB and execute that just by pressing the Enter Key from the keyboard. Both the MATLAB Command Prompt and its menu bar provide means of getting into simulink. Figure 1.1(a) shows the upper portion of a typical MATLAB Command Window. Either you type simulink in the command prompt and press enter or click the simulink icon as shown in the MATLAB menu bar in figure 1.1(a). Simulink is an aggregation of functional blocks arranged in a tree structure as presented in figure 1.1(b).

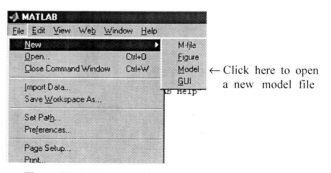

Figure 1.1(c) *Location of a new simulink model file*

2

1.3 Where can I build a simulink model?

Like other software package, there should be a file where we can build our model. The model we intend to build is problem dependent. First of all, we need to decide what we want to model. Referring to the MATLAB Command Window of figure 1.1(a), if you click File under MATLAB menu bar and then New, you see the pull down menu as shown in figure 1.1(c). Click at Model in the Pull down menu and you will see the untitled model file as shown in figure 1.1(d). The simulink library browser icon is also found in the menu bar in the untitled model file as you have seen in figure 1.1(a). The reader should have the knowledge about the blocks' function and their input-output descriptions before you bring them from the simulink library to your model file. A model can be defined as the interconnected blocks found in simulink library to describe a particular problem. Let us bring a block from the simulink library in your untitled model file. To perform such action, click the simulink icon located in the menu bar of the untitled model file (the simulink library of figure 1.1(b) appears and the cursor is residing in simulink) and click continuous down the simulink to see various blocks available under the subclass continuous (the library browser response is shown in figure 1.1(e)). Referring to the figure, we see a block called Derivative displayed by a shaded area. Let us bring the Derivative block in our untitled model file. Bring your mouse pointer on the Derivative block, move the mouse pointer keeping your finger pressed in the left button of the mouse to any convenient area of the untitled model and release the left button of the mouse. Now you see the Derivative block in your untitled model file as shown in figure 1.1(f). Another way of bringing the block is rightclick the mouse, see the *Add to the current model* in the dialog window, click that, and find the block in your model file. So we say that the link for the Derivative block is *'Simulink → Continuous → Derivative'*. We maintain this style of locating a block throughout the text. Microsoft Windows Start

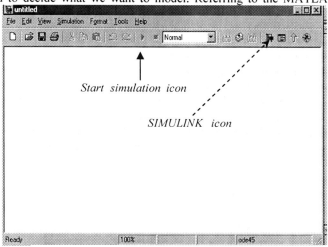

Start simulation icon

SIMULINK icon

Figure 1.1(d) *An untitled simulink model file*

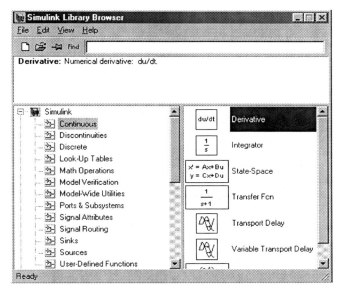

Figure 1.1(e) *Blocks under the subclass continuous of simulink library*

bar keeps the open minimized windows side by side. You may work in your model file keeping other windows open. When you bring some block from the simulink library in your model file, make sure you first click the model file and then click the library browser minimized window. Suppose you worked in MATLAB Command Window just now and switched to simulink library to bring a block in your model. You will not be successful in bringing the block. However, you have been successful in bringing the Derivative block in your model file. Now you can save your model in any convenient name in your working directory. By the way, MATLAB script file has the extension .m but the simulink model file does .mdl. *Keep in mind that one should know the link of a block to bring it from the simulink library to the model file.*

1.4 Block manipulation

A model consisting of interconnected components requires a number of blocks to be inserted. Each block has its own input-output characteristics. Not only the simulink library contains hundreds of blocks, but also the toolbox blocks enhance the application to a further extent. During a simulink model building process, you need some manipulations of the blocks to construct a seemly, well-placed, and well-devised model. We mention the most frequently encountered ones in the following.

☞ *How can I select a block?*

Let us say you brought the Derivative block in an untitled model file following the link *'Simulink → Continuous → Derivative'*. Bring the mouse pointer on the Derivative block and click the left button of the mouse to see the selection as shown in figure 1.2(a).

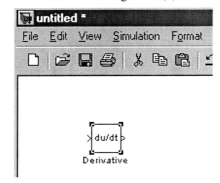

Figure 1.1(f) *Derivative block in the untitled model file*

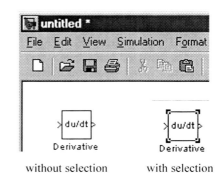

without selection   with selection

Figure 1.2(a) *Derivative block with and without selection*

Figure 1.2(b) *Derivative block mentioning the input and output ports*

Real-Imag to   Complex to
Complex   Magnitude-Angle

Figure 1.2(c) *Blocks with multiple input and output ports*

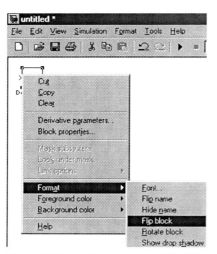

Figure 1.2(d) *Pull down menu of the block Derivative following the right click on the block*

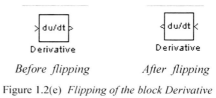

Before flipping   After flipping

Figure 1.2(e) *Flipping of the block Derivative*

Before rotating   After rotating

Figure 1.2(f) *Rotating the block Derivative by 90⁰ clockwise*

☞ *How can I detect the input and output ports of a block?*

Referring to the Derivative block of figure 1.1(f), we see that the left and right sides of the block contain the

Figure 1.2(g) *270⁰ clockwise rotation of the Derivative block*

Figure 1.2(h) *The Derivative block without the block name*

symbol '>'. One can identify the input and output ports of the block as presented in figure 1.2(b). Not necessarily the number of the input and output ports will be unity, the blocks Real-Imag to Complex and Complex to Magnitude-Angle of the figure 1.2(c) have two input ports and two output ports respectively.

⬚ *How can I delete a block?*

Let us say you brought the Derivative block in an untitled model file but you want to delete that. There are several options for this. First select the block, then press the Delete button from the keyboard, click the Cut icon in the model menu bar, or click Cut followed by the rightclick of the mouse.

⬚ *How can I flip a block?*

Let us say we have the Derivative block in our model file. We want to flip the block according to the figure 1.2(e). First, bring the mouse pointer on the block, click the right button of the mouse, and then click Flip via Format. The action is shown in the figure 1.2(d).

⬚ *How can I rotate a block?*

Suppose we want to rotate previous mentioned Derivative block. Bring the mouse pointer on the block, click the right button of the mouse, and click the Rotate block via Format. You see the change as shown in figure 1.2(f). Figure 1.2(d) also indicates the operation by which a block is rotated clockwise 90^0 at a time. If you want to rotate the block 270^0 clockwise (indicated in figure 1.2(g)), you need the operation three times.

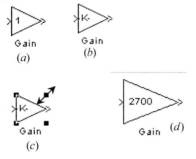

Figure 1.3 *The gain block a) with default gain, b) with gain 2700, c) with selection, and d) after enlargement*

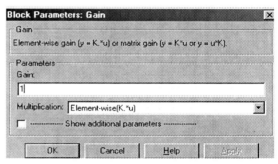

Figure 1.4(a) *Block parameter window for the Gain block*

⬚ *How can I remove the block name?*

The name of the Derivative block can be removed as presented in figure 1.2(h) by the Hide name of the figure 1.2(d). A clumsy model might give better look by removing the block name if the reader is well acquainted with the blocks.

Figure 1.4(b) *Derivative block with the name D*

⬚ *How can I enlarge or contract a block?*

Let us bring the block Gain in your simulink model file as shown in figure 1.3(a) following the link '*Simulink → Math Operations → Gain*' which has the default gain 1. If we have five or six digits gain, the default size will not allow to display that. Doubleclick the block to see the dialog window of figure 1.4(a), let us change the gain from default 1 to 2700, and click OK. The block displays the inside gain as shown in figure 1.3(b). Select the block, bring your mouse pointer on the upper right square target to see the figure 1.3(c), move the mouse pointer to the right keeping the left button of the mouse pressed, and release the left button of the mouse. You should see the Gain block of figure

Derivative block in a model

Figure 1.4(c) *Derivative block with the annotation*

1.3(d). In a similar way for the oversize block, you can reduce the block size by moving towards the inside of the block after selection. Resizing might be necessary for other kinds of blocks such as Transfer Fcn, Fcn, MATLAB Fcn, ... etc specifically where we enter some user defined code or expression inside the block.

⬚ *How can I rename a block?*

During the simulink model building process, it may be necessary to use the same kind of block twice or more. Then it requires renaming the block for identification. Let us say you brought a Derivative block in your model file as shown in figure 1.1(f). You want to write just D as the block name instead of the Derivative (like the figure 1.4(b)). Bring the mouse pointer on the word Derivative, click the left button of the mouse (word is selected), and delete the other letters except D using the delete button from the keyboard, bring the mouse pointer outside the block, and click it.

☞ *How can I give annotation to a block?*

Let us say you have the Derivative block in your model file as shown in figure 1.1(f) and want to write 'Derivative block in a model' down the block in the model file. Bring the mouse pointer at the desired position in your model file, doubleclick the mouse to see the blinking cursor, type 'Derivative block in a model' from the keyboard, bring mouse pointer out of the block, and click the left button

1.4(d) *Without drop shadow*   1.4(e) *With drop shadow*

Figures 1.4(d)-(e) *Derivative block without and with the drop shadow*

of the mouse. Figure 1.4(c) shows the action we performed. Once typed, you can even drag the whole text to move anywhere in your model.

☞ *How can I add drop shadow to my block?*

Some reader might be interested to see the drop shadow form of the simulink block rather than the plain form. Let us say we have the derivative block is a new simulink model file. Rightclick on the block to see the pull down menu, go to the Show drop shadow via Format, and click it. The necessary change is depicted in the figures 1.4(d)-(e). To remove the shadow, again rightclick on the block and click the Hide drop shadow via format.

☞ *How can I change simulink model file background color?*

Rightclick anywhere in the model file and see the Screen color in the prompt menu. From there, you can choose any background color you like for the model file.

☞ *How can I copy a block within a simulink model?*

Select the block, click Copy icon in the model menu bar, and paste it as many times as you want.

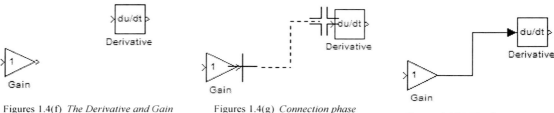

Figures 1.4(f) *The Derivative and Gain blocks are residing in a simulink model file*

Figures 1.4(g) *Connection phase of the two blocks*

Figures 1.4(h) *The Derivative and Gain blocks are connected*

☞ *How can I connect two blocks?*

This manipulation is very important in the sense that we need to connect frequently blocks in a model in the subsequent chapters while working in simulink. The reader is familiar with the Gain and Derivative blocks from previous discussion. Let us say that the two blocks are residing in a simulink model file as shown in the figure 1.4(f). We want to connect the two blocks. The connection of blocks must be correct syntactically. The output port of any block can be connected to the input port of any other block but not to the output port of others. The same syntax is also true for the input port. However, bring the mouse pointer on the output port of the Gain block, see the single cross target as shown in the figure 1.4(g), press the left button of the mouse, move the single cross target anywhere in the model you like keeping your finger pressed, bring the single cross target close to the input port of the Derivative block, see the double cross target as shown in the figure 1.4(g), and release the left button of the mouse. You should find the two blocks connected as shown in the figure 1.4(h).

☞ *What is a parameter or block parameter window?*

In the following chapters, we are going to mention many times the term 'parameter window'. After bringing any block in your model file, you can doubleclick the block. Due to the action, always you see a dialog window in which you find one or more slots for value taking depending on the purpose of the block. That dialog window is termed as the parameter window, for example, the figure 1.4(a).

1.5 How to get started in simulink?

This is the most important startup for the beginners. In previous sections the reader has seen how one can bring and connect blocks in a simulink model file. Here in this section we present simple modeling lessons so that the novice in simulink feels how simulink functions. Whatever operations such as manipulations, computations, assignments, or comparisons are carried out in conventional software can be conducted in simulink through various blocks and functional lines. Since most algorithms are hidden in the functional lines, initially one might feel it complicated. Due to the fact that most of the model building happens through the mouse operation rather than writing the source codes, one would feel little by little with the increase in use how convenient simulink is.

Let us go through the following three tutorials as a quick start in simulink.

✦✦ Tutorial one

Two numbers are to be added – 4 and 6. The output should be 10. The question is where we should keep the numbers. In simulink every programming aspect happens through blocks. There are in general two kinds of blocks – constant and dependent on simulink independent variable generation. You find a block called Constant in the link *'Simulink → Sources → Constant'*. Open a new simulink model file and bring the Constant block in your model file as we did before. The default value in the block is 1. Doubleclick the block and enter the constant value in the parameter window as 4 from the keyboard by deleting the default 1. Similarly bring another Constant block from the same link and enter 6 in the block. You see the latter block by the name Constant1. If you bring one more Constant block, simulink names that as Constant2. This style of naming is followed for all other blocks. However, we need a Sum block to add the two numbers that can be reached via *'Simulink → Math Operations → Sum'*. Bring the Sum block in your model file. To see the computation, we need a Display block whose link is *'Simulink → Sinks*

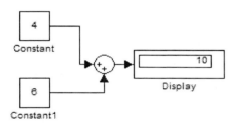

Figure 1.5(a) *Adding two constants and displaying the result*

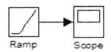

Figure 1.5(b) *Ramp block connected with the Scope block*

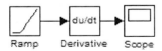

Figure 1.5(d) *Derivative block differentiates the output generated by the Ramp and Scope shows the Derivative output*

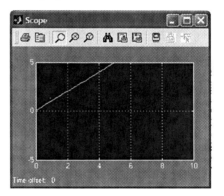

Figure 1.5(c) *Scope block shows the ramp function*

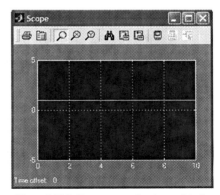

Figure 1.5(e) *Scope output for the model of the figure 1.5(d)*

→ Display' and also bring the block in the model file. Place the four blocks relatively and connect them according to the figure 1.5(a). Referring to the figure 1.1(d), you find there inactivated start simulation icon (inactivated because no blocks are present). Click the Start simulation icon and simulink responds showing the summation 10 in the display block. You can also run the model file from the menu bar by clicking Simulation and then Start.

The MATLAB Command window provides another alternative for running a simulink model from its command prompt. Let us say we have a simulink model by the name test.mdl. To run it from the command prompt we carry out the following:

MATLAB Command
>>sim('test') ⏎

✦ ✦ Tutorial two

Let us say we want to generate a function $y(t) = t$. The function is a straight line passing through the origin and has a slope of unity, which is also known as the ramp function. Bring a Ramp and a Scope blocks in a new simulink model file following the links *'Simulink → Sources → Ramp'* and *'Simulink → Sinks → Scope'* respectively. Connect the two blocks as shown in the figure 1.5(b) and run the model by clicking the start simulation icon. Doubleclick the Scope and you see the figure 1.5(c) in which our required straight line is plotted. So the Ramp block generated the function and the Scope just displayed that.

✦ ✦ Tutorial three

If you differentiate the ramp function $y(t) = t$ with respect to t, you should get 1. The reader is familiar with the Derivative block from earlier discussion. Insert the Derivative block between the Ramp and Scope blocks of the figure 1.5(b) so that we have the model in figure 1.5(d). Select the Ramp or Scope block, use the left or right arrow key from your keyboard so that the space is enough between the two blocks to accommodate the Derivative block. When you bring the Derivative block, drop the block keeping its input and output ports in line with the connection line of the Ramp and Scope. Simulink is so smart that it connects the block the way you want. On forming the model, click the start simulation icon and then doubleclick the Scope in the model to view the output like the figure 1.5(e) which is essentially a straight line parallel to the horizontal axis and located at 1 in the vertical axis of the Scope – that is what we expect from simulink.

1.6 Importance of the Display and Scope blocks in simulink

A practical model contains dozens of blocks. The whole simulation occurs in simulink with the agency of blocks and functional lines. When a model is being run, the blocks Display and Scope can display how the functions are changing or flowing. The functional flow or computation may or may not be seen during the run time because it happens so rapid - in the fraction of a second. It also depends on your model whether it is time consuming.

However, the Display block is convenient only for showing a single scalar output and a row, column, or rectangular matrix output at the end of the simulation. The block is designed to show the instantaneous value flowing through the functional line in which it is connected to. Once simulink has finished the simulation, the block shows the last value. The default size of the block is for a single scalar. For matrix, one needs to enlarge the block to view all in it.

If the turnout of a simulink model block is in the form of a long matrix in which hundreds of elements are associated with, it is not feasible to see the results through the Display block. Instead the graphical plot is the better way to observe the output. The Scope in all sense mimics the Oscilloscope that essentially displays the signal flow with time. The Scope has two axes – horizontal and vertical. The horizontal and vertical axes simulate the independent and dependent variables respectively. To all extent, the horizontal axis does not have to be time even though it is originated in that name, any physical quantity such as displacement, frequency, speed or other can be assigned to the horizontal axis. In the subsequent chapters, the Scope axis usage will be more illustrative.

Modeling Mathematical Functions and Waves

We outline simulink approach of managing the common mathematical function generation problems for the periodic and non-periodic context in this chapter. Primarily, we highlighted the built-in functional codes and blocks for the mathematical expressions in simulink. Waveform or function generation comfort is one of many appreciative features of simulink that just happens through the dialog window without rigorous programming. Our focus also comprises the complex functional and data-based modeling. The function generation approach of simulink is entirely numerical. Even though simulink is an extension of MATLAB, but the package itself can be regarded as an independent one. The package is very efficient and convenient in modeling various functions and waves in a smarter way without too much complicity often found in practice.

2.1 SIMULINK coding of functions

Simulink codes the functions in terms of a string, which is a set of characters placed consecutively. One distinguishing feature of simulink is that the variable itself is a matrix. The strings adopted for computation can be divided into two classes – scalar and vector. The scalar computation results the order of the output matrix same as that of the variable matrix. On the contrary, the order for the vector computation is determined in accordance with the matrix algebra rules. A list of symbolic functions and their simulink counterparts are presented in table 2.A. The operators for the arithmetic computations are as follows:

addition by	+
subtraction by	−
multiplication by	*
division by	/
power by	^

The operation sequence of different operators in a scalar or vector string observes the following order:

enclosing braces	( )	first,
power operator	^	then,
division operator	/	next,
multiplication operator	*	after that,
addition operator	+	then, and
subtraction operator	−	finally.

The syntax of the scalar computation urges to use .*, ./, and .^ in lieu of *, /, and ^ respectively. The operators *, /, or ^ are never preceded by . for the vector computation. This string is the simulink code of any symbolic expression or function often found in mathematics. Starting from the simplest one, we present some examples for writing the long expressions in simulink for the scalar form and for the vector as well.

Write the SIMULINK code in scalar and vector forms for the following functions:

$$A.\ \sin^3 x \cos^5 x \qquad B.\ 2+\ln x \qquad C.\ x^4+3x-5 \qquad D.\ \frac{x^3-5}{x^2-7x-7} \qquad E.\ \sqrt{|x^3|+\sec^{-1} x}$$

$$F.\ (1+e^{\sin x})^{x^2+3} \qquad G.\ \frac{\cosh x+3}{\sqrt{\dfrac{x+4}{\log_{10}(x^3-6)}}} \qquad H.\ \frac{1}{(x-3)(x+4)(x-2)} \qquad I.\ \frac{u^2 v^3 w^9}{x^4 y^7 z^6}$$

$$J.\ \frac{a}{x+a}+\frac{b}{y+b}+\frac{c}{z+c} \qquad K.\ \frac{1}{1+\dfrac{1}{1+\dfrac{1}{x}}}$$

In tabular form, they are coded as follows:

Example	String for scalar computation	String for vector computation
A	sin(x).^3.*cos(x).^5	sin(x)^3*cos(x)^5
B	2+log(x)	2+log(x)
C	x.^4+3*x-5	x^4+3*x-5
D	(x.^3-5)./(x.^2-7*x-7)	(x^3-5)/(x^2-7*x-7)
E	sqrt(abs(x.^3)+asec(x))	sqrt(abs(x^3)+asec(x))
F	(1+exp(sin(x))).^(x.^2+3)	(1+exp(sin(x)))^(x^2+3)
G	(cosh(x)+3)./sqrt((x+4)./log10(x.^3-6))	(cosh(x)+3)/sqrt((x+4)/log10(x^3-6))
H	1./(x-3)./(x+4)./(x-2)	1/(x-3)/(x+4)/(x-2)
I	u.^2.*v.^3.*w.^9./x.^4./y.^7./z.^6	u^2*v^3*w^9/x^4/y^7/z^6
J	a./(x+a)+b./(y+b)+c./(z+c)	a/(x+a)+b/(y+b)+c/(z+c)
K	1./(1+1./(1+1./x))	1/(1+1/(1+1/x))

Table 2.A   Some mathematical functions and their simulink counterparts

Mathematical notation	Simulink notation	Mathematical notation	Simulink notation	Mathematical notation	Simulink notation
$\sin x$	sin(x)	$\sin^{-1} x$	asin(x)	π	pi
$\cos x$	cos(x)	$\cos^{-1} x$	acos(x)	A+B	A+B
$\tan x$	tan(x)	$\tan^{-1} x$	atan(x)	A−B	A−B
$\cot x$	cot(x)	$\cot^{-1} x$	acot(x)	A×B	A*B
$\mathrm{cosec} x$	csc(x)	$\sec^{-1} x$	asec(x)	e^x	exp(x)
$\sec x$	sec(x)	$\mathrm{cosec}^{-1} x$	acsc(x)	A^B	A^B
$\sinh x$	sinh(x)	$\sinh^{-1} x$	asinh(x)	$\ln x$	log(x)
$\cosh x$	cosh(x)	$\cosh^{-1} x$	acosh(x)	$\log_{10} x$	log10(x)
$\mathrm{sec}\, hx$	sech(x)	$\mathrm{sec}\, h^{-1} x$	asech(x)	$\sqrt{x}$	sqrt(x)
$\mathrm{cosec}\, hx$	csch(x)	$\mathrm{cosec}\, h^{-1} x$	acsch(x)	Σ	sum
$\tanh x$	tanh(x)	$\tanh^{-1} x$	atanh(x)	Π	prod
$\coth x$	coth(x)	$\coth^{-1} x$	acoth(x)	$|x|$	abs(x)

2.2 Modeling waveforms, functions, or signals

The generation of the waveforms is a part and parcel of many functional analyses. The simulink family of waves is opulent and easy to implement without rigorous programming. Most of the waveforms possess certain associated parameters or properties. Prior to the generation of the waves in simulink, selection of the waveform parameters is essential. A number of commonly practiced waveform generations are addressed in the following.

2.2.1 Modeling the unit step and its derived functions

The unit step function or Heaviside function $u(t) = \begin{cases} 1 & \text{for } t \geq 0 \\ 0 & \text{for } t < 0 \end{cases}$ which has the plot 2.1(a) can be

simulated in simulink by the block Step found in the link *'Simulink* $\rightarrow$ *Sources* $\rightarrow$ *Step'*. Bring the block in a new simulink model file, connect the block with the Scope block as shown in figure 2.1(b), run the model, and doubleclick the Scope block with the autoscale setting (click the indicated icon of the figure 2.1(c)) to see the output

Figure 2.1(a) *The unit step function*

Figure 2.1(b) *Step block connected with Scope*

Figure 2.1(e) *The unit step function shifted at 1.5 and of final value 12.5*

autoscale icon

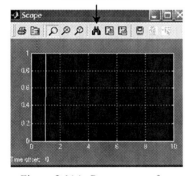

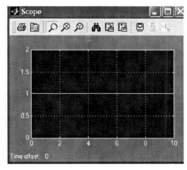

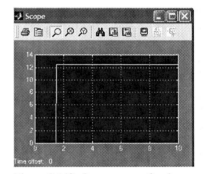

Figure 2.1(c) *Scope output for the default Step block*

Figure 2.1(d) *Scope output for the* $u(t)$

Figure 2.1(f) *Scope output for the* $12.5u(t-1.5)$

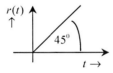

Figure 2.2(a) *The ramp function*

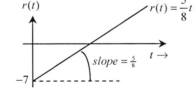

Figure 2.2(c) *The ramp function with different slope and initial output*

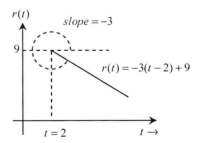

Figure 2.2(d) *The ramp function* $r(t) = -3(t-2)+9$ *with different starting time*

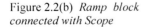

Figure 2.2(b) *Ramp block connected with Scope*

like the figure 2.1(c). The point is the displayed output has the step value at 1 or the function returned is $u(t-1)$. Doubleclick the Step block to see its parameter window like the figure 3.7(e), change the step time from 1 to 0 in the parameter window, run the model, doubleclick the Scope block to see the correct unit step function with autoscale setting like the figure 2.1(d). As another example, the function $12.5u(t-1.5)$, which has the plot in figure 2.1(e), has the final value 12.5 and is shifted at 1.5 to the right on the time axis. We need to enter the setting in the parameter window of the figure 3.7(e) as $\begin{cases} \text{Step value to 1.5} \\ \text{Final value to 12.5} \end{cases}$ and the Scope output for the setting is shown in figure 2.1(f) with the autoscale. A finite duration constant value can be generated with the help of two Step blocks. The reader is referred to the second example of the section 3.7 for the simulation of a finite duration constant value.

2.2.2 Modeling the ramp and its derived functions

The ramp function is defined as $r(t) = t$ which is a straight line with 45^0 slope and is plotted in figure 2.2(a). The block Ramp found in *'Simulink $\rightarrow$ Sources $\rightarrow$ Ramp'* can generate the function $r(t)$. Let us bring one Ramp and one Scope blocks in a new simulink model file, connect them according to the figure 2.2(b), run the model, and doubleclick the Scope block to see the output as shown in figure 2.2(e) with the autoscale setting. Ramp

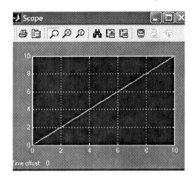

Figure 2.2(e) *Scope output for the default Ramp block i.e.* $r(t) = t$

Figure 2.2(f) *Scope output for the function* $r(t) = \dfrac{5}{8}t - 7$

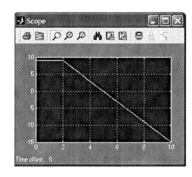

Figure 2.2(g) *Scope output for the function* $r(t) = -3(t-2) + 9$

Figure 2.2(h) *The model for the function*
$r(t) = [-3(t-2)+9]u(t-2)$

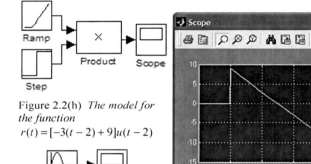

Figure 2.3(a) *The model for the sine wave generation for example 1*

Figure 2.2(i) *Scope output for the model in the figure 2.3(h)*

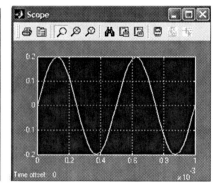

Figure 2.3(b) *The Scope output of the model for the example 1*

derived functions can also be generated with a slight change in the parameter window of Ramp. Let us doubleclick the Ramp block to see the parameter window like the figure 3.2(b). You find there the parameters as $\begin{Bmatrix} \text{Slope} \\ \text{Start time} \\ \text{Initial output} \end{Bmatrix}$.

The function $r(t) = t$ has the slope 1, the start time 0, and the initial output 0.

Let us see another Ramp whose equation is $r(t) = \dfrac{5}{8}t - 7$ and the plot is the figure 2.2(c). Doubleclick the

Ramp block, enter the setting in the parameter window of the Ramp as $\begin{Bmatrix} \text{Slope} = 5/8 \\ \text{Start time} = 0 \\ \text{Initial output} = -7 \end{Bmatrix}$, run the model, display

the output as shown in the figure 2.2(f) with the autoscale setting.

The function $r(t) = -3(t-2)+9$ is also a ramp which has the parameters $\begin{cases} \text{Slope} = -3 \\ \text{Start time} = 2 \\ \text{Initial output} = 9 \end{cases}$ and which is

detailed in figure 2.2(d). The corresponding Scope output with the autoscale setting is depicted in figure 2.2(g). Referring to the figure, the block assumes the value of $t = 2$ for the interval $0 \le t \le 2$. If the reader wants the function to be strictly 0 for $0 \le t \le 2$, the functional description had better be $r(t) = [-3(t-2)+9]u(t-2)$ and you need the model of the figure 2.2(h) whose Scope output is the figure 2.2(i).

2.2.3 Modeling the sine wave and its derived functions

Probably the sine wave is the most frequent one used in various engineering disciplines. The sine wave can be defined as $y(t) = A\sin(2\pi f t + \theta)$, where A, f, θ, and t are the amplitude, frequency, phase angle, and the time interval over which the wave is to be generated respectively. Also the time period of the wave is given by $T = \dfrac{1}{f}$.

The block Sine Wave (link: '*Simulink* $\rightarrow$ *Sources* $\rightarrow$ *Sine Wave*') can help us generate different waves derived from the sinusoidal for which a number of examples are presented in the following:

⊟ Example 1

Generate a sine wave of frequency $2KHz$ and amplitude ± 0.2. The wave should exist for 1 millisecond.

Solution:

In practice, the amplitude of the wave can represent any physical quantity such as displacement, voltage, or current. The wave has the time period $T = \dfrac{1}{2 \times 10^3}$ sec or 0.5 milliseconds hence in the given time interval we expect the wave to have two cycles. The wave and the model for implementation are shown in figures 2.3(c) and 2.3(a) respectively. Doubleclick the block Sine Wave, enter the settings as $\begin{cases} \text{Amplitude}: 0.2 \\ \text{Frequency (rad/sec)}: 2*\text{pi}*2\text{e}3 \end{cases}$ in the parameter window keeping the

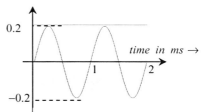

Figure 2.3(c) *Plot of the sine wave of the example 1*

others as default ($\omega = 2\pi f$ is used for the frequency). For the time interval, let us click the '*Simulation* $\rightarrow$ *Simulation parameters* $\rightarrow$ *Solver*' from the model menu bar and enter the stop time as 1e-3 (for 1 millisecond). Run the model and the Scope output should look like the figure of 2.3(b) with the autoscale setting.

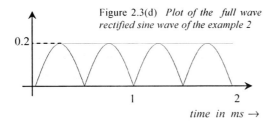

Figure 2.3(d) *Plot of the full wave rectified sine wave of the example 2*

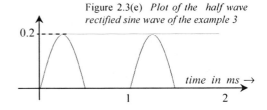

Figure 2.3(e) *Plot of the half wave rectified sine wave of the example 3*

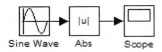

Figure 2.3(f) *Simulink model for the full wave rectified sine wave of the example 2*

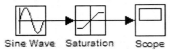

Figure 2.3(g) *Simulink model for the half wave rectified sine wave of the example 3*

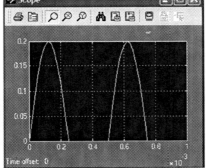

Figure 2.3(h) *Scope output for the half wave rectified sine wave of the example 3*

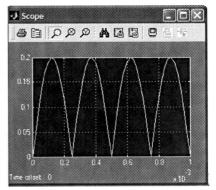

Figure 2.3(i) *Scope return for the full wave rectified sine wave*

13

⌗ Example 2

Generate a full wave rectified sine wave of frequency $2KHz$, amplitude ± 0.2, and duration 1 millisecond.

Solution:

This is basically the example 1 but with the exception that the negative peaks become positive as shown in figure 2.3(d). What we do is we bring one Abs (link: '*Simulink $\rightarrow$ Math Operations $\rightarrow$ Abs*') block (the block finds the absolute vale of the function) and insert that between the Sine Wave and the Scope as presented in figure 2.3(f). After running the model, the reader should see the Scope output as shown in figure 2.3(i) with the autoscale setting.

⌗ Example 3

Generate a half wave rectified sine wave of frequency $2KHz$ and amplitude ± 0.2. The wave should exist for 1 millisecond.

Solution:

The wave we are having is the wave of the example 1 but the negative portion of the wave is set to zero like the figure 2.3(e). To model the wave, we bring the block Saturation (link: '*Simulink $\rightarrow$ Discontinuities $\rightarrow$ Saturation*') and connect that as shown in figure 2.3(g). The block has two saturation limits: lower and upper. The block sets the wave to the saturation limit if the wave reaches below or above the saturation value. Referring to the example 1, if we set the value of the wave to 0 for any negative value, we can have the half wave rectified wave but we do not want to change the positive portion of the wave. So let us doubleclick the block, enter the settings as $\left\{\begin{array}{l}\text{Upper limit}:0.2\\\text{Lower limit}:0\end{array}\right\}$, and run the model. The simulink Scope response with the autoscale setting is shown in figure 2.3(h).

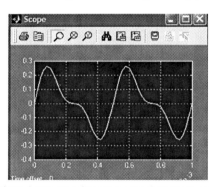

Figure 2.4(a) *Two frequency wave from the Scope*

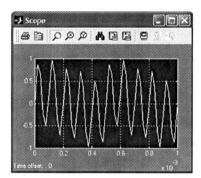

Figure 2.4(b) *Scope output for the three frequency wave*

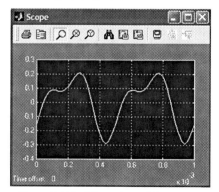

Figure 2.4(c) *Model for the two frequency wave*

Figure 2.4(d) *Model for the three frequency wave*

⌗ Example 4

Let us generate the two frequency wave $y(t) = 0.2\sin 2\pi f t + 0.1\sin 4\pi f t$ for $0 \le t \le 1ms$, where $f = 2KHz$.

Solution:

The given $y(t)$ has two sine components: the first one with the amplitude 0.2 and frequency $2KHz$ and the second one with the amplitude 0.1 and frequency $4KHz$. Figures 2.4(c) and 2.4(a) show the model and its scope output with autoscale setting respectively. The necessary parameter settings for the Sine Wave and Sine Wave1 are $\left\{\begin{array}{l}\text{Amplitude}:0.2\\\text{Frequency (rad/sec)}:2*pi*2e3\end{array}\right\}$ and $\left\{\begin{array}{l}\text{Amplitude}:0.1\\\text{Frequency (rad/sec)}:2*pi*4e3\end{array}\right\}$ respectively.

⌗ Example 5

In this example we generate three frequency wave $y(t) = 0.2\sin 2\pi f t + 0.1\sin 4\pi f t + 0.08\sin 10\pi f t$ for $0 \le t \le 1ms$, where $f = 2KHz$.

Figure 2.4(e) *Waves formed from two different phase angle sine waves*

Solution:

Obviously one needs three Sine Wave blocks and a three input Sum (doubleclick the block Sum and change its list of signs to +++) block as depicted in the model 2.4(d) for the design. The settings for the Sine Wave, Sine Wave1, and Sine Wave2 are $\left\{\begin{array}{l}\text{Amplitude}: 0.2 \\ \text{Frequency (rad/sec)}: 2*\text{pi}*2e3\end{array}\right\}$, $\left\{\begin{array}{l}\text{Amplitude}: 0.1 \\ \text{Frequency (rad/sec)}: 2*\text{pi}*4e3\end{array}\right\}$, and $\left\{\begin{array}{l}\text{Amplitude}: 0.08 \\ \text{Frequency (rad/sec)}: 2*\text{pi}*10e3\end{array}\right\}$ respectively. Figure 2.4(b) is the Scope output with the autoscale setting for the wave.

⊟ Example 6

Adding some phase in each wave which is $y(t) = 0.2\sin(2\pi f t - 60^0) + 0.1\sin(4\pi f t + 10^0)$ modifies the wave of the example 4. In the parameter window of the Sine Waves just append the phases as –pi/3 and 10*pi/180 respectively. The Scope returns the output with the autoscale setting as depicted in figure 2.4(e).

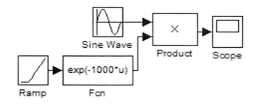

Figure 2.4(f) *Model for the damped sine wave generation*

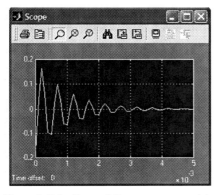

Figure 2.4(g) *Damped Sine wave returned by simulink with the adaptive setting*

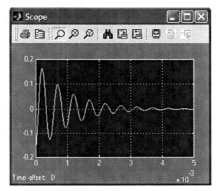

Figure 2.4(h) *Damped sine wave when the solver is set for the fixed step*

⊟ Example 7

An example of the damped sine wave can be $y(t) = 0.2e^{-1000t}\sin(2\pi f t - 60^0)$, where $f = 2KHz$. Let us generate the wave for $0 \le t \le 5m\sec$.

Solution:

One can separate the function $y(t)$ as e^{-1000t} and $0.2\sin(2\pi f t - 60^0)$ then the latter part is like the sine wave of the example 1. The exponent part can be modeled through the blocks Ramp and Fcn (link: *'Simulink → User-Defined Functions → Fcn'*) since the Ramp simulates t. Let us bring one Ramp, one Fcn, one Sine Wave, one Product (link: *'Simulink → Math Operations → Product'*), and one Scope blocks in a new simulink model file. Connect the blocks as shown in figure 2.4(f). Doubleclick the block Fcn, enter the code of e^{-1000t} as exp(-1000*u) considering the independent variable u, and resize the block to display its contents. Doubleclick the block Sine Wave, enter its settings as $\left\{\begin{array}{l}\text{Amplitude}: 0.2 \\ \text{Frequency (rad/sec)}: 2*\text{pi}*2e3 \\ \text{Phase (rad)}: -\text{pi/3}\end{array}\right\}$, change the solver stop time to 5e-3 for $0 \le t \le 5m\sec$ (as we did before), run the model, doubleclick the Scope to see the output as shown in figure 2.4(g) with the autoscale setting. Looking into the figure, the successive maximums of the wave are not so smooth. The reason for this is the simulink solver adaptively selects the time step of the wave which is non uniform. Simulink offers us the provision for changing that. Since the wave existence is in the millisecond range, let us choose the fixed step size as 0.01 $m\sec$. Change the Solver option –Type from the Variable step to the Fixed step, enter the Fixed step size as 0.01e-3, and run the model. The output is depicted in the figure 2.4(h) with the autoscale setting.

⊡ Example 8

A clipped sine wave can be generated employing the saturation block like the model of the figure 2.3(g). Clipping means that the wave is constant below and after some user defined value. Considering the sine wave of the example 1, let us clip the wave to 0.15 toward the positive peak and to –0.05 toward the negative peak. Doubleclick the block Saturation and enter the settings as $\begin{Bmatrix} \text{Upper limit}: 0.15 \\ \text{Lower limit}: -0.05 \end{Bmatrix}$. The output with the autoscale setting is shown in figure 2.4(i).

⊡ Example 9

A sine wave can be raised up or down from the horizontal axis by some specific value. This operation is called adding a bias to the sine wave. Let us consider the wave of the example 1. We wish to raise the sine wave by 0.1 that is the equation of the wave becomes $y(t) = 0.2 \sin 2\pi f t +$ 0.1. So the swing from the +0.2 to –0.2 should be from

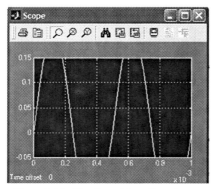

Figure 2.4(i) *The clipped Sine Wave for the example 8*

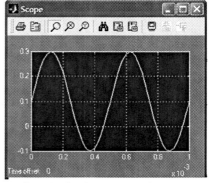

Figure 2.4(j) *The sine wave of the example 1 with a bias of 0.1*

the +0.3 to –0.1 due to the bias. However, let us doubleclick the block Sine Wave in the model for the example 1 and enter the bias 0.1 in the parameter window. The output with the autoscale adjustment is shown in the figure 2.4(j).

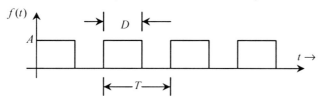

Figure 2.5(a) *The general rectangular pulse*

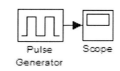

Figure 2.5(b) *The model for the rectangular pulse of the example 1*

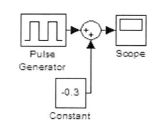

Figure 2.5(d) *The model for the equal positive and negative swing*

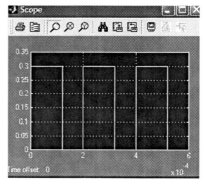

Figure 2.5(c) *The Scope output for the model of the example 1 with the autoscale setting*

2.2.4 Modeling the rectangular wave and its derived functions

A rectangular wave is the one whose wave value is constant over some interval. Even though the wave value is constant, different applications need different kinds of the wave generation. We ensemble some of the wave generations in the following.

⊡ Example 1

Figure 2.5(a) presents the general rectangular pulse $f(t)$, where A is the amplitude of the wave, D is the duty cycle (or the time when it exists) of the wave measured as the percentage of the period, and t is the desired time interval through which we want to see the square wave. Let us generate a rectangular pulse of amplitude 0.3, frequency $5KHz$, duty cycle 60%, and duration 0.6 $m$sec. The shape of the wave is similar to that of the figure 2.5(a). The model of the figure 2.5(b) can be constructed in this regard in which the block Pulse Generator generates the rectangular wave and has the link '*Simulink* → *Sources* → *Pulse Generator*'. The time period of the wave is $T = \dfrac{1}{f} = \dfrac{1}{5KHz} = 0.2\,m\sec$. Doubleclick the block Pulse Generator and enter its settings as $\left\{\begin{array}{l} \text{Amplitude}: 0.3 \\ \text{Period(secs)}: 0.2e-3 \\ \text{Pulse Width (\% of period)}: 60 \end{array}\right\}$ in the parameter window keeping the other as default. For the duration $0.6\,m\sec$, change the solver stop time to 0.6e-3. Figure 2.5(c) shows the simulation output.

☐ Example 2

In the last example, the pulse amplitude is 0.3 and the minimum is 0. What if we would like to have a wave of the same frequency and duration but the swing from –0.3 to 0.3. The difference between the maximum and minimum is 0.6. We doubleclick the block Pulse Generator of the figure 2.5(b), change its amplitude to 0.6 and add a constant value of –0.3 to

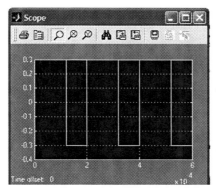

Figure 2.5(e) *The Scope output for the model of the example 2 with the autoscale setting*

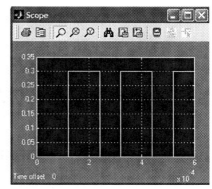

Figure 2.5(f) *The shifted pulse of the example 4 with the autoscale setting*

obtain the expected swing. The model of the figure 2.5(d) depicts the implementation, and its output is presented in the figure 2.5(e).

☐ Example 3

Generate a square wave of the same frequency and duration as the wave of the example 1 does. The wave has the duty cycle 50% and equal positive and negative swing. The last two examples can help us simulate the wave.

☐ Example 4

In the examples 1 and 2, the wave started exactly at the time $t = 0$. Let us say we want to shift the wave of the example 1 to the right exactly by the duty

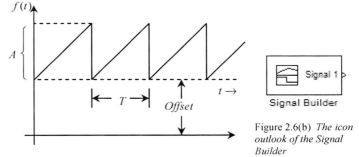

Figure 2.6(a) *A triangular wave of period T seconds and amplitude swing from 0 to A*

Figure 2.6(b) *The icon outlook of the Signal Builder*

cycle. In second, the time shift becomes $0.12\,m\sec$. Doubleclick the block Pulse Generator of the model 2.5(b) and enter the Phase delay (secs) as 0.12e-3 in the parameter window of the block. Figure 2.5(f) is the outcome of the simulation.

2.2.5 Modeling the triangular wave and its derived functions

Like the rectangular and sine waves, the triangular wave also has a lot of varieties. Let us explore few variants of the wave. A triangular wave is basically a ramp function over one period with different slop, swing, and offset. Figure 2.6(a) shows one typical triangular wave whose equation is given by $f(t) = \left\{ \dfrac{At}{T} \; for \; 0 \le t \le T \right.$ excluding the offset. The offset just shifts the wave up or down.

☐ Example 1

Generate a triangular wave which has the frequency $500\,Hz$, the amplitude swing from –0.05 to 0.05, and duration 0.01 sec.

Solution:

We wish to introduce here the Signal Builder facility of simulink. Pertaining to the example, the time period of the wave is $T = \dfrac{1}{f} = 0.002$ sec indicating five cycles in the given duration. Let us bring the block Signal Builder following the link '*Simulink → Sources → Signal Builder*' in a new simulink model file whose icon outlook is presented in the figure 2.6(b). Simulink responds with the design window of the figure 2.6(c) on doubleclicking the block in which a finite duration rectangular pulse exists as the default. Since the wave frequency or time period and the duration must be consistent in the wave modeling, we first enter the wave duration in the Signal Builder design window. To do so, click the Axes from the menu of the design window, and you see the option Change time range in the pull down menu. Let us click that and enter $\left\{\begin{array}{l}\text{Min time}:0\\\text{Max time}:0.01\end{array}\right\}$ in the prompt window for entering the wave duration 0.01 sec (the design window is updated on account of the change). Again the required amplitude

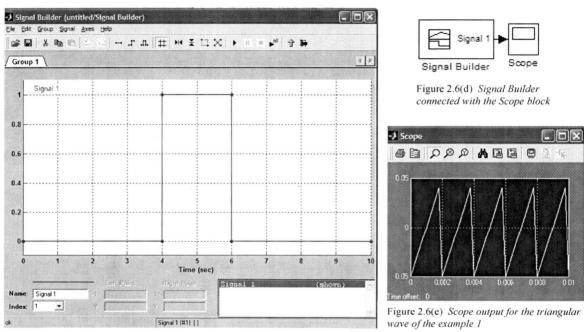

Figure 2.6(d) *Signal Builder connected with the Scope block*

Figure 2.6(e) *Scope output for the triangular wave of the example 1*

Figure 2.6(c) *Design window of the Signal Builder block*

swing is from –0.05 to 0.05 hence we click again the Axes in the menu bar of the design window and then click Set Y Display limits in the pull down menu. We enter $\left\{\begin{array}{l}\text{Minimum}:-0.05\\\text{Maximum}:0.05\end{array}\right\}$ in the prompt window. In the design window menu bar, you also see the menu for Signal, click the Signal menu, and you find Replace with option in the pull down menu and Triangle in the second stage pull down menu. Click the Triangle and enter $\left\{\begin{array}{l}\text{Frequency}:500\\\text{Amplitude}:0.05\\\text{Offset}:0\end{array}\right\}$ in the prompt window. With that action, the updated wave appears in the design window and that is what we require. Our wave design is finished and let us save it from the Save icon or from the File menu of the design window. We can close the Signal Builder window. Let us jump into the simulink model file. Simulink does not have any information about the wave duration. Enter it by changing the solver stop time to 0.01 from the simulink model menu bar as we did before. The Signal Builder block behaves as the triangular wave generator, and it is connected with the Scope as shown in the figure 2.6(d). Run the model and see the Scope output as shown in figure 2.6(e) with the autoscale setting. Referring to the Scope output, the vertical edge of each wave does not seem to be vertical. This is because of the variable step or adaptive setting of the solver. If we can make the solver setting fixed step and small within the duration for sure the wave becomes as expected.

⌗ Example 2

Generate a triangular wave which has the frequency 500 *Hz*, the amplitude swing from 0 to 0.1, and duration 0.01 sec. This is basically the wave of the example 1 but with little modification. The previous swing is from –0.05 to 0.05 but now we need from 0 to 0.1 so just adding an offset of 0.05 will have our simulation done. Referring to the design window of the figure 2.6(c), click the *Signal →*

Replace With → Triangle, enter $\begin{bmatrix} Frequency: 500 \\ Amplitude: 0.05 \\ Offset: 0.05 \end{bmatrix}$ in the prompt

window, save the design, and run the simulink model of the figure 2.6(d). The output is shown in figure 2.6(f) with Fixed Step solver option, the Fixed step size 0.00001 (*Simulation → Simulation parameters → Solver* from the simulink menu bar), and the autoscale setting.

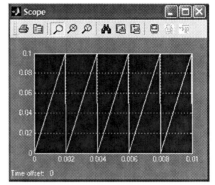

Figure 2.6(f) *Scope output for the triangular wave of the example 2*

⌗ Example 3

The periodic wave of the figure 2.6(g) is to be generated for $0 \leq t \leq 0.01$ sec which has the time period 4 *m*sec but the wave is off from 2 to 4 *m*sec considering the first cycle. So what we perform is we form a triangular wave of the swing from 0 to 0.1 and the time period 2 *m*sec (means frequency 500 Hz) and then set the alternate cycle to zero. Let us imagine the pulse whose value is 1 from 0 to 2 *m*sec and 0 from 2 to 4 *m*sec and the pulse continues this way for the other cycles. Our purpose will be

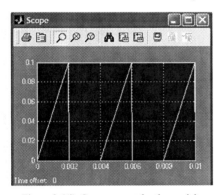

Figure 2.6(i) *Scope output for the model of the figure 2.6(h)*

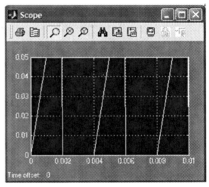

Figure 2.6(k) *Scope output for the model of the figure 2.6(l)*

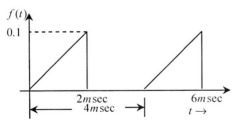

Figure 2.6(g) *A triangular wave with some off interval*

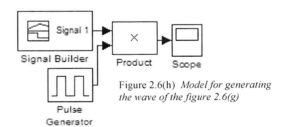

Figure 2.6(h) *Model for generating the wave of the figure 2.6(g)*

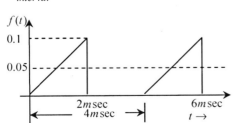

Figure 2.6(j) *The triangular wave of the figure 2.6(g) is cropped at 0.05*

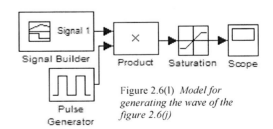

Figure 2.6(l) *Model for generating the wave of the figure 2.6(j)*

served if the pulse is multiplied with the triangular wave. However, first you generate the wave of the example 2 employing the Signal Builder and connect the Pulse Generator as shown in the figure 2.6(h). The parameter window settings for the Pulse Generator should be
$$\begin{cases} \text{Amplitude: } 1 \\ \text{Period(secs): } 4e-3 \\ \text{Pulse Width (\% of period): } 50 \end{cases}$$
. The Scope output is shown in the figure 2.6(i) with the autoscale setting.

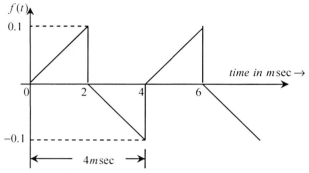

Figure 2.7(a) *A symmetric positive and negative halves triangular wave*

⬚ Example 4

Suppose the wave of the figure 2.6(g) is cropped at 0.05 so that we only get the lower portion of the wave under the dotted line at 0.05 (shown in figure 2.6(j)). The solution is just insert a Saturation block as indicated in the figure 2.6(l) with the settings
$$\begin{cases} \text{Upper limit: } 0.05 \\ \text{Lower limit: } 0 \end{cases}$$
in the parameter window whose output is the figure 2.6(k) with the autoscale setting.

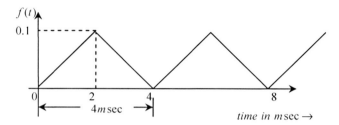

Figure 2.7(b) *The negative half of the wave in 2.7(a) is made positive*

⬚ Example 5

Figure 2.7(a) represents a wave in which the symmetric positive and negative halves are located. Once we generate the wave of the figure 2.6(g), it is easy to go through this wave. The wave with the positive half is exactly the same as that of the figure 2.6(g). We shift the wave of the figure 2.6(g) by $2\,m\sec$ to the right and then flip the wave about the horizontal axis. That is how we form the negative halves. The shifting operation can be done through the block Transport Delay (link: '*Simulink → Continuous → Transport Delay*') and the flipping about the t axis can happen just by multiplying -1 (carried out by a Gain block of gain -1). At last we add these two halves to form the complete wave by a Sum block. Anyhow extending the model of the figure 2.6(h), we have the model for the problem as shown in figure 2.7(d). Once you bring the Transport Delay and Gain blocks, doubleclick them to enter the Time Delay: 2e-3 and Gain: -1 in the parameter windows respectively. Figure 2.7(e) is the Scope result on simulation with the autoscale setting.

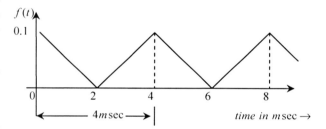

Figure 2.7(c) *The wave of the figure 2.7(b) is shifted by 2 msec*

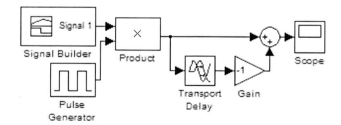

Figure 2.7(d) *Model for generating the wave in the figure 2.7(a)*

⬚ Example 6

In this example we wish to simulate the wave of the figure 2.7(b) in a slight different approach for $0 \le t \le 10\,m\sec$. The first cycle of the wave has two ramps or straight line functions, the first and second of which have the slopes $\dfrac{0.1}{2\,m\sec}$ and $-\dfrac{0.1}{2\,m\sec}$ respectively. The differential coefficient of a straight line is the slop of the line so from the slope if we want to have the line, we need to integrate the slope. Now the differentiation of the wave in figure 2.7(b) takes the shape of a square wave (like the figure 2.5(e)) whose amplitude swings are from $-\dfrac{0.1}{2\,m\sec}$ to

$\dfrac{0.1}{2m\sec}$. The block Pulse Generator generates pulse from 0 to some value A. So if we want to generate the swing from –A to A, we generate the wave swing from 0 to 2A and then add a constant value –A to obtain the required swing. The slope $\dfrac{0.1}{2m\sec}$ has the code 0.1/2e-3 and $2 \times \dfrac{0.1}{2m\sec}$ can be written as 2*0.1/2e-3. For the problem at hand, the Pulse Generator generates a pulse of swing 0 to 2*0.1/2e-3 and we add –0.1/2e-3 with the generator output

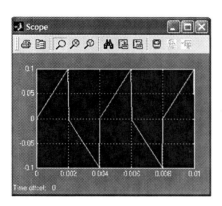

Figure 2.7(e) *Scope output for the model in the figure 2.7(d)*

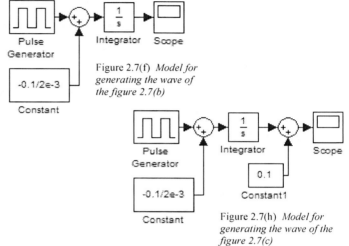

Figure 2.7(f) *Model for generating the wave of the figure 2.7(b)*

Figure 2.7(h) *Model for generating the wave of the figure 2.7(c)*

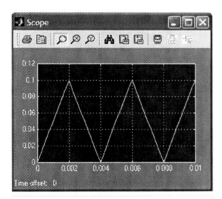

Figure 2.7(g) *Scope output for the model of the figure 2.7(f) with autoscale setting*

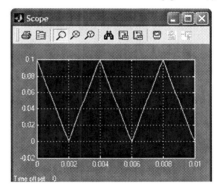

Figure 2.7(i) *Scope output for the model of the figure 2.7(h) with autoscale setting*

to obtain the required swing from $-\dfrac{0.1}{2m\sec}$ to $\dfrac{0.1}{2m\sec}$. At last we integrate the resultant output to have the wave of the figure 2.7(b). The model of the figure 2.7(f) illustrates the simulation in which you need the settings for the Pulse Generator and Constant blocks as $\begin{Bmatrix} \text{Amplitude}: 2*0.1/2\text{e}-3 \\ \text{Period(secs)}: 4\text{e}-3 \end{Bmatrix}$ and Constant Value: $-0.1/2\text{e}-3$ in the parameter windows respectively keeping the others as default. Also resize the block Constant to display its contents, and the block Integrator has the link: *'Simulink $\to$ Continuous $\to$ integrator'*. However, the Scope output is presented in the figure 2.7(g) with the autoscale setting.

🗗 Example 7

The wave of the figure 2.7(c) is the shifted version of the wave in 2.7(b) by $2\,m\sec$. The model in the figure 2.7(f) needs slight modification to implement the wave. Doubleclick the block Pulse Generator and change its Phase Delay (secs) to 2e-3 (for $2\,m\sec$ shifting). But the point is the swing becomes –0.1 to 0. We just add a constant value of 0.1 to attain the swing from 0 to 0.1. The modified model is shown in the figure 2.7(h) whose Scope output is the figure 2.7(i) with the autoscale setting.

With this example we close the section of the triangular wave generation.

21

2.2.6 Modeling the triggered waves and user-defined non periodic functions

Once we have a wave of specific frequency, the wave can be turned on or off at any instant of the cycle depending on the user requirement – this is termed as the triggering a wave. Let us say we have a sine wave $y = A\sin 2\pi f t$ whose frequency and amplitude are $f =50\ Hz$ and $A =0.8$ respectively. The wave is off in one quarter of the period and the rest three quarters of the period the wave is on. Or it can be rephrased that the wave is triggered at 25% of the period. Figure 2.8(a) illustrates the triggering over four cycles or 0.08 seconds. To mention about the simulink solution, we generate a rectangular pulse of the same period as that of the sine wave but with the amplitude swing from 0 to 1. The rectangular pulse will have the duty cycle which is exactly the on time of the sine wave.

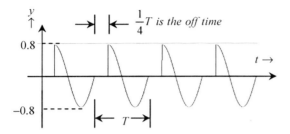

Figure 2.8(a) *A sine wave is triggered at 25% of its period*

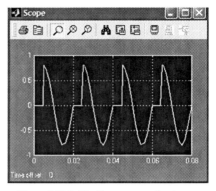

Figure 2.8(c) *The Scope output for the triggered sine wave*

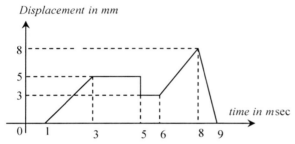

Figure 2.8(d) *A displacement versus time function*

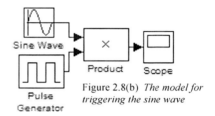

Figure 2.8(b) *The model for triggering the sine wave*

Since the duty cycle of the pulse starts at zero, we give a shift to the pulse of exactly (1–duty cycle) where the duty cycle is expressed as the percentage of the time period. So to simulate the current problem, let us bring one Sine Wave, one Pulse Generator, one Product, and one Scope blocks in a new simulink model file and connect them as presented in figure 2.8(b). The time period of the given wave is $\dfrac{1}{50Hz}=0.02\sec$. Hence the duty cycle is either 75% of the period or 0.015 sec. With that the off period is 25% of the period or 0.005 sec. Enter the settings of the Sine Wave and Pulse Generator as

$$\begin{cases}\text{Amplitude}: 0.8 \\ \text{Frequency(rad/sec)}: 2*pi*50\end{cases} \text{ and } \begin{cases}\text{Period(secs)}: 0.02 \\ \text{Pulse Width (\% of period)}: 75 \\ \text{Phase Delay (secs)}: 0.005\end{cases}$$

in the parameter windows keeping the others as default and set the solver stop time to 0.08 for the four cycles. In the event on running the model, we obtain the output from the Scope as depicted in the figure 2.8(c) with the autoscale setting.

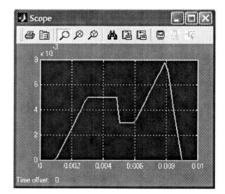

Figure 2.8(e) *The Scope output for the figure 2.8(d) with the autoscale setting*

This sort of triggering can occur for any other wave for example the wave of the example 2.7(b) or 2.7(c).

Not all functions are periodic, we also come across the finite duration function or signal of varying shapes. If the function is non periodic and defined by straight line segments and edges, the Signal Builder is the appropriate block to generate it which we introduced before. Just to address the problem specifically, let us design the finite displacement versus time variation curve of the figure 2.8(d) in simulink employing the block Signal Builder.

Let us bring the block in a new simulink model file and doubleclick it. The Design Window of the figure 2.6(c) appears before you with the default setting of a finite duration pulse. Referring to the figure 2.8(d), the function has the vertical and horizontal axes ranges as 0 to 8 mm and 1 to 9 msec respectively but we design the

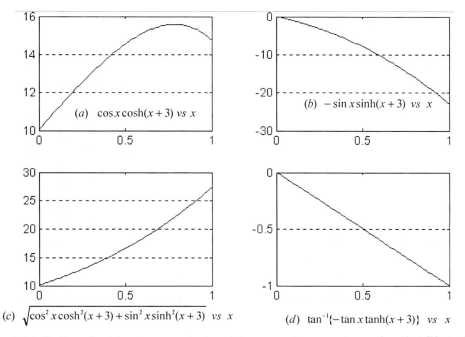

(a) $\cos x \cosh(x+3)$ *vs* x

(b) $-\sin x \sinh(x+3)$ *vs* x

(c) $\sqrt{\cos^2 x \cosh^2(x+3) + \sin^2 x \sinh^2(x+3)}$ *vs* x

(d) $\tan^{-1}\{-\tan x \tanh(x+3)\}$ *vs* x

Figure 2.9(a)-(d) *Plots of real, imaginary, magnitude, and phase parts of the expression* $\cos\{x+i(x+3)\}$ *for* $0 \le x \le 1$

simulink code of
$\cos\{x+i(x+3)\}$ *is here*

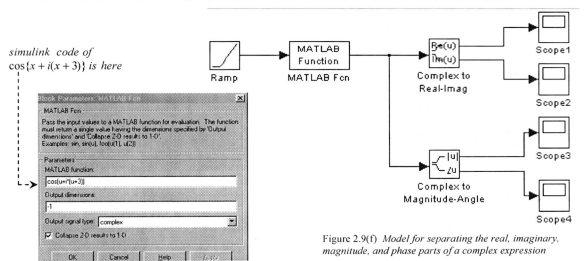

Figure 2.9(f) *Model for separating the real, imaginary, magnitude, and phase parts of a complex expression*

Figure 2.9(e) *MATLAB Fcn block parameter window*

function in the standard units - meter and second respectively. The line segments of the figure 2.8(d) have the coordinates (0msec, 0mm), (1msec, 0mm), (3msec, 5mm), (5msec, 5mm), (5msec, 3mm), (6msec, 3mm), (8msec, 8mm), and (9msec, 0mm). Collecting consecutive time and displacement coordinates, we have $[0\ 1\ 3\ 5\ 5\ 6\ 8\ 9] \times 10^{-3}$ sec and $[0\ 0\ 5\ 5\ 3\ 3\ 8\ 0] \times 10^{-3}$ meter respectively. Considering the deign window of the figure 2.6(c), click the *'Signal $\rightarrow$ Replace with $\rightarrow$ Custom'* from the menu bar, enter $\begin{Bmatrix} \text{Time Values: } [0\ 1\ 3\ 5\ 5\ 6\ 8\ 9] * 1e - 3 \\ \text{Y Values: } [0\ 0\ 5\ 5\ 3\ 3\ 8\ 0] * 1e - 3 \end{Bmatrix}$ in the prompt window for entering the horizontal and vertical coordinates in standard units, click the *'Axes $\rightarrow$ Change time range'* from the menu bar, enter $\begin{Bmatrix} \text{Min time: } 0 \\ \text{Max time: } 9e - 3 \end{Bmatrix}$ in the prompt window for the horizontal range, and save your function design from the design window menu bar. In doing so, the design window updates help us see the function designed in accordance with the figure 2.8(d). So the Signal Builder will generate the wave you wanted. Just to verify, let us connect the Signal Builder in conjunction with the Scope as shown in the figure 2.6(d), change the solver stop time to 9e-3, and run the model. The Scope output of the figure 2.8(e) confirms the design.

However, we just highlighted generic real functions or signals formation. In the next section we introduce the implementation style for the complex functions.

2.3 Modeling the complex functions

Complex expressions are also accepted in simulink. Assume that the expression $\cos\{x + i(x + 3)\}$ is to be simulated for $0 \le x \le 1$ and we wish to see the real, imaginary, magnitude, and phase components of the expression following the simulation.

Employing the analytical approach, one would obtain that the real, imaginary, magnitude, and phase components of $\cos\{x + i(x + 3)\}$ are $\cos x \cosh(x + 3)$, $-\sin x \sinh(x + 3)$, $\sqrt{\cos^2 x \cosh^2(x + 3) + \sin^2 x \sinh^2(x + 3)}$, and $\tan^{-1}\{-\tan x \tanh(x + 3)\}$, and they get the graphical shapes of the figures 2.9(a), 2.9(b), 2.9(c), and 2.9(d) respectively over the domain $0 \le x \le 1$. So our objective of the simulation is to obtain them in simulink.

☞ Simulink Solution

We can code the expression $\cos\{x + i(x + 3)\}$ in simulink as cos(u+i*(u+3)), considering the independent variable u (the dialog window does not accept other independent variables). The necessary model is presented in the figure 2.9(f) and for which following procedure can be adopted:

⇒ *bring a Ramp, a MATLAB Fcn (link: 'Simulink → User-Defined Functions → MATLAB Fcn'), a Complex to Real-Imag (link: 'Simulink → Math Operations → Complex to Real-Imag'), a Complex to*

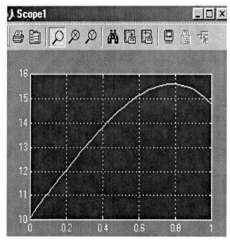

(g) *Scope1 output for the real part*

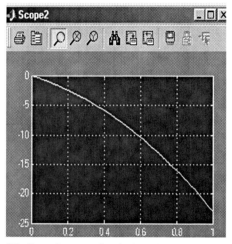

(h) *Scope2 output for the imaginary part*

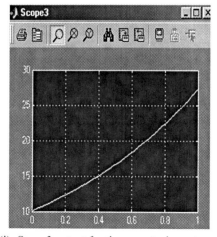

(i) *Scope3 output for the magnitude part*

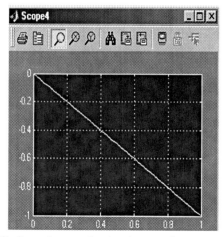

(j) *Scope4 output for the phase angle part*

Figure 2.9(g)-(j) *Simulink scopes' output for the model of figure 2.9(f)*

24

Magnitude-Angle (link: 'Simulink $\rightarrow$ Math Operations $\rightarrow$ Complex to Magnitude-Angle'), and a Scope blocks in a new simulink model file

$\Rightarrow$ rename the Scope as Scope1, select the Scope1, click the copy icon of the model file, click the paste icon of the model file three times to get the other three Scope blocks in the untitled model file

$\Rightarrow$ place various blocks relatively and connect them according to the figure 2.9(f)

$\Rightarrow$ set the solver start and stop time as 0 and 1 respectively for the insertion of $0 \leq x \leq 1$

$\Rightarrow$ doubleclick the MATLAB Fcn block, insert the functional code of the complex expression in the dialog window as shown in figure 2.9(e), and change the output signal type from auto to complex in the same window

$\Rightarrow$ run the model and doubleclick each Scope to see the output

With the autoscale setting, figures 2.9(g), 2.9(h), 2.9(i), and 2.9(j) present the simulink Scopes' output for the real, imaginary, magnitude, and phase (in radians) components which can be authenticated comparing to the figures of 2.9(a)-(d) respectively. If you select the block Complex to Real-Imag and doubleclick it, you see all three output options, complex, real, and imaginary, in the output parameter slot. The Complex to Magnitude-Angle block also has three output options. The reader is referred to chapter 8 for displaying the phase angle in degrees. The Ramp block by default generates the function like t versus t which can simulate the independent variable x. Even though the horizontal axes in all graphs show time as the independent variable, that can be assumed as the independent variable x without the loss of generosity.

We illustrate few examples on simulating the complex expressions in simulink in the following.

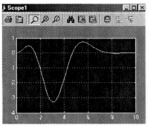
2.10(a)

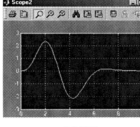
2.10(b)

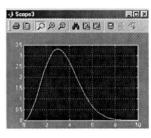
2.10(c)

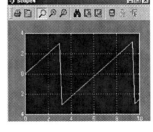
2.10(d)

Figure 2.10(a)-(d) *Simulink scope outputs for the expression* $x^2 e^{-\frac{x^2}{9}+jx}$ *over* $0 \leq x \leq 10$:
(a) *Scope1 for the real,*
(b) *Scope2 for the imaginary,*
(c) *Scope3 for the magnitude, and*
(d) *Scope4 for the phase parts of the expression*

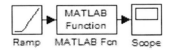

Figure 2.10(e) *Model for simulating the complex expression of the example 2*

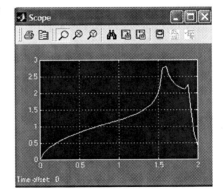

Figure 2.10(f) *Simulink Scope output for the model of the figure 2.10(e)*

✦ ✦ Example 1

Construct a model for simulating the real, imaginary, magnitude, and phase parts of the complex expression $x^2 e^{-\frac{x^2}{9}+jx}$ over the interval $0 \leq x \leq 10$.

☞ Simulink Solution

The string that can code $x^2 e^{-\frac{x^2}{9}+jx}$ is u^2*exp(-u^2/9+i*u) assuming u as the independent variable. The model presented in figure 2.9(f) is still operational for the problem at the hand. Doubleclick the MATLAB Fcn block (see figure 2.9(e)), change its MATLAB function to u^2*exp(-u^2/9+i*u), and change the solver start and stop time to 0 and 10 respectively. The outputs of the model for each scope with the autoscale setting are furnished in figures 2.10(a)-(d).

❖ ❖ **Example 2**

Model the expression $|\{\sin^2 x\cos^2 x + j\ln(1 + x + \tan x)\}^{\frac{2}{3}}|$ over $0 \le x \le 2$ in simulink.

🗗 *Simulink Solution*

Inside the second braces, the real and the imaginary parts of the expression are $\sin^2 x\cos^2 x$ and $\ln(1 + x + \tan x)$ respectively. Assuming the independent variable u, one can write their MATLAB codes as sin(u)^2*cos(u)^2 and log(1+u+tan(u)) respectively. With that, the complete coding of the given function is abs((sin(u)^2*cos(u)^2+i*log(1+u+tan(u)))^(2/3)). The model is presented in figure 2.10(e) in which the functional setting of the Fcn block should contain the last code. Because of the absolute value, we need to set the Fcn output data type as real in the same parameter window. However, insert the stop time as 2 for the solver and run the model. Figure 2.10(f) depicts the simulink scope output for the model with the autoscale setting.

❖ ❖ **Example 3**

Let us construct a simulink model for the complex function $F(\theta) = (1 + \cos\theta)e^{j\theta} + \left(1 + \dfrac{\theta}{9}\right)e^{-j2\theta}$ from which

we are interested to see the responses for the magnitude and phase angle in degrees over $0 \le \theta \le 2\pi$.

🗗 *Simulink Solution*

There are several ways to simulate this problem, one of which is entering the u independent variable code for the expression as (1+cos(u))* exp(j*u)+(1+u/9)*exp(-j*2 u) into the MATLAB Fcn block of the model in figure 2.10(e) or employ the model we presented in the figure 2.10(g). Referring to the figure, both Clocks generate the variable θ or Ramp like functions. All block links are attached at the end of the chapter in the table 2.B. Bring all blocks associated with the figure in a new simulink model file, place the blocks relatively, connect them, and run the

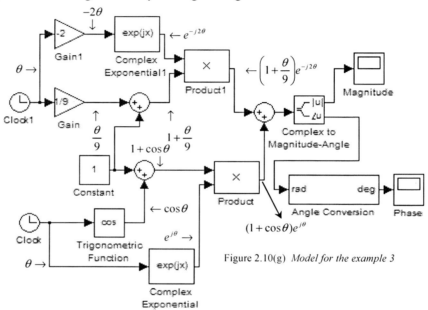

Figure 2.10(g) *Model for the example 3*

model. We renamed two Scope blocks as Magnitude and Phase respectively. The block Complex Exponential takes input θ as radians and returns $e^{j\theta}$ or $\cos\theta + j\sin\theta$. In the solver stop time, we enter the stop time as 2*pi for $0 \le \theta \le 2\pi$. Figures 2.10(h) and 2.10(i) show the magnitude and phase variations of the expression $F(\theta)$ respectively with the autoscale setting and the horizontal axes for both figures share common θ in radians.

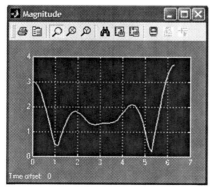

Figure 2.10(h) *Magnitude of $F(\theta)$*

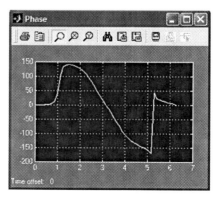

Figure 2.10(i) *Phase angle of $F(\theta)$ in degrees*

2.4 Modeling some standard functions

Simulink can import and export many of the standard mathematical functions often found in pure and applied sciences. In previous sections we addressed both the code and block oriented generations of functions yet a lot are to be explored. We point out here more functions both to the context of the code and block.

In the table 2.A, we mainly emphasized on the codes but some of those functions are implementable through blocks as well. The reader has already employed one block (Trigonometric Function of the figure 2.10(g)) in the last section. On doubleclicking the block, you find most standard trigonometric and hyperbolic functions including their inverses as in the drop down menu of the figure 2.11(a). For instance, the function asinh in the drop down menu means $\sinh^{-1} x$, so does others. But the important point is you must supply some input to the input port of the block to model the function.

In general the reader can generate the standard functions in two ways: constant generation and model dependent generation.

Let us consider the standard sine integral function, which is defined as $Si(x) = \int_{t=0}^{t=x} \frac{\sin t}{t} dt$ and has the simulink code

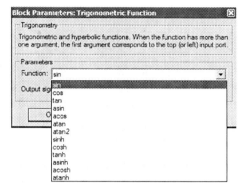

Figure 2.11(a) *Drop down menu of the Trigonometric Function block*

sinint. It is given that $Si(2) = 1.6054$, we take the help of the Constant and Display blocks to model that as shown in

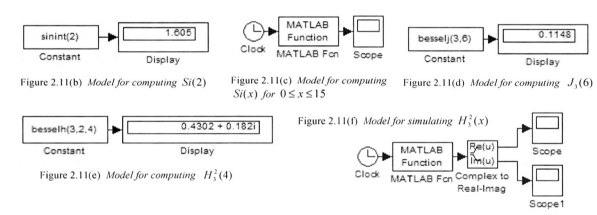

Figure 2.11(b) *Model for computing* $Si(2)$

Figure 2.11(c) *Model for computing* $Si(x)$ *for* $0 \le x \le 15$

Figure 2.11(d) *Model for computing* $J_3(6)$

Figure 2.11(e) *Model for computing* $H_3^2(4)$

Figure 2.11(f) *Model for simulating* $H_3^2(x)$

the figure 2.11(b). You need to enlarge the Constant block while modeling. The generation is entirely constant because the function has no dependency or interaction with the time or other blocks.

Again assume that the same function is to be computed for $0 \le x \le 15$. Now we can assign the interval to the independent variable of the simulink that is simulated by either the Ramp or the Clock block. But the matter is there is no specific block for the sine integral. So what we do is we send the horizontal data to the MATLAB Fcn (doubleclick the block and enter MATLAB Fcn: sinint in the parameter window) block for the computation. To pass the information about the interval $0 \le x \le 15$, we change the stop time of the solver to 15. There is no constancy in the generation because simulink now decides the independent variable generation whether it is fixed step or variable step. Furthermore, other blocks presented in the model also share the same clock. However, the model is presented in the figure 2.11(c). While entering the code in the parameter window of the MATLAB Fcn block, there is no need to write the input argument for the single input argument function. This way you can generate or pass any other functions placed in the table 2.B.

Let us explore a two input argument function, Bessel function of the first kind of order α, from the table 2.B. We want to compute $J_3(6) = 0.1148$ for which one can employ the model of the figure 2.11(d). Again let us compute and plot the $J_3(x)$ for $0 \le x \le 15$. In the model of the figure 2.11(c), doubleclick the block MATLAB Fcn and enter besselj(3,u) in the slot of MATLAB function.

To involve a three input argument function, let us consider the Hankel function of the second kind of order α from the same table $\{ H_\alpha^2(x) \}$. The function is a complex one and it is given that $H_3^2(4) = 0.4302 + j\, 0.1820$ whose implementation is shown in the figure 2.11(e). You need to enlarge both blocks to display the contents. How if we simulate $H_3^2(x)$ for $0 \le x \le 15$ is the employment of the model of the figure 2.11(c). But as far as the complex return is associated with, some alteration is required. Doubleclick the block MATLAB Fcn in the model of the figure 2.11(c), enter besselh(3,2,u) in the slit of MATLAB function, and change the output signal type from the default auto to complex. The complex output needs to be separated because the Scope can not display the complex numbers. Hence we bring the block Complex to Real-Imag and connect two Scopes to see the real and imaginary parts of the $H_3^2(x)$ separately. Again at $x=0$, the $H_3^2(x)$ is not defined on account of that you encounter some error until you change the start time. Let us change the start time of

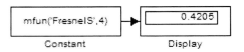

Figure 2.11(g) *Model for computing* $S(4)$

the solver to 1 so that we simulate $H_3^2(x)$ for $1 \le x \le 15$. Anyhow the figure 2.11(f) presents the modeling.

Table 2.B does not present all available functions that can be invoked or simulated. Over fifty functions can be accessed in a similar fashion. Go to MATLAB command window and execute help mfunlist, you see a long list of the functions. But they reside in the specific name mfun. From the MATLAB execution, let us choose Fresnel sine integral, which is defined as $S(x) = \int\limits_{t=0}^{t=x} \sin\left(\frac{\pi}{2}t^2\right) dt$ and which has the function name FresnelS (the function name is case sensitive). It is given that $S(4)$ =0.4205, we implement the computation by writing the code inside the constant block as mfun('FresnelS',4) for which the model of the figure 2.11(g) is shown. The computation on an interval can happen by entering mfun('FresnelS',u) in the MATLAB Fcn block parameter window as we did before for the other functions. We bring to an end of the section with this.

Table 2.B   Some standard functions that can be invoked or coded in simulink

Function name	Symbolic notation or definition	Simulink counterpart	Argument needed
Cosine integral	$Ci(x) = \int\limits_{t=\infty}^{t=x} \frac{\cos t}{t} dt$	cosint(x)	one
Exponential integral	$Ei(x) = \int\limits_{t=x}^{t=\infty} \frac{e^{-t}}{t} dt$	expint(x)	one
Error function	$erf(x) = \frac{2}{\sqrt{\pi}} \int\limits_{t=0}^{t=x} e^{-t^2} dt$	erf(x)	one
Complementary error function	$erfc = \frac{2}{\sqrt{\pi}} \int\limits_{t=x}^{t=\infty} e^{-t^2} dt$	erfc(x)	one
Gamma function	$\Gamma x = \int\limits_{t=0}^{t=\infty} t^{x-1} e^{-t} dt$	gamma(x)	one
Sinc function	$\sin c(x) = \frac{\sin \pi x}{\pi x}$	sinc(x)	one
Beta function	$B(z,w) = \int\limits_{t=0}^{t=1} t^{z-1}(1-t)^{w-1} dt$	beta(z , w)	two
Bessel function of the first kind of order α	$J_\alpha(x) = \sum\limits_{n=0}^{\infty} \frac{(-1)^n}{2^{2n+\alpha}\Gamma(n+1)\Gamma(n+\alpha+1)} x^{2n+\alpha}$	besselj(α , x)	two
Bessel function of the second kind of order α	$Y_\alpha(x) = \frac{J_\alpha(x)\cos(\alpha\pi) - J_{-\alpha}(x)}{\sin(\alpha\pi)}$	bessely(α , x)	two
Modified Bessel function of the first kind of order α	$I_\alpha(x) = J_\alpha(ix)$	besseli(α , x)	two
Modified Bessel function of the second kind of order α	$K_\alpha(x) = Y_\alpha(ix)$	besselk(α , x)	two
Hankel function of the first kind of order α	$H_\alpha^1(x) = J_\alpha(x) + jY_\alpha(x)$	besselh(α ,1, x)	three
Hankel function of the second kind of order α	$H_\alpha^2(x) = J_\alpha(x) - jY_\alpha(x)$	besselh(α ,2, x)	three

2.5 Data based function or signal generation

So far we focused primarily the modeling of the standard functions or their composites. In practice the reader may have experimental or observational data to simulate or analyze in simulink. A practical data set may have hundreds or thousands of elements in it. The entered data can reside in other software package like Microsoft Word,

Text file, or Excel data sheet. Reference [2] can be mentioned for exporting or importing data to and from MATLAB. In this section we assume that the data is available in MATLAB Workspace. We present data handling in simulink engaging simplistic set in the following.

❖ ❖ Example 1

Let us say we have the data set $x = \begin{bmatrix} -2 \\ -6 \\ 2 \\ 8 \\ 17 \\ 45 \end{bmatrix}$ and $y = \begin{bmatrix} 0 & -1 \\ 1 & -3 \\ 2 & 0 \\ 3 & -4 \end{bmatrix}$. We want to import

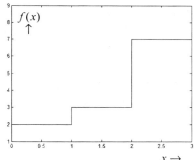

Figure 2.12(a) *Constant block holds the matrix x and y*

the sets in simulink. First we enter them into MATLAB workspace from the command prompt as follows:

MATLAB Command
>>x=[-2 -6 2 8 17 45]'; y=[0 -1;1 -3;2 0;3 -4]; ↵

Now bring a Constant block in a new simulink model file, doubleclick the block, and enter x in the slit of the constant value in the parameter window of the block (block appearance is in the figure 2.12(a)). So the matrix x is ready to enter in any simulation carried out by simulink. It is important to mention that the generation happens through out the solver timing. For instance, the default solver start and stop timings are 0 and 10 respectively. The block Constant provides matrix x at all time of the solver starting from 0 through 10. That is why if you connect a Scope block with the Constant block you see constant lines as the Scope output. Similarly you can also import the matrix y as in figure 2.12(a). The matrices x and y now can be regarded as the one and two dimensional functions or signals respectively. In practice you will have many elements in x (may be thousands) and y (may be hundreds of rows and columns). So the elements in x and y are just like the functional values of $f(x)$ and $h(x, y)$ respectively as happens in mathematics.

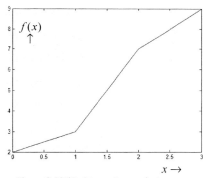

Figure 2.12(b) *Linear interpolation of the intervals of the function*

Figure 2.12(c) *Constancy between the intervals of the function*

❖ ❖ Example 2

In the example 1, the entered values only consider the functional values, for instance, if we think about $f(x)$ versus x, only the information about the $f(x)$ is associated and no information about the x is present. This example presents how both the $f(x)$ and x are invoked in simulink. Let us say that we have the functional values as

$f(x) = \begin{bmatrix} 2 \\ 3 \\ 7 \\ 9 \end{bmatrix}$ for $0 \le x \le 3$ where x is an integer. When we enter data into computer, basically we enter the samples or

the discrete values of the continuous $f(x)$. Theoretically speaking, a continuous curve representation needs infinite number of samples at infinite observations on x. However, the x is 0, 1, 2, and 3 within the given limit. The functional values of $f(x)$ can be plotted in two ways: either plot of the figures 2.12(b) and 2.13(c). How is one different from the other? The first two sample coordinates are $(x, f(x)) = (0, 2)$ and $(1, 3)$. In the figure 2.12(a) the two points are linearly interpolated that is how they are continuous. On the other hand the last sample value is held in between the interval in the figure 2.12(c). Which style you need depends on which analysis you perform —continuous or discrete. To feed the values of x and $f(x)$ in simulink, we form a rectangular matrix in which the first and second columns contain the x and $f(x)$ values respectively whence we obtain the matrix $\begin{bmatrix} 0 & 2 \\ 1 & 3 \\ 2 & 7 \\ 3 & 9 \end{bmatrix}$. Now we assign

the last matrix to some variable simin in the MATLAB command window as follows:

MATLAB Command
>>simin=[0 2;1 3;2 7;3 9]; ↵

The reason we chose the variable name as simin is simulink has the same variable name for the functional data import. Let us bring a From Workspace (link: *'Simulink → Sources → From Workspace'*) and a Scope blocks in a new simulink model file and connect them as presented in the figure 2.12(d). Our x changes from 0 to 3 hence we set the solver start and stop timings as 0 and 3 respectively. On running he model, the Scope returns you the output like the figure 2.12(e) with the autoscale setting which is identical with the earlier plot of the figure 2.12(b). Even though the horizontal axis is exhibiting time, that can be assumed as the x axis without the loss of generosity. So to note, we passed our $f(x)$ versus x data to simulink. Now for the discrete analysis we may not need the interpolated curve instead we prefer the curve of the figure 2.12(c). To implement so, insert a Zero-Order Hold block (link: *'Simulink → Discrete → Zero-Order Hold'*) in the model of the figure 2.12(d) so that the modified model outlook becomes the figure 2.12(f). One would obtain the Scope output of the figure 2.12(g) with the autoscale setting, which is verified by the figure 2.12(c).

Figure 2.12(d) *Model for importing the $f(x)$ versus x data in simulink*

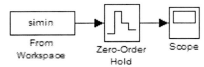

Figure 2.12(f) *Model for keeping constancy between the intervals of the function*

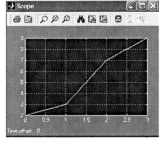

Figure 2.12(e) *Scope output for the model of the figure 2.12(d)*

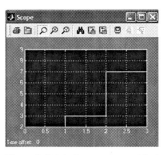

Figure 2.12(g) *Scope output with the Zero-Order Hold block*

❖ ❖ Example 3

In the example 2, we have only one functional value. Let us say now we hold multiple functional values like $f_1(x) = \begin{bmatrix} 2 \\ 3 \\ 7 \\ 9 \end{bmatrix}$, $f_2(x) = \begin{bmatrix} 3 \\ 5 \\ 6 \\ 0 \end{bmatrix}$,

and $f_3(x) = \begin{bmatrix} 8 \\ 5 \\ 1 \\ 4 \end{bmatrix}$ for the same range of x. The procedure is very similar.

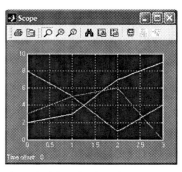

Figure 2.12(h) *The three functional data on a common horizontal axis*

We form a rectangular matrix in which the x values occupy the first column and the $f_1(x)$, $f_2(x)$, and $f_3(x)$ values occupy the second, third, and fourth columns respectively

wherefrom assignee simin should have $\begin{bmatrix} 0 & 2 & 3 & 8 \\ 1 & 3 & 5 & 5 \\ 2 & 7 & 6 & 1 \\ 3 & 9 & 0 & 4 \end{bmatrix}$. Let us enter that in MATLAB as follows:

MATLAB Command
>>simin=[0 2 3 8;1 3 5 5;2 7 6 1;3 9 0 4]; ↵

After entering the matrix, the reader can run the model in 2.12(d) that exhibits the output of the figure 2.12(h) with the autoscale setting.

♦ ♦ **Example 4**

Not necessarily the x data should be uniformly separated. Their spacing can be non-uniform as well. For

example, the simin formed from the data $x = \begin{bmatrix} 0 \\ 0.1 \\ 0.2 \\ 3 \end{bmatrix}$ and $f(x) = \begin{bmatrix} 2 \\ 3 \\ 7 \\ 9 \end{bmatrix}$ would keep the model of the figure 2.12(d)

operational too.

The solver setting is also adaptive regardless of the input data spacing. Variable step option in the solver considers taking the non uniform steps. If the output is not satisfactory, the variable step can be changed to fixed step with smaller step size within the given range. Whether it is variable or fixed step option, simulink still employs interpolation until you set it from your input and solver mode. Once our functional data is imported in simulink in the form of functional representation that can be manipulated for analysis. However, we bring an end to the section with the implementation of this example.

2.6 Block links used in this chapter

Our main approach in this chapter is to attain the mathematical functions simulated in simulink both from the expression and data oriented cases for which we employed the blocks of the table 2.C.

Table 2.C   Necessary blocks for the functional modeling as found in simulink library
(not in the alphabetical order)

Block name	Representative Symbol/Function	Icon Outlook	Block name	Representative Symbol/Function	Icon Outlook
Step	Generates unit step function, $u(t-1)$		Scope	It displays the function of the connected line	
Link: *Simulink → Sources → Step*			Link: *Simulink → Sinks → Scope*		
Ramp	It generates straight line functions of various slopes and intercepts		Product	It multiplies two functions on the common time or horizontal scale	
Link: *Simulink → Sources → Ramp*			Link: *Simulink → Math Operations → Product*		
Sine Wave	It generates sine waves of various frequencies, amplitudes, and phases		Abs	It takes the absolute value of the signal or functions	
Link: *Simulink → Sources → Sine Wave*			Link: *Simulink → Math Operations → Abs*		
Saturation	It crops a curve on user defined specific upper and lower limits		Fcn	The user can define any function being operated on its input considering the independent variable u	
Link: *Simulink → Discontinuities → Saturation*			Link: *Simulink → User-Defined Functions → Fcn*		
Sum	It sums two or more functional values, constants, or combinations of them		Pulse Generator	It generates pulses of swing from 0 to some amplitude of different frequencies and phases	
Link: *Simulink → Math Operations → Sum*			Link: *Simulink → Sources → Pulse Generator*		
Constant	It generates constant value (s) or constant built-in functions		Signal Builder	It generates signals of standard form or gives the provision for user defined signal	
Link: *Simulink → Sources → Constant*			Link: *Simulink → Sources → Signal Builder*		
Transport Delay	It shifts a function to the right of the horizontal axis at any user defined point		Gain	It multiples the input function by a specified constant	
Link: *Simulink → Continuous → Transport Delay*			Link: *Simulink → Math Operations → Gain*		

Continuation of the previous table:

Block name	Representative Symbol/Function	Icon Outlook	Block name	Representative Symbol/Function	Icon Outlook
Integrator	It integrates the function entering to its input port	$\frac{1}{s}$ Integrator	MATLAB Fcn	It executes any built-in MATLAB or user defined MATLAB coded function	MATLAB Function MATLAB Fcn
Link: *Simulink → Continuous → Integrator*			Link: *Simulink → User-Defined Functions → MATLAB Fcn*		
Clock	It generates solver times or independent variable of the functions	Clock	Trigonometric Function	It performs most trigonometric or hyperbolic functions including their inverses	sin Trigonometric Function
Link: *Simulink → Sources → Clock*			Link: *Simulink → Sources → Trigonometric Function*		
Complex to Real-Imag	It separates the complex functional values into real and imaginary parts	Re{u} Im{u} Complex to Real-Imag	Complex to Magnitude-Angle	It separates the complex functional values into the magnitude and phase angle parts	|u| ∠u Complex to Magnitude-Angle
Link: *Simulink → Math Operations → Complex to Real-Imag*			Link: *Simulink → Math Operations → Complex to Magnitude-Angle*		
Complex Exponential	It takes θ as the input and returns $\cos\theta + j\,\sin\theta$	exp(jx) Complex Exponential	Angle Conversion	It converts the angle from radian to degrees or vice versa	deg rad Angle Conversion
Link: *DSP Blockset → Math Functions → Math Operations → Complex Exponential*			Link: *Aerospace Blockset → Transformations → Units → Angle Conversion*		
From Workspace	It imports functional data from the workspace of MATLAB	simin From Workspace	Display	It displays the instantaneous functional value of the concern functional line	0 Display
Link: *Simulink → Sources → From Workspace*			Link: *Simulink → Sinks → Display*		
Zero-Order Hold	It retains the functional values as constant in the successive interval of the samples without interpolation	Zero-Order Hold			
Link: *Simulink → Discrete → Zero-Order Hold*					

Modeling Ordinary Differential Equations

The chapter highlights the implementation of simulink models comprising the ordinary differential equations largely to the context of pure mathematics. We deposited the common problems of the differential equations to grasp the concept of the modeling and simulation and pay unified attention. Solving or simulating the differential equations is often practiced in many branches of engineering and sciences. Presence of analytical solutions for some examples assists us gaining the consistency and trustworthiness on simulink. We devoted to the modeling and simulating of various order differential equations throughout the chapter. Differential equations are the equations that comprise from the derivatives of some unknown quantity, parameter, or variable with respect to some other parameter, variable, time, or space. The whole purpose of simulating such equation is to examine the response of a system under different circumstances. Very often, the differential equations come into view in the mathematical or computer models that endeavor to describe the real-life problems or govern the system dynamics with regards to the continuous time systems. The analytical solution is pedagogically preferable yet a large number of continuous time systems may not reserve the analytical solution and they demand the numerical solution. Simulink can be an appropriate tool for solving the differential equations numerically which models the continuous time systems of various order.

3.1 Ordinary differential equations and the integrator operator

To forecast the behavior of a physical system, we often construct the mathematical models comprising the differential equations. The derivatives measure the rates of change of a system's parameter with respect to some other parameter such as time or space. The mathematical models or equations involving the derivatives are called the differential equations. Presence of one dependent and one independent variables makes the differential equation as the ordinary differential equation. For instance, $\frac{dy}{dx} + y = 9$ and $\frac{d^4 y}{dx^4} + y = 9$ are the ordinary differential equations, where y is the dependent and x is the independent variables. The x can represent time, space, or other parameter. The order of a differential equation is the highest order of derivatives present in the equation. Thus, a first order ordinary differential equation has the form $f\left(x, y, \frac{dy}{dx}\right) = 0$, a second order ordinary differential equation has the form

$f\left(x, y, \dfrac{dy}{dx}, \dfrac{d^2y}{dx^2}\right) = 0$…so forth. A differential equation can also be catalogued as regards with and without the input or excitation, for example,

$\dfrac{dy}{dt} + 2y = 0$ input function is zero, the independent variable is t

$\dfrac{d^2y}{dt^2} + 2\dfrac{dy}{dt} + 8y = \cos t + \sin t$ input function is $\cos t + \sin t$, the independent variable is t

$u\dfrac{d^3v}{du^3} + \dfrac{dy}{du} - 34y\cos u = 1 + u^2 + \log u$ input function is $1 + u^2 + \log u$, the independent variable is u

The solution of a differential equation is a function that satisfies the differential equation over some domain of the independent variable. Method of finding the solution of a differential equation can be analytical or numerical. The analytical solution has two components – complementary function and particular integral. To acquire more experience and mastery of techniques, we modeled and simulated various differential equations in the following sections.

To find the solution of a differential equation in simulink, we always employ the integrator operator not the differential one. There are some reasons why the differential operator is not hired to find the solution of a differential equation: the input-output functional flow can not be maintained by the differential operator, the initial conditions of the dependent variable is lost by the differential operator, and a sudden change of input function or step input to the operator gives rise spike like output. However, to find the integrator operator in simulink, the clicking sequence is *'Simulink → Continuous → Integrator'*. When you bring the block in your model file, it appears as follows:

$$\boxed{\dfrac{1}{s}}$$
Integrator

The function $\dfrac{1}{s}$ in Laplace domain is equivalent to the time domain integration i.e. $L\left[\displaystyle\int_{\tau=0}^{\tau=t} y\,d\tau\right] = \dfrac{Y(s)}{s}$. Our main

objective of simulating the differential equations is to find the wave shape of the dependent y. If we have $\dfrac{dy}{dt}$, y

can be obtained as $y = \displaystyle\int \dfrac{dy}{dt}\,dt$ and whose simulink operation can be presented as follows:

$$\dfrac{dy}{dt} \rightarrow \boxed{\dfrac{1}{s}} \rightarrow y$$
Integrator

y from the second derivative $\dfrac{d^2y}{dt^2}$ is obtained as $y = \displaystyle\int\int \dfrac{d^2y}{dt^2}\,dt\,dt$ and the simulink operation is as follows:

$$\dfrac{d^2y}{dt^2} \rightarrow \boxed{\dfrac{1}{s}} \xrightarrow{\;\;\downarrow\dfrac{dy}{dt}\;\;} \boxed{\dfrac{1}{s}} \rightarrow y$$
Integrator1   Integrator2

Similarly, the recovery of y from $\dfrac{d^3y}{dt^3}$ i.e. $y = \displaystyle\int\int\int \dfrac{d^3y}{dt^3}\,dt\,dt\,dt$ can be cited as follows:

$$\dfrac{d^3y}{dt^3} \rightarrow \boxed{\dfrac{1}{s}} \xrightarrow{\;\;\downarrow\dfrac{d^2y}{dt^2}\;\;} \boxed{\dfrac{1}{s}} \xrightarrow{\;\;\downarrow\dfrac{dy}{dt}\;\;} \boxed{\dfrac{1}{s}} \rightarrow y$$
Integrator1   Integrator2   Integrator3

The differential equation is an algebraic equation as well. When various blocks are connected to simulate a differential equation, the equality of the equation and appropriate input-output functional flow of the blocks employed must have to be maintained. Laplace transform of integration says that the initial conditions are always necessary and they start at $t = 0$. Since the concept of past and present is relative, one can onset the initial conditions to $t = 0$ without the loss of generosity.

3.2 Modeling a first order differential equation

The highest derivative in a first order differential equation has the order one, for example, $\frac{dy}{dt} + 2y = 0$, $t\frac{dy}{dt} + 2\ln t = \sin t$, $\frac{dy}{dt} + 2y = e^{-2t}$, ... etc. To solve this kind of differential equation, one needs to know one initial condition preferably the value of the dependent variable at the zero value of the independent variable.

Let us start with the example of the homogeneous equation $\frac{dy}{dt} + 2y = 0$ with the initial condition $y = 10$ at $t = 0$.

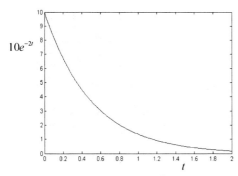

Figure 3.1(a) *Plot of $10e^{-2t}$ versus t over $0 \le t \le 2$*

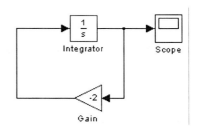

Figure 3.1(b) *Simulink model for solving the differential equation $\frac{dy}{dt} + 2y = 0$*

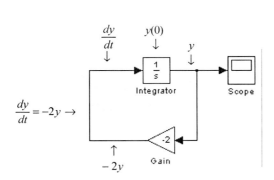

Figure 3.1(c) *Simulink model of figure 3.1(b) mentioning the functional flow*

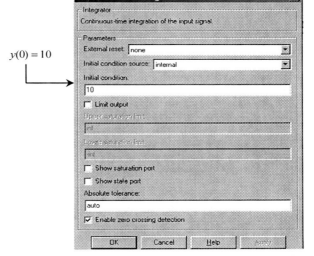

Figure 3.1(d) *Integrator block parameters of simulink*

To solve the differential equation analytically, we find the auxiliary equation by assuming $y = e^{mt}$ which provides $m + 2 = 0$ and the general solution is $y = Ae^{-2t}$. Application of the initial condition gives us $y = 10e^{-2t}$, plot of which is shown in figure 3.1(a) for $0 \le t \le 2$. In a nutshell, we can summarize that we have

$$\underbrace{\frac{dy}{dt} + 2y = 0 \ with \ y = 10 \ at \ t = 0}_{problem \ space} \ and \ \underbrace{y = 10e^{-2t} \ plot \ like \ figure \ 3.1(a) \ for \ 0 \le t \le 2}_{solution \ space}.$$

Our objective is to achieve the solution like the figure 3.1(a) by simulink. We rearrange the differential equation $\frac{dy}{dt} + 2y = 0$ so that the highest order derivative is on the left side of the equation, i.e., $\frac{dy}{dt} = -2y$ and the simulink model must hold the equality between $\frac{dy}{dt}$ and $-2y$. To obtain y from $\frac{dy}{dt}$, one integrator is needed where the output of the integrator is y. On the right side of the equation what we have is $-2y$. Multiplying the y by -2 provides $-2y$, so, a gain block of gain -2 is required. To mention the simulink procedure,

⇒ *open a new simulink model file, save the file in your working path, and click the library browser icon*

⇒ *bring an integrator block in the model file following the link 'Simulink → Continuous → Integrator' and doubleclick it to insert the initial condition $y(0) = 10$ as shown in figure 3.1(d)*

⇒ *bring a gain block in the model file following the link 'Simulink → Math Operations → Gain'*

$t = 0$ $\qquad t = 2$

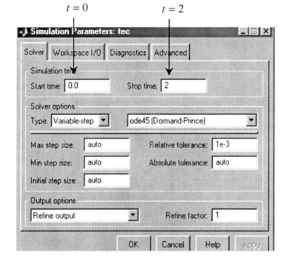 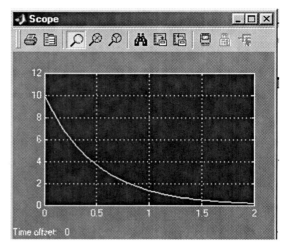

Figure 3.1(e) *Solver settings for $\frac{dy}{dt} + 2y = 0$* Figure 3.1(f) *Output of simulink scope for $\frac{dy}{dt} + 2y = 0$*

⇒ *doubleclick the gain block to change its gain to –2 and flip the block (select → click the right button of the mouse → format → flip block)*

⇒ *bring a scope block in the model file following the link 'Simulink → Sinks → Scope'*

⇒ *place the three blocks relatively and connect them according to the figure 3.1(b)*

⇒ *click 'Simulation → Simulation parameters → Solver' from the menu bar in your model file and enter start time as 0 and stop time as 2 to insert $0 \leq t \leq 2$ in the solver parameter window as shown in the figure 3.1(e)*

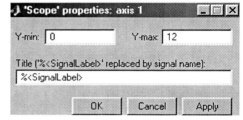

Figure 3.1(g) *Scope y axis properties*

⇒ *click the start simulation icon*

⇒ *doubleclick the Scope block, bring the mouse pointer in the plot area of the scope output, click 'the right button of the mouse → Axes properties' to see the Y axes properties like figure 3.1(g), and enter there 0 and 12 as Y min and Y max axes values respectively (to be consistent with the figure 3.1(a))*

⇒ *these all steps finally result the scope output of the figure 3.1(f) which is exactly the same as the figure 3.1(a)*

How different functions are flowing in the simulink model are shown in the figure 3.1(c). *We advise you to keep the other settings of the dialog windows of figures 3.1(d), 3.1(e), and 3.1(g) leave as they are.* Sometimes a dialog box may contain many block parameters. We address only the essential parameters to solve the particular problem keeping the other block parameters unchanged. A number of examples for modeling the first order differential equations are illustrated in the following.

♦ ♦ Example 1

The differential equation example we started with does not hold any input function. Now we address a differential equation containing polynomial input function, for example, $\frac{dy}{dt} + 7y = 1 + t - t^2$. Let us simulate the equation for $0 \leq t \leq 5$ with the initial condition $y(0) = -4$.

⎙ Analytical solution

The trial solution, auxiliary equation, root of the auxiliary equation, the complementary function, the particular integral, and the complete solution of the non homogeneous equation are $y = e^{mt}$, $m + 7 = 0$, $m = -7$,

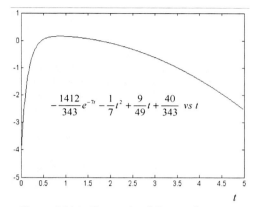

$$-\frac{1412}{343}e^{-7t}-\frac{1}{7}t^2+\frac{9}{49}t+\frac{40}{343}\ \ vs\ t$$

Figure 3.2(a) *First order differential equation with the polynomial excitation*

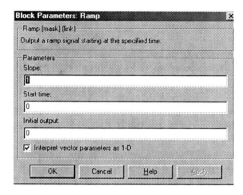

Figure 3.2(b) *Ramp block parameters*

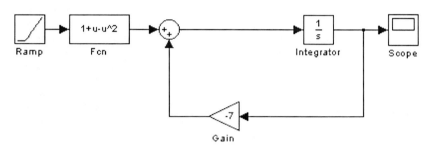

Figure 3.2(c) *Simulink model for solving the differential equation* $\dfrac{dy}{dt}+7y=1+t-t^2$

Equality $\dfrac{dy}{dt}=-7y+1+t-t^2$ *is here*

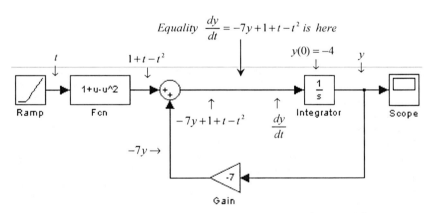

Figure 3.2(d) *Simulink model of figure 3.2(c) mentioning the functional flow*

$y_{CF}=Ae^{-7t}$, $y_{PI}=\dfrac{1+t-t^2}{D+7}$ (where D is the operator $\dfrac{d}{dt}$) $=-\dfrac{1}{7}t^2+\dfrac{9}{49}t+\dfrac{40}{343}$, and $y=Ae^{-7t}-\dfrac{1}{7}t^2+\dfrac{9}{49}t+\dfrac{40}{343}$

respectively. Insertion of the initial condition $y(0)=-4$ yields $y=-\dfrac{1412}{343}e^{-7t}-\dfrac{1}{7}t^2+\dfrac{9}{49}t+\dfrac{40}{343}$ which when graphed over $0\le t\le 5$ results in the figure 3.2(a). Based on the analytical computation, the inference can be cited as

$$\overbrace{\dfrac{dy}{dt}+7y=1+t-t^2\ \ with\ \ y(0)=-4}^{\textit{problem space}}\ \ \text{and}$$

37

$$\underbrace{y = -\frac{1412}{343}e^{-7t} - \frac{1}{7}t^2 + \frac{9}{49}t + \frac{40}{343}}_{solution\ space}\ plot\ like\ figure\ 3.2(a)\ for\ 0 \le t \le 5\ .$$

⑤ Simulink solution

Rearranging the differential equation, one obtains $\frac{dy}{dt} = -7y + 1 + t - t^2$ whose right side can be thought as the summation of $-7y$ and $1 + t - t^2$ (can be added by a two input sum block). The polynomial $1 + t - t^2$ can be generated from the ramp function (which simulates t) by passing through a user defined block (which contains the functional code of the polynomial $1 + t - t^2$). The design of the model will be such that the equality between $\frac{dy}{dt}$ and $-7y + 1 + t - t^2$ is maintained. The simulink model is depicted in figure 3.2(c). For convenience of understanding, the model with the functional flow is also presented in figure 3.2(d). The simulink procedure we apply is

⇒ *open a new simulink model file and click the library browser icon*

⇒ *bring an Integrator block in the model file and doubleclick it to insert the initial condition $y(0) = -4$*

⇒ *bring one Gain block in the model file and doubleclick the Gain block to change its gain to -7 and flip the block (right button of the mouse → format →flip block)*

⇒ *bring one Scope and one Sum blocks in the model file*

⇒ *bring a Fcn (simulates function) block in the model file following the link 'Simulink → User-Defined Functions → Fcn' and doubleclick the block to change the parameters expression to $1+u-u^2$ in the parameter window which is the code for $1 + t - t^2$ considering the independent variable as u (see figure 3.2(e) for Fcn block window)*

⇒ *bring a Ramp block in the model file following the link 'Simulink → Sources → Ramp' whose parameter window appears as in the figure 3.2(b) on doubleclicking*

⇒ *place various blocks relatively in the model file according to the figure 3.2(c) and connect them*

⇒ *click the 'Simulation → Simulation Parameters → Solver' in the model menu bar and set the start and stop time of the solver as 0 and 5 respectively for the insertion of $0 \le t \le 5$*

⇒ *click the Start simulation icon, doubleclick the Scope, bring the mouse pointer in the plot area of the Scope, press 'the right button of the mouse → Axes properties' to set the Ymin and Ymax to -5 and 1 respectively*

By performing all these steps, the Scope displays the output as shown in the figure 3.2(f) which is identical with the foreseen figure 3.2(a).

Figure 3.2(e) *Fcn block expression*

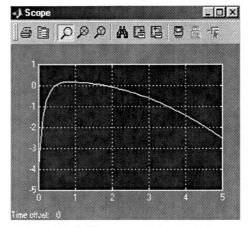

Figure 3.2(f) *The output of the simulink Scope for* $\frac{dy}{dt} + 7y = 1 + t - t^2$

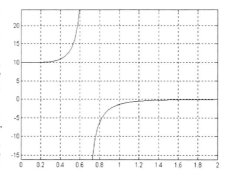

Figure 3.3(a) *Plot of* $-\frac{10}{8x^5 - 1}$ *versus* x *over* $0 \le x \le 2$

❖ ❖ Example 2

Find the solution for $\frac{dy}{dx} = 4y^2x^4$ with the initial condition $y(0) = 10$ over $0 \le x \le 2$.

38

☞ Analytical solution

Trial solution does not apply in the equation instead one employs the separation of variable method from which $y = -\dfrac{5}{4x^5 + A}$. The arbitrary constant is removed by the initial condition $y(0) = 10$ and we obtain $y = -\dfrac{10}{8x^5 - 1}$, plot of which is traced in figure 3.3(a) for $0 \le x \le 2$.

☞ Simulink solution

We assume the independent variable x as t and obtain the dependent y by passing $\dfrac{dy}{dx}$ through the integrator operator. The y so found can be passed through a square function block (to perform y^2) and later we can multiply that by a Gain block of gain 4. In a similar way, we can generate x^4 and then multiply $4y^2$ and x^4 by a Product block as far as the equality of $\dfrac{dy}{dx}$ and $4y^2x^4$ is concern. However, the simulink model and its functional flow description are presented in figures 3.3(b) and 3.3(c) respectively.

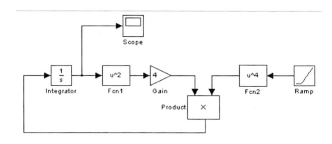

Figure 3.3(b) *Simulink model for solving the differential equation* $\dfrac{dy}{dx} = 4y^2x^4$

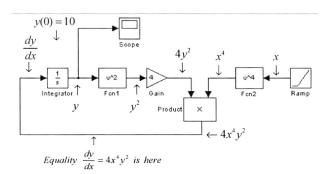

Figure 3.3(c) *Simulink model of figure 3.3(b) mentioning the functional flow*

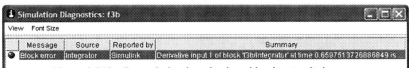

Figure 3.3(d) *Error dialog box displayed by the simulink*

Open a new simulink model file. The preceding examples mentioned the links for bringing the Integrator, Gain, Ramp, Scope, and Fcn blocks in your working model file. Now two Fcn blocks are necessary. Bring one Fcn block, rename it as Fcn1, set the functional expression as u^2 for y^2, then bring the other Fcn block, rename it as Fcn2, and set its expression as u^4 for x^4. Apart from these blocks, you need one two-input Product block which can be reached through the link *'Simulink → Math Operations → Product'*. Bring all necessary blocks, place them relatively, and connect them in your simulink model file as presented in the figure 3.3(b). The Ramp and Fcn2 blocks need to be flipped and the Product one needs to be rotated *(select → right button of the mouse → format → rotate block)* by 90^0 clockwise. Doubleclick the Integrator and Gain blocks to set the initial condition $y(0) = 10$ and the gain 4 respectively. Click *'the 'Simulation → Simulation parameter → Solver'* in the menu bar to insert the start (0) and stop time (2) for the simulation as we did before. Now click the start simulation icon. An error dialog box appears as shown in figure 3.3(d). Let us investigate the analytical solution $y = -\dfrac{10}{8x^5 - 1}$, the function becomes undefined

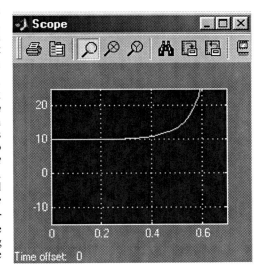

Figure 3.3(e) *The output of the simulink Scope for* $\dfrac{dy}{dx} = 4y^2x^4$ *over* $0 \le x \le 0.6$

at $8x^5 - 1 = 0$ or $x = \dfrac{1}{\sqrt[5]{8}} = 0.6598$. The asymptotic behavior of the function is also displayed in the figure 3.3(a). The simulation fails at $x = 0.6598$ because computer can not numerically deal with the infinity. Let us change the start and stop time via *'Simulation $\rightarrow$ Simulation parameter $\rightarrow$ Solver'* to 0 and 0.6 respectively and click the start simulation icon. Simulink responded with the Scope output (with the Ymin and Ymax settings as −15 and 25 respectively) as shown in figure 3.3(e). The reader can compare that the output of the simulink and the analytical plot presented in figure 3.3(a) are identical within the interval $0 \le x \le 0.6$. The remedy for handling the asymptotic circumstances is provoided in section 3.8.

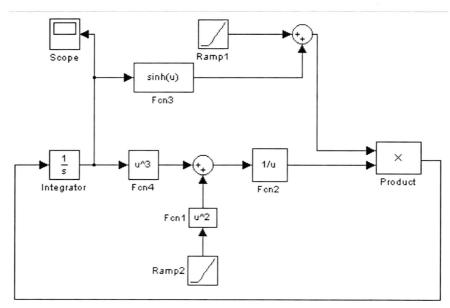

Figure 3.4(a) *Simulink model for solving the differential equation* $(x^2 + y^3)\dfrac{dy}{dx} = x + \sinh y$

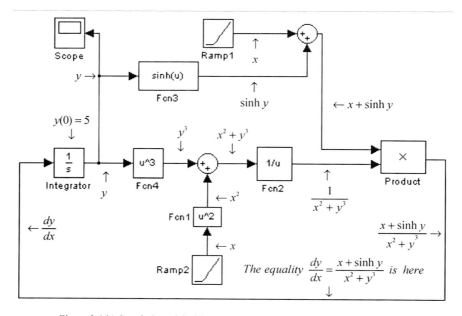

Figure 3.4(b) *Simulink model of figure 3.4(a) mentioning the functional flow*

40

❖ ❖ **Example 3**

Some differential equations may not have close form analytical solution but to some extent one can obtain the numerical or simulink solution. For the numerical solution of the differential equation in MATLAB, reference [2] can be considered. One example can be the simulation of the differential equation $(x^2 + y^3)\frac{dy}{dx} = x + \sinh y$ with the initial condition $y(0) = 5$ over $0 \le x \le 3$.

🖉 *Simulink solution*

Figures 3.4(a) and 3.4(b) delineate the simulink model and the model with the functional flow respectively. We briefly mention some detail of the model building process:

⇒ *the rearranged equation is* $\frac{dy}{dx} = \frac{x + \sinh y}{x^2 + y^3}$

⇒ *four function blocks are necessary: Fcn1, Fcn2, Fcn3, and Fcn4 for* x^2, $\frac{1}{x^2 + y^3}$, $\sinh y$, *and* y^3 *respectively*

⇒ *two Ramp function blocks are necessary: Ramp1 and Ramp2. The Ramp2 is rotated by 90 degrees counter clockwise. Both Ramp settings must be according to the figure 3.2(b)*

⇒ *the Scope needs to be flipped because of the relative position in the model*

⇒ *inject the initial condition* $y(0) = 5$ *into the Integrator*

⇒ *set the start and stop time as 0 and 3 respectively for* $0 \le x \le 3$

Successful building of the model and the simulation should return the Scope output as shown in figure 3.4(c) (with Ymin and Ymax setting as 5 and 10 respectively).

3.3 Modeling a second order differential equation

A second order ordinary differential equation possesses the highest order derivative as two. Two integrator blocks and two initial conditions are necessary to simulate the second order ordinary differential equation. Let us choose the equation $4\frac{d^2y}{dt^2} + 12\frac{dy}{dt} + 8y = 0$ with the initial conditions $y(0) = -5$ and $y'(0) = 3$ for the simulation and which has to be plotted for $0 \le t \le 9$.

🖉 *Analytical solution*

Employing the trial solution $y = e^{mt}$ provides the roots of the auxiliary equation as $m = -2$ and -1, and where from the general equation is $y = Ae^{-2t} + Be^{-t}$. We apply the initial conditions $y(0) = -5$ and $y'(0) = 3$

to have $\begin{Bmatrix} A + B = -5 \\ -2A - B = 3 \end{Bmatrix}$ whose solution is $\begin{Bmatrix} A = 2 \\ B = -7 \end{Bmatrix}$

thereby providing the final solution as $y = 2e^{-2t} - 7e^{-t}$.

The plot of the function is depicted in figure 3.5(a) so our requirement is

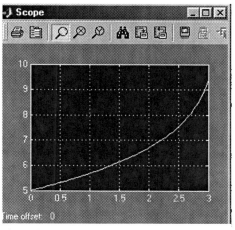

Figure 3.4(c) *Output of the simulink Scope for* $(x^2 + y^3)\frac{dy}{dx} = x + \sinh y$

$2e^{-2t} - 7e^{-t}$
↑

→ t

Figure 3.5(a) *Plot of* $y = 2e^{-2t} - 7e^{-t}$ *versus* t *for* $0 \le t \le 9$

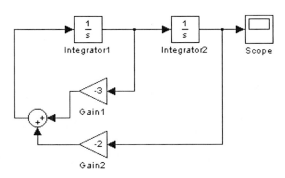

Figure 3.5(b) *Simulink model for solving the differential equation* $4\frac{d^2y}{dt^2} + 12\frac{dy}{dt} + 8y = 0$

41

problem space

and

solution space

$$4\frac{d^2y}{dt^2}+12\frac{dy}{dt}+8y=0 \; with \quad y'(0)=3 \quad and \quad y(0)=-5$$

$$y=2e^{-2t}-7e^{-t} \quad plot \; like \; figure \; 3.5(a)$$

.

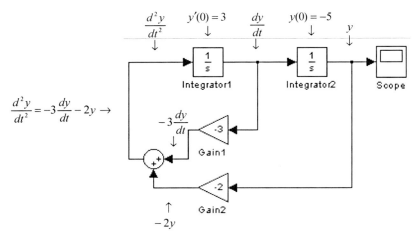

$$\frac{d^2y}{dt^2}=-3\frac{dy}{dt}-2y \rightarrow$$

Figure 3.5(c) *Simulink model of figure 3.5(b) mentioning the functional flow*

⎙ *Simulink solution*

We rearrange the equation so that the highest order derivative is on the left side where from $\frac{d^2y}{dt^2}=-3\frac{dy}{dt}-2y$. Two Integrator blocks are required to obtain y from $\frac{d^2y}{dt^2}$. The gain associated with $\frac{dy}{dt}$ and y are -3 and -2 respectively. A two input adder can sum the $-3\frac{dy}{dt}$ and $-2y$. The simulink model is presented in figure 3.5(b) followed by the functional flow description in figure 3.5(c). We described the links for bringing various blocks displayed in the figure 3.5(b) in previous examples. The procedural sequence for the simulation is

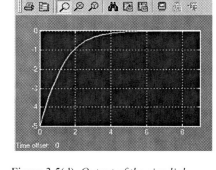

Figure 3.5(d) *Output of the simulink Scope for* $4\frac{d^2y}{dt^2}+12\frac{dy}{dt}+8y=0$

> ⇒ *open a new simulink model file*
> ⇒ *bring one Integrator block, rename it as Integrator1, set the initial condition $y'(0)=3$ in Integrator1, then bring the other Integrator block, rename it as Integrator2, and set the initial condition $y(0)=-5$ in it*
> ⇒ *bring one Gain block, rename it as Gain1, set its gain to -3, then bring the other Gain block, rename it as Gain2, set its gain to -2, and flip both of them*
> ⇒ *bring one Sum block and flip it, and then bring one Scope block*
> ⇒ *place various blocks relatively in the model file according to the figure 3.5(b) and connect them*
> ⇒ *set the start and stop time as 0 and 9 respectively for the insertion of $0 \le t \le 9$*
> ⇒ *click the Start simulation icon, the Scope displays the output as shown in figure 3.5(d) (with the Ymin and Ymax settings as -5 and 0 respectively)*

One can compare that the plots in the figures 3.5(a) and 3.5(d) are identical as it is expected. We present three more examples on simulating the second order differential equations in the following.

❖ ❖ Example 1

Let us choose a second order differential equation which has the input function, for example, $3\frac{d^2y}{dt^2}+30\frac{dy}{dt}+63y=-8+3t-e^{-5t}$ with the initial conditions $y(0)=-2$ and $y'(0)=-900$ over $0 \le t \le 2$.

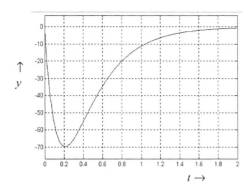

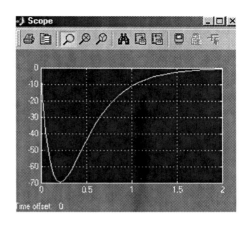

Figure 3.6(a) *Plot of y =*

$$\frac{266197}{1176}e^{-7t} - \frac{5479}{24}e^{-3t} - \frac{22}{147} + \frac{1}{21}t + \frac{1}{12}e^{-5t}$$

vs t over $0 \le t \le 2$

Figure 3.6(b) *Output of the simulink scope for*

$$3\frac{d^2y}{dt^2} + 30\frac{dy}{dt} + 63y = -8 + 3t - e^{-5t}$$

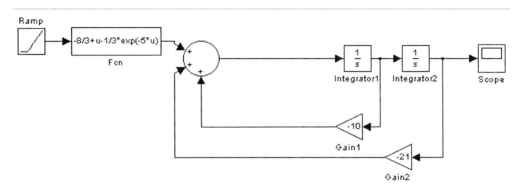

Figure 3.6(c) *Simulink model for solving the differential equation $3\frac{d^2y}{dt^2} + 30\frac{dy}{dt} + 63y = -8 + 3t - e^{-5t}$*

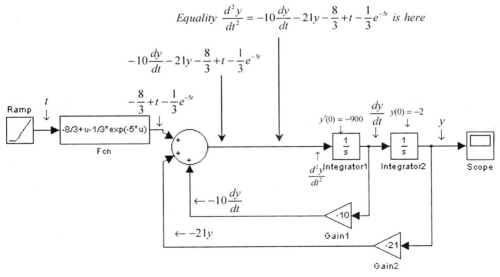

Figure 3.6(d) *Simulink model of figure 3.6(c) mentioning the functional flow*

43

☞ Analytical solution

The roots of the auxiliary equation, the complementary function, the particular integral, the general equation, and the complete solution considering the initial conditions are -7 and -3, $y_{CF} = Ae^{-7t} + Be^{-3t}$,

$y_{PI} = -\dfrac{22}{147} + \dfrac{1}{21}t + \dfrac{1}{12}e^{-5t}$, $y = Ae^{-7t} + Be^{-3t} - \dfrac{22}{147} + \dfrac{1}{21}t + \dfrac{1}{12}e^{-5t}$, and $y = \dfrac{266197}{1176}e^{-7t} - \dfrac{5479}{24}e^{-3t} - \dfrac{22}{147} + \dfrac{1}{21}t + \dfrac{1}{12}e^{-5t}$

respectively. The plot of the function takes the form as shown in figure 3.6(a). Our issue statement is

$\overbrace{}^{\text{problem space}}$ and $\overbrace{}^{\text{solution space}}$.

$3\dfrac{d^2y}{dt^2} + 30\dfrac{dy}{dt} + 63y = -8 + 3t - e^{-5t}$ with $y'(0) = -900$ and $y(0) = -2$ and y vs t plot like figure 3.6(a)

☞ Simulink solution

In order to construct the simulink model for the differential equation as depicted in figure 3.6(c), the important elements to be considered are as follows:

⇒ *on rearranging, the equation becomes* $\dfrac{d^2y}{dt^2} = -10\dfrac{dy}{dt} - 21y - \dfrac{8}{3} + t - \dfrac{1}{3}e^{-5t}$

⇒ *two Integrator blocks are required to obtain* y *from* $\dfrac{d^2y}{dt^2}$ *and the gains for* $\dfrac{dy}{dt}$ *and* y *are* -10 *and* -21 *respectively*

⇒ *the input functional block Fcn should contain the code for* $-\dfrac{8}{3} + t - \dfrac{1}{3}e^{-5t}$ *which is -8/3+u-exp(-5*u)/3 considering the independent variable as u and the block needs to be enlarged horizontally otherwise the functional expression may not be displayed*

⇒ *a three input Sum block is necessary to add* $-10\dfrac{dy}{dt}$, $-21y$, *and* $-\dfrac{8}{3} + t - \dfrac{1}{3}e^{-5t}$

⇒ *open a new simulink model file, bring all associated blocks, rename or flip the blocks if it is necessary, place various blocks relatively, and connect them as indicated in the figure 3.6(c)*

⇒ *for the three input Sum block, first bring the default two-input Sum block, doubleclick the block, type one more + beside ++ in 'list of signs' in the parameter window of the block, and then click OK*

⇒ *the three input Sum block so brought can now be enlarged for convenience*

⇒ *however, set the start and stop time as 0 and 2 respectively for the insertion of* $0 \le t \le 2$

⇒ *the two initial conditions* $y'(0) = -900$ *and* $y(0) = -2$ *are set to the Integrator1 and Integrator2 respectively*

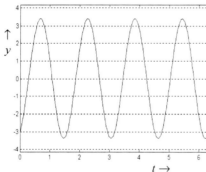

Figure 3.7(a) *Plot of* $y = $

$\dfrac{1}{63}u(t) - \dfrac{1}{63}u(t)\cos\dfrac{3\sqrt{7}}{2}t + \dfrac{4}{\sqrt{7}}\sin\dfrac{3\sqrt{7}}{2}t - 3\cos\dfrac{3\sqrt{7}}{2}t$

versus t *over* $0 \le t \le 2\pi$

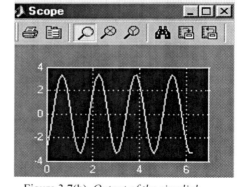

Figure 3.7(b) *Output of the simulink Scope for* $4\dfrac{d^2y}{dt^2} + 63y = u(t)$

We also presented the functional flow diagram for the equation in figure 3.6(d). The outturn of the simulink scope should be like the figure 3.6(b) (with the autoscale setting for the Y axes) on successful construction and simulation of the model, which is indistinguishable with the analytical plot of the figure 3.6(a).

❖ ❖ Example 2

A differential equation may not contain all successive derivatives, some derivative term might be missing, for example, $4\dfrac{d^2y}{dt^2} + 63y = u(t)$. Still one needs two initial conditions (let us say, $y(0) = -3$ and $y'(0) = 6$) as far as the numerical approach of the simulink is concern. By the way, the function $u(t)$ on the right side of the equation is the unit step function whose step starts at $t = 0$. The equation does not hold the derivative $\frac{dy}{dt}$ that means we do not have to employ the gain block associated with the $\frac{dy}{dt}$. However, the analytical and simulink solutions of the equation over $0 \le t \le 2\pi$ are presented in the following.

⌗ Analytical solution

By employing the trial solution approach, one obtains in due course the complete solution of the equation as $y = \frac{1}{63}u(t) - \frac{1}{63}u(t)\cos\frac{3\sqrt{7}}{2}t + \frac{4}{\sqrt{7}}\sin\frac{3\sqrt{7}}{2}t - 3\cos\frac{3\sqrt{7}}{2}t$ whose plot is presented in figure 3.7(a). Our objective

can be revealed as $\overbrace{4\frac{d^2y}{dt^2} + 63y = u(t) \text{ with } y'(0) = 6 \text{ and } y(0) = -3}^{problem\ space}$ and $\overbrace{y \text{ vs } t \text{ plot like figure } 3.7(a)}^{solution\ space}$.

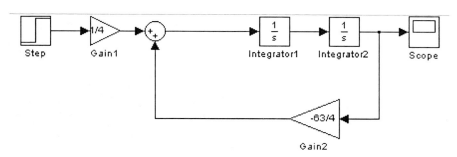

Figure 3.7(c) *Simulink model for solving the differential equation* $4\frac{d^2y}{dt^2} + 63y = u(t)$

Equality $\frac{d^2y}{dt^2} = -\frac{63}{4}y + \frac{1}{4}u(t)$ is here

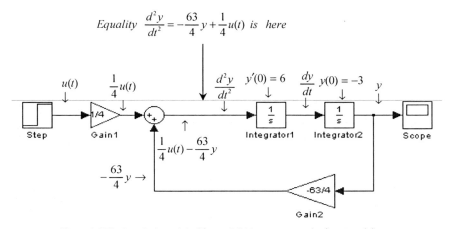

Figure 3.7(d) *Simulink model of figure 3.7(c) mentioning the functional flow*

⌗ Simulink solution

Depicted figure 3.7(c) is the simulink model for the differential equation. To model the equation, following considerations are necessary:

⇒ *the rearranged equation is* $\frac{d^2y}{dt^2} = -\frac{63}{4}y + \frac{1}{4}u(t)$

⇒ *a two input Sum block is needed to add* $-\frac{63}{4}y$ *and* $\frac{1}{4}u(t)$

⇒ *the unit step function (the default block parameters' settings are shown in figure 3.7(e)) can be reached following the link 'Simulink* → *Sources* → *Step' and change its step time to 0 for* $u(t)$

45

$\Rightarrow$ *two Integrator blocks are required to obtain* y *from* $\dfrac{d^2y}{dt^2}$

$\Rightarrow$ *the gains for* y *and* $u(t)$ *are* $-\dfrac{63}{4}$ *and* $\dfrac{1}{4}$ *respectively*

$\Rightarrow$ *open a new simulink model file, bring all associated blocks, rename or flip the blocks if it is necessary, place various blocks relatively, and connect them as indicated in figure 3.7(c)*

$\Rightarrow$ *the Gain2 block might need to be enlarged horizontally to show the gain of* y

$\Rightarrow$ *set the start and stop time as 0 and 2*pi respectively for the insertion of* $0 \le t \le 2\pi$

$\Rightarrow$ *set the two initial conditions* $y'(0) = 6$ *and* $y(0) = -3$ *to the Integrator1 and Integrator2 respectively*

Figure 3.7(d) shows the functional flow diagram for the model in the figure 3.7(c) and whose simulated solution can be seen in figure 3.7(b) (with the autoscale setting) upon successful simulation.

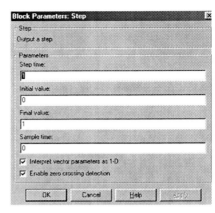

Figure 3.7(e) *Step block parameter window*

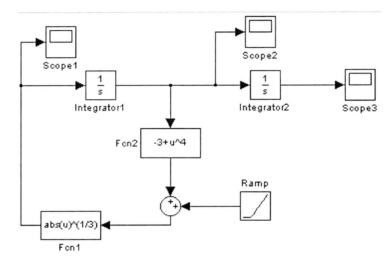

Figure 3.8(a) *Simulink model for solving the differential equation* $\left(\dfrac{d^2y}{dt^2}\right)^3 - \left|\left(\dfrac{dy}{dt}\right)^4 + t - 3\right| = 0$

❖ ❖ Example 3

Construct a simulink model to solve the differential equation $\left(\dfrac{d^2y}{dt^2}\right)^3 - \left|\left(\dfrac{dy}{dt}\right)^4 + t - 3\right| = 0$ under the initial conditions $y'(0) = -4$ and $y(0) = -5$ on the interval $0 \le t \le 4$. We also want to see the response or variation of $\dfrac{d^2y}{dt^2}$ and $\dfrac{dy}{dt}$ separately on the same interval.

⌨ Simulink solution

Figure 3.8(a) presents the simulink model for solving the differential equation. Some details about the design are the following:

$\Rightarrow$ *keeping the highest order derivative on the left side, we have* $\dfrac{d^2y}{dt^2} = \left|\left(\dfrac{dy}{dt}\right)^4 + t - 3\right|^{\frac{1}{3}}$

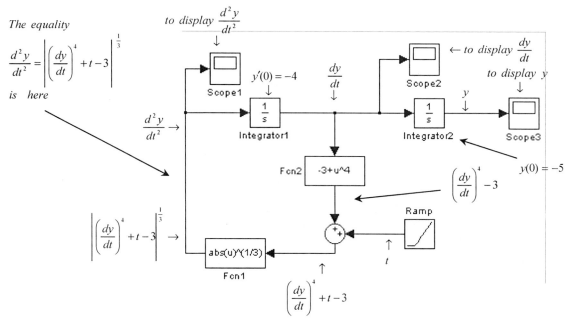

The equality

$$\frac{d^2 y}{dt^2} = \left| \left(\frac{dy}{dt} \right)^4 + t - 3 \right|^{\frac{1}{3}}$$

is   here

Figure 3.8(b) *Simulink model of figure 3.8(a) mentioning the functional flow*

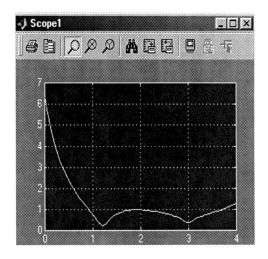

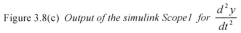

Figure 3.8(c) *Output of the simulink Scope1 for* $\frac{d^2 y}{dt^2}$

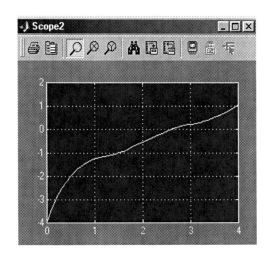

Figure 3.8(d) *Output of the simulink Scope2 for* $\frac{dy}{dt}$

$\Rightarrow \left(\dfrac{dy}{dt} \right)^4 - 3$ *can be obtained by passing* $\dfrac{dy}{dt}$ *through a functional block (which has the functional expression* $u^4 - 3$ *) and then necessary addition with the ramp can be performed by a two input Sum block*

$\Rightarrow$ *to display* y *,* $\dfrac{dy}{dt}$ *, and* $\dfrac{d^2 y}{dt^2}$ *separately, we require three Scope blocks as shown by Scope3, Scope2, and Scope1 respectively*

$\Rightarrow$ *open a new simulink model file, bring all associated blocks, rename, rotate, or flip the blocks if it is necessary, place various blocks relatively, and connect them as presented in figure 3.8(a)*

47

⇒ *the functional blocks Fcn1 and Fcn2 might need to be stretched to show the functional expressions in them*

⇒ *set the start and stop time as 0 and 6 respectively for the insertion of $0 \le t \le 4$*

⇒ *set the two initial conditions $y'(0) = -4$ and $y(0) = -5$ to the Integrator1 and Integrator2 respectively*

⇒ *save the model file, run the simulation from the start simulation icon, and finally select and doubleclick each scope block to see various outputs*

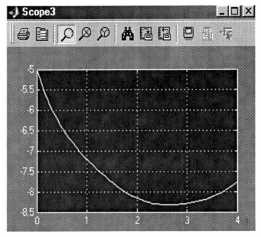

Figure 3.8(b) points out the functional flow in various path of the model. The outputs exhibited by Scope1, Scope2, and Scope3 of the simulink model with autoscale setting are shown in figures 3.8(c), 3.8(d), and 3.8(e) for $\dfrac{d^2 y}{dt^2}$, $\dfrac{dy}{dt}$, and y respectively. It goes without saying that the time axes of all three Scope blocks share the common variation, which is $0 \le t \le 4$ for the example at hand. A Scope block can be placed anywhere in the functional flow path to see the functional variation in the path over the specific time domain.

Figure 3.8(e) *Output of the simulink Scope3 for y*

3.4 Modeling a third order differential equation

It is quite obvious now that we have to deal with three integrator blocks and three initial conditions as far as the order of a third order differential equation is concern. Let us find the output response for the third order differential equation $\dfrac{d^3 y}{dt^3} - 4\dfrac{d^2 y}{dt^2} + 14\dfrac{dy}{dt} - 20y = 0$ without the input excitation and with the initial conditions

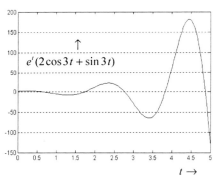

Figure 3.9(a) *Plot of $y = e^t(2\cos 3t + \sin 3t)$ versus t over $0 \le t \le 5$*

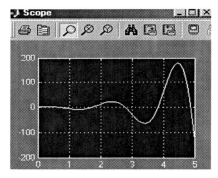

Figure 3.9(b) *Output of the simulink Scope for $\dfrac{d^3 y}{dt^3} - 4\dfrac{d^2 y}{dt^2} + 14\dfrac{dy}{dt} - 20y = 0$*

$y(0) = 2$, $y'(0) = 5$, and $y''(0) = -10$ over the time domain $0 \le t \le 5$. The trial solution, auxiliary equation, roots of the auxiliary equation, the general solution, and the solution subject to the initial conditions are given by $y = e^{mt}$,

$m^3 - 4m^2 + 14m - 20 = 0$, $m = \begin{Bmatrix} 2 \\ 1 \pm j3 \end{Bmatrix}$, $y = Ae^{2t} + e^t(B\cos 3t + C\sin 3t)$, and $y = e^t(2\cos 3t + \sin 3t)$ respectively. The plot of the function can be seen in figure 3.9(a). The agenda we are dealing with is

problem space

$$\overbrace{\dfrac{d^3 y}{dt^3} - 4\dfrac{d^2 y}{dt^2} + 14\dfrac{dy}{dt} - 20y = 0 \ \text{with} \ y(0) = 2, \ y'(0) = 5 \ \text{and} \ y''(0) = -10}$$ and

solution space

$$\overbrace{y = e^t(2\cos 3t + \sin 3t) \ \text{plot like figure } 3.9(a)}$$.

☞ Simulink solution

Figure 3.9(c) shows the necessary simulink model to implement the differential equation. As we rearranged before, the equation can be written as $\frac{d^3y}{dt^3} = 4\frac{d^2y}{dt^2} - 14\frac{dy}{dt} + 20y$. To find the dependent y from the $\frac{d^3y}{dt^3}$, three integrator blocks are required which are labeled as Integrator1, Integrator2, and Integrator3 in the model. The output of these blocks are $\frac{d^2y}{dt^2}$, $\frac{dy}{dt}$, and y respectively. The initial conditions $y''(0) = -10$, $y'(0) = 5$, and $y(0) = 2$ are injected to the three integrator blocks respectively. The gains associated with the three Integrators are 4, -14, and 20

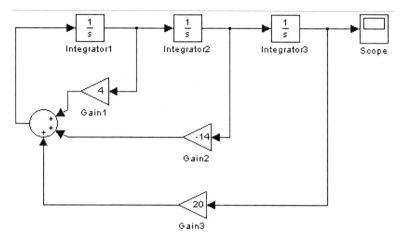

Figure 3.9(c) *Simulink model for solving the differential equation* $\frac{d^3y}{dt^3} - 4\frac{d^2y}{dt^2} + 14\frac{dy}{dt} - 20y = 0$

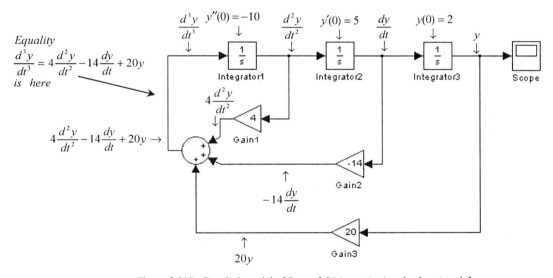

Figure 3.9(d) *Simulink model of figure 3.9(c) mentioning the functional flow*

respectively. From the rearranged equation, a three input Sum block is required to sum $4\frac{d^2y}{dt^2}$, $-14\frac{dy}{dt}$, and $20y$.

The simulink model along with the functional flow is given in figure 3.9(d). Open a new simulink model file, bring all associated blocks, rename, rotate, or flip the blocks if it is necessary, place various blocks relatively, and connect them as placed in figure 3.9(c). To enter the interval $0 \le t \le 5$, we set the start and stop time as 0 and 5 respectively in the solver settings. Successful modeling and simulation of the equation should return you the scope output with the autoscale setting as shown in figure 3.9(b).

We switch to the next section followed by two more examples on the third order differential equation.

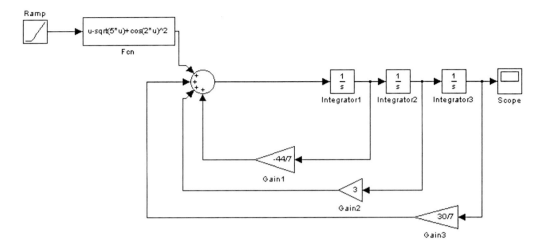

Figure 3.10(a) *Simulink model for solving the differential equation* $7\dfrac{d^3v}{dt^3} + 44\dfrac{d^2v}{dt^2} - 21\dfrac{dv}{dt} - 30v = 7t - 7\sqrt{5t} + 7\cos^2 2t$

$$Equality \quad \frac{d^3v}{dt^3} = -\frac{44}{7}\frac{d^2v}{dt^2} + 3\frac{dv}{dt} + \frac{30}{7}v + t - \sqrt{5t} + \cos^2 2t \quad is \quad here$$

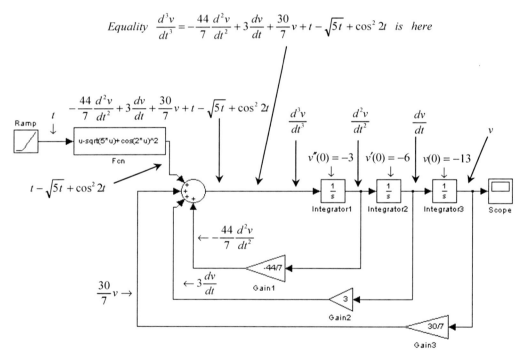

Figure 3.10(b) *Simulink model of figure 3.10(a) mentioning the functional flow*

✦ ✦ Example 1

The third order differential equation so simulated does not possess any input function. This example introduces some input function so that the reader achieves mastery in dealing with and without the input excitation functions in a third order differential equation. Let us consider the equation $7\dfrac{d^3v}{dt^3} + 44\dfrac{d^2v}{dt^2} - 21\dfrac{dv}{dt} - 30v =$ $7t - 7\sqrt{5t} + 7\cos^2 2t$ with the initial conditions $v''(0) = -3$, $v'(0) = -6$, and $v(0) = -13$ over the domain $0 \le t \le 4$.

Since the reader is familiar with bringing the associated blocks and insertion of initial conditions from the beginning examples, we present only the essential elements for the design process in the following:

⇒ *the useful rearranged form for the model is*

$$\frac{d^3v}{dt^3} = -\frac{44}{7}\frac{d^2v}{dt^2} + 3\frac{dv}{dt} + \frac{30}{7}v + t - \sqrt{5t} + \cos^2 2t ,$$ *where the dependent variable is* v

⇒ *the gains pertaining to* $\frac{d^2v}{dt^2}$, $\frac{dv}{dt}$, v *are* $-\frac{44}{7}$, 3, and $\frac{30}{7}$ *respectively*

⇒ *the input excitation* $t - \sqrt{5t} + \cos^2 2t$ *can be generated by passing through the functional code u-sqrt(5*u)+cos(2*t)^2 bearing in mind that the independent variable is u and which simulates the ramp*

⇒ *open a new simulink model file, bring all associated blocks, rename, rotate, or flip the blocks if it is necessary, place various blocks relatively, and connect them as indicated in figure 3.10(a)*

⇒ *the input excitation and the three Gain blocks' output require to use a four input Sum block (bring the two-input Sum block and doubleclick the block to include two more plus in 'The list of sign' in the parameter window)*

⇒ *stretch the functional block Fcn to show the functional expressions*

⇒ *set the start and stop time as 0 and 4 respectively for the insertion of* $0 \le t \le 4$

⇒ *set the three initial conditions* $v''(0) = -3$, $v'(0) = -6$, *and* $v(0) = -13$ *to the Integrator1, Integrator2, and Integrator3 respectively*

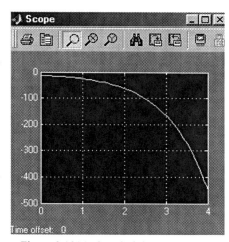

Figure 3.10(c) *Simulink Scope output for* $7\frac{d^3v}{dt^3} + 44\frac{d^2v}{dt^2} - 21\frac{dv}{dt} - 30v =$
$7t - 7\sqrt{5t} + 7\cos^2 2t$

Presented in figure 3.10(b) is the functional flow diagram of the model and whose Scope output resembles to the figure 3.10(c) (with the autoscale setting). The vertical and horizontal axes of the Scope correspond to v and t variations respectively.

✦✦ Example 2

Most examples illustrated so far largely contain the constant gains. The gain associated with the derivative or the dependent variable might be a function of dependent or independent variable. The equation

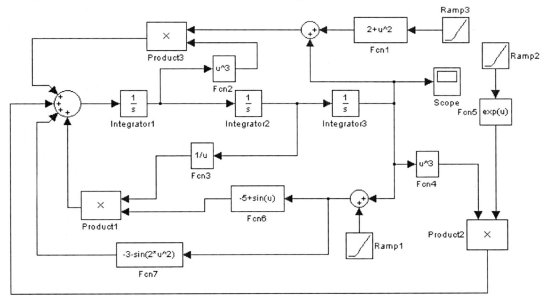

Figure 3.11(a) *Simulink model for solving the differential equation*

$$\frac{dy}{dt}\frac{d^3y}{dt^3} + 5\sin(y+t) + [3 + \sin\{2(y+t)^2\} - y^3e^t]\frac{dy}{dt} - \left(\frac{d^2y}{dt^2}\right)^3\frac{dy}{dt}(y+2+t^2) = 0$$

$\dfrac{dy}{dt}\dfrac{d^3y}{dt^3} + 5\sin(y+t) + [3 + \sin\{2(y+t)^2\} - y^3 e^t]\dfrac{dy}{dt} - \left(\dfrac{d^2y}{dt^2}\right)^3\dfrac{dy}{dt}(y+2+t^2) = 0$ is a third order differential equation

whose gains are not constants. Let us model the equation contingent to the initial conditions $y''(0) = -4$, $y'(0) = -2$, and $y(0) = -12$ over the interval $0 \le t \le 1$.

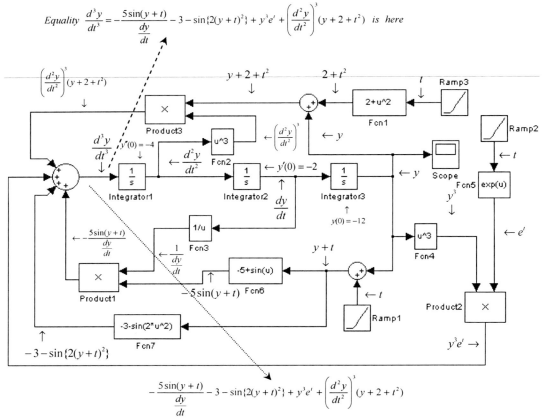

Equality $\dfrac{d^3y}{dt^3} = -\dfrac{5\sin(y+t)}{\dfrac{dy}{dt}} - 3 - \sin\{2(y+t)^2\} + y^3 e^t + \left(\dfrac{d^2y}{dt^2}\right)^3 (y+2+t^2)$ *is here*

Figure 3.11(b) *Simulink model of figure 3.11(a) mentioning the functional flow*

☞ Simulink solution

The model for solving the equation is depicted in figure 3.11(a) followed by the functional flow descriptive diagram in figure 3.11(b). The important elements regarding the equation modeling that need to be addressed are as follows:

$\Rightarrow$ *by placing the highest order derivative on the left side, one gets* $\dfrac{d^3y}{dt^3} = -\dfrac{5\sin(y+t)}{\dfrac{dy}{dt}} - 3 - \sin\{2(y+t)^2\}$

$+ y^3 e^t + \left(\dfrac{d^2y}{dt^2}\right)^3 (y+2+t^2)$

$\Rightarrow$ *to the context of the dependent variable and its derivatives, the right side of the equation is above all the*

sum of four components: $\left(\dfrac{d^2y}{dt^2}\right)^3 (y+2+t^2)$, $-\dfrac{5\sin(y+t)}{\dfrac{dy}{dt}}$, $-3 - \sin\{2(y+t)^2\}$, *and* $y^3 e^t$

$\Rightarrow$ *the four components in the right side of the equation necessitate to use a four input Sum block*

$\Rightarrow$ *three Integrator blocks are needed to obtain* y *from* $\dfrac{d^3y}{dt^3}$ *as shown in the model 3.11(a)*

⇒ *the Integrator1, Integrator2, and Integrator3 in the model produce* $\dfrac{d^2y}{dt^2}$, $\dfrac{dy}{dt}$, *and y and whose initial conditions are* $y''(0) = -4$, $y'(0) = -2$, *and* $y(0) = -12$ *respectively*

⇒ *the Integrator2 of the model outturns* $\dfrac{d^2y}{dt^2}$ *which can be passed through a functional block* u^3 *to obtain*

$\left(\dfrac{d^2y}{dt^2}\right)^3$ *(performed by the block Fcn2 in the model)*

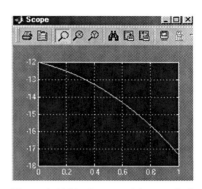

⇒ *the function* $(y + 2 + t^2)$ *mixes (can be obtained by adding) the dependent y with* $2 + t^2$ *which again is a function of the independent t and can be generated by passing through a functional block with the expression* $2 + u^2$ *(done by block Fcn1 in the model)*

⇒ *the component* $\left(\dfrac{d^2y}{dt^2}\right)^3 (y + 2 + t^2)$ *in the right side of the equation is the product of* $\left(\dfrac{d^2y}{dt^2}\right)^3$ *and* $(y + 2 + t^2)$

Figure 3.11(c) Output of the simulink Scope for the differential equation of example 2

(performed by the Product3 block in the model)

⇒ *the part* $\dfrac{1}{\dfrac{dy}{dt}}$ *of the second component in the right side of the equation can be formed by passing the output of the Integrator2 through a functional block with expression* $\dfrac{1}{u}$ *(indicated by Fcn3 in the model)*

⇒ *the part* $-5\sin(y + t)$ *of the second component is a function of* $y + t$ *which can be generated by summing y with t (indicated by Ramp1 in the model)*

⇒ *the second component* $-\dfrac{5\sin(y + t)}{\dfrac{dy}{dt}}$ *is in a division form which turns to the multiplied form as*

$-5\sin(y + t) \times \dfrac{1}{\dfrac{dy}{dt}}$ *(performed by the block Product1 in the model)*

⇒ *the third component* $-3 - \sin\{2(y + t)^2\}$ *is again a function of* $y + t$ *whose functional flow is available from the second component (performed by block Fcn7)*

⇒ *the fourth component* $y^3 e^t$ *is the product of* y^3 *(obtained by passing y through* u^3 *and performed by Fcn4 in the model) and* e^t *(obtained by passing t through* e^u *and performed by Fcn5 in the model)*

⇒ *however, open a new simulink model file, bring all associated blocks, rename, rotate, or flip the blocks if it is necessary, place various blocks relatively, and connect them as indicated in figure 3.11(a)*

⇒ *set the start and stop time as 0 and 1 respectively for the insertion of* $0 \leq t \leq 1$

⇒ *set the three initial conditions as indicated in the figure 3.11(b)*

⇒ *finally save your model file and run it*

With all these successful steps, connections, modeling, and specific settings, the reader is supposed to see the Scope output with the autoscale setting for y versus t as shown in figure 3.11(c). Notice that we brought three ramp generators (indicated by Ramp1, Ramp2, and Ramp3 in the model file), all of which generate the same function (plot like t vs t) over the common time interval $0 \leq t \leq 1$. We could have taken the t functional flow from one ramp to the others. Yet, the model we presented may not be the exclusively one as regards the orientation of the blocks. The reader may systematize his own fashion of designing the model but the important point is the appropriate functional flow must be maintained in the model during the design process.

3.5 Modeling a system of differential equations

A system of differential equations can be expressed as $\begin{cases} \dfrac{dy_1}{dt} = f_1(t, y_1, y_2, \ldots, y_n) \\ \dfrac{dy_2}{dt} = f_2(t, y_1, y_2, \ldots, y_n) \\ \qquad\vdots \\ \dfrac{dy_n}{dt} = f_n(t, y_1, y_2, \ldots, y_n) \end{cases}$, where y_1, y_2, ..., and

y_n are the n dependent variables and t is the independent variable. Our intention is to find the functions y_1, y_2, ..., and y_n satisfying simultaneously the differential equations of the system. For a linear system, the system of the

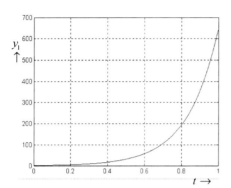

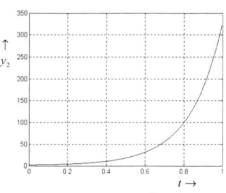

Figure 3.12(a) *Plot of* $y_1 = -\dfrac{3}{5} e^t + \dfrac{8}{5} e^{6t}$ *vs* t Figure 3.12(b) *Plot of* $y_2 = \dfrac{6}{5} e^t + \dfrac{4}{5} e^{6t}$ *vs* t

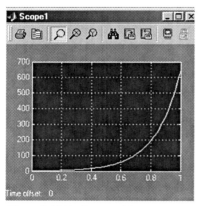

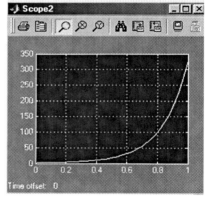

Figure 3.12(c) *Output of the simulink*   Figure 3.12(d) *Output of the simulink*
Scope1 for y_1 *versus* t *Scope2 for* y_2 *versus* t

equations can be written as $\begin{cases} \dfrac{dy_1}{dt} = A_{11}y_1 + A_{12}y_2 + \cdots + A_{1n}y_n + b_1 \\ \dfrac{dy_2}{dt} = A_{21}y_1 + A_{22}y_2 + \cdots + A_{2n}y_n + b_2 \\ \qquad\vdots \\ \dfrac{dy_n}{dt} = A_{n1}y_1 + A_{n2}y_2 + \cdots + A_{nn}y_n + b_n \end{cases}$, where A_{ij}'s are the functions of t and b's are the

forcing or input functions.

All the while our approach has been exemplary so let us find the solution of the system of differential

equations $\left\{\begin{array}{l} \dfrac{dy_1}{dt} = 5y_1 + 2y_2 \\ \dfrac{dy_2}{dt} = 2y_1 + 2y_2 \end{array}\right\}$ over the domain $0 \le t \le 1$ that is homogeneous and that has the constant coefficients

contingent to the initial conditions $\left\{\begin{array}{l} y_1(0) = 1 \\ y_2(0) = 2 \end{array}\right\}$.

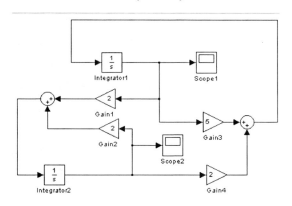

Figure 3.12(e) *Simulink model for*

solving the system $\left\{\begin{array}{l} \dfrac{dy_1}{dt} = 5y_1 + 2y_2 \\ \dfrac{dy_2}{dt} = 2y_1 + 2y_2 \end{array}\right\}$

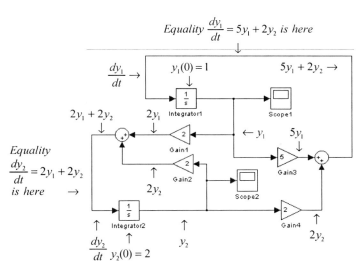

Figure 3.12(f) *Simulink model of figure 3.12(e) mentioning the functional flow*

There are several methods to find the solution of a system of differential equations, namely, fundamental matrix method, Laplace transform method, and eigenvector method. Reference [1] can be considered in this regard. We do not wish to make the reader overwhelmed with too much mathematics even though sound mathematical background is very essential for understanding the very basics of modeling and simulation. However, employing the fundamental matrix method and applying the two initial conditions, one would obtain the solution of the system of

the equations as $\begin{bmatrix} y_1 \\ y_2 \end{bmatrix} = \begin{bmatrix} -\dfrac{3}{5}e^t + \dfrac{8}{5}e^{6t} \\ \dfrac{6}{5}e^t + \dfrac{4}{5}e^{6t} \end{bmatrix}$. The system possesses two dependent variables – y_1 and y_2 with the

independent variable t. Figures 3.12(a) and 3.12(b) show the plots of y_1 versus t and y_2 versus t respectively. The number of initial conditions is always determined by the order of each differential equation in the system. Since each equation is of order one, the initial values of y_1 and y_2 are sufficient to solve the system. Let us list our expectation as follows:

problem space　　　　　　　　　　*solution space*

$$\left.\begin{matrix}\dfrac{dy_1}{dt} = 5y_1 + 2y_2 \\[2mm] \dfrac{dy_2}{dt} = 2y_1 + 2y_2\end{matrix}\right\} \text{ with } \left\{\begin{matrix}y_1(0) = 1 \\ y_2(0) = 2\end{matrix}\right\} \text{ and } \left.\begin{matrix}y_1 = -\dfrac{3}{5}e^t + \dfrac{8}{5}e^{6t} \text{ plot like figure } 3.12(a) \\[2mm] y_2 = \dfrac{6}{5}e^t + \dfrac{4}{5}e^{6t} \text{ plot like figure } 3.12(b)\end{matrix}\right\} \text{ for } 0 \le t \le 1 \text{ .}$$

☞ *Simulink solution*

We presented the model for solving the system in figure 3.12(e) followed by the functional flow representation in figure 3.12(f). Much of the terminology and concepts employed in the model are very similar to the ones we simulated before but now we have more than one differential equation and the equations are interconnected or coupled either by derivatives, dependent variables, independent variables, and/or their functions. However, we describe some elements of the modeling as follows:

⇒ *both equations are already in rearranged form*

⇒ y_1 *and* y_2 *can be generated from* $\dfrac{dy_1}{dt}$ *and* $\dfrac{dy_2}{dt}$ *via the Integrator1 and Integrator2 respectively as shown in the model as if we are dealing with two different differential equations*

⇒ *the right side of the first differential equation (* $5y_1 + 2y_2$ *) not only has the first dependent variable but it also has the second dependent variable in sum form hence we pick up the second dependent variable from the output of the second Integrator and sum that with the first one followed by the appropriate gain*

⇒ *equality of each equation must be observed in accordance with the rearranged form of the equation*

⇒ *however, open a new simulink model file, bring all associated blocks, rename, rotate, or flip the blocks if it is necessary, place various blocks relatively, and connect them as indicated in figure 3.12(e)*

⇒ *set the start and stop time as 0 and 1 respectively for the insertion of* $0 \le t \le 1$

⇒ *set the two initial conditions as indicated in the figure 3.12(f)*

⇒ *finally save your model file and run it*

Figures 3.12(c) and 3.12(d) depict the Scope outputs for y_1 and y_2 (doubleclick the Scope1 and Scope2) with the autoscale setting which are identical with the analytical plots as shown in figures 3.12(a) and 3.12(b) respectively.

We bring an end to the section presenting two more examples on modeling and simulating of two systems of differential equations thereafter.

❖ ❖ Example 1

Let us find the simulink model and solution for the nonhomogeneous system of equations

$$\left.\begin{matrix}\dfrac{dx}{dt} = 31x - 21y + 9z - e^{-3t} \\[2mm] \dfrac{dy}{dt} = 44x - 30y + 12z + 2t \\[2mm] \dfrac{dz}{dt} = -22x + 14y - 8z + \sin t\end{matrix}\right\} \text{ over the domain } 0 \le t \le 0.5 \text{ satisfying the initial conditions } \left\{\begin{matrix}x(0) = -2 \\ y(0) = 1 \\ z(0) = 0\end{matrix}\right\}.$$

☞ *Simulink solution*

The simulink model that can solve the system of the differential equations is shown in figure 3.12(g) followed by the functional flow indicatory figure 3.12(h). The important elements in the modeling are as follows:

⇒ *there are three dependent variables –* x *,* y *, and* z *and the independent variable is* t

⇒ *each equation has four components so the use of four input Sum block before the Integrator block of each equation is obvious*

⇒ *assume that we are modeling three different first order equations, for instance, modeling the dependent* x *in* $\dfrac{dx}{dt} = 31x - 21y + 9z - e^{-3t}$ *but we acquire the* $-21y$ *and* $9z$ *components of the right side in the equation from the second and the third differential equations respectively, and the procedure applies evenly to the other two differential equations*

⇒ *the Integrator1, Integrator2, and Integrator3 in the model are employed to solve the first, second, and third differential equations respectively*

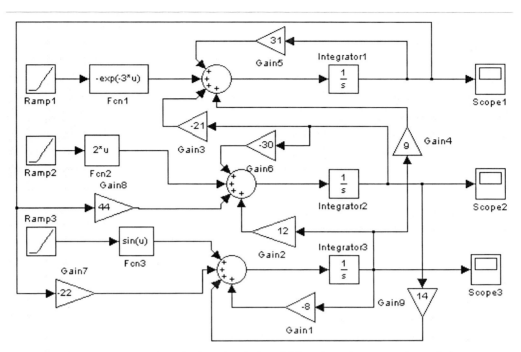

Figure 3.12(g) *Simulink model for solving the system of example 1*

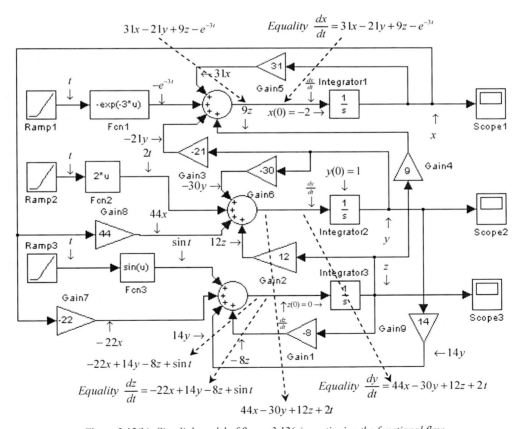

Figure 3.12(h) *Simulink model of figure 3.12(g) mentioning the functional flow*

$\Rightarrow$ *however, open a new simulink model file, construct the model as indicated in figure 3.12(g), set the specified start and stop timings, set the three initial conditions to each Integrator, save your model file, and finally run it*

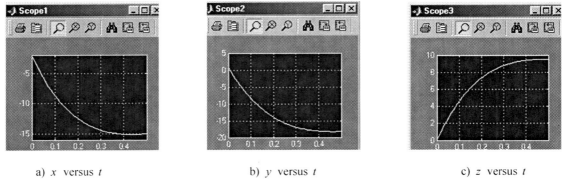

a) x versus t b) y versus t c) z versus t

Figure 3.13 *Scope1, Scope2, and Scope3 outputs of the model for x , y , and z respectively*

Successful construction and simulation of the model should provide the Scope1, Scope2, and Scope3 turnouts as shown in figures 3.13(a), 3.13(b), and 3.13(c) for x vs t , y vs t , and z vs t respectively that is what we are after.

Let us focus our attention in the model file surrounding the region of the Scope 2 and Scope3 as shown in figure 3.12(g). Depending on the model's complicity, planar connection of the model blocks may not always be possible. Overlapping lines do not mean that they are connected. A connection between two lines is only assured by a node (bold dot) as shown in the figure 3.13(d).

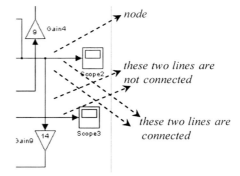

Figure 3.13(d) *Line connections in a model*

❖ ❖ Example 2

Implement the simulink model for the system

$$\begin{cases} \dfrac{d^2 y_1}{dt^2} + \dfrac{dy_2}{dt} - y_1 y_2 \left(1 + \dfrac{t^2}{1000}\right) = 0 \\ \dfrac{d^3 y_2}{dt^3} - \left(\dfrac{dy_1}{dt}\right)^3 \left(\dfrac{dy_2}{dt}\right)^2 - y_1 - 6\cos 9t = 0 \end{cases}$$ to find the solution

of y_1 and y_2 over the domain $0 \le t \le 0.4$ subject to the initial conditions $\begin{cases} y_1'(0) = 0, \ y_1(0) = 7 \\ y_2''(0) = 95, \ y_2'(0) = 11, \ y_2(0) = -2 \end{cases}$.

☞ Simulink solution

Figures 3.14(a) and 3.14(b) show the simulink model and its functional flow indicatory representation. We discuss few steps in the modeling very briefly as follows:

⇒ *the first phase of the modeling is to rearrange the system so that the highest order derivatives of both*

equations are on the right side whence $\dfrac{d^2 y_1}{dt^2} = -\dfrac{dy_2}{dt} + y_1 y_2 \left(1 + \dfrac{t^2}{1000}\right)$ and $\dfrac{d^3 y_2}{dt^3} = \left(\dfrac{dy_1}{dt}\right)^3 \left(\dfrac{dy_2}{dt}\right)^2 +$

$y_1 + 6\cos 9t$

⇒ *the system possesses different order derivatives, the first and second equations of which are of order 2 and 3 respectively*

⇒ *the Integrator1 and Integrator2 in the model simulate the first differential equation, so do the Integrator3, Integrator4, and Integrator5 for the second differential equation*

Figures 3.14(c) and 3.14(d) exhibit the required output for y_1 versus t and y_2 versus t respectively. Notice that the

sum block before the Integrator1 is holding a negative sign. The first rearranged equation contains $-\dfrac{dy_2}{dt}$ which

requires using a Gain block of gain -1. One can avoid the negative unity gain block by altering the input functional line of the Sum block negative. Doubleclick the Sum block, delete one plus sign in the 'List of signs' in the parameters window, type one negative sign in the same window, and click okay to see the change.

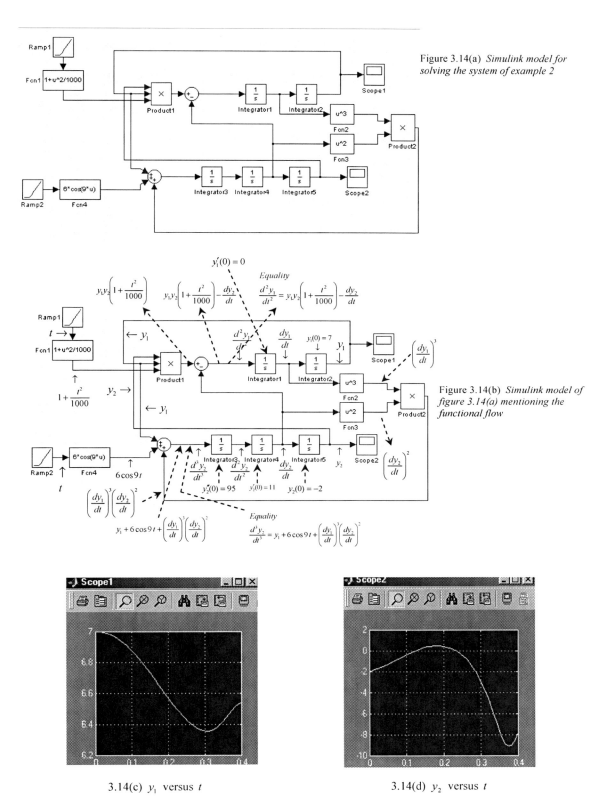

Figure 3.14(a) *Simulink model for solving the system of example 2*

Figure 3.14(b) *Simulink model of figure 3.14(a) mentioning the functional flow*

3.14(c) y_1 versus t

3.14(d) y_2 versus t

Figure 3.14(c)-(d) *Scope1 and Scope2 outputs of the model for y_1 and y_2*

59

Parameters icon
of scope

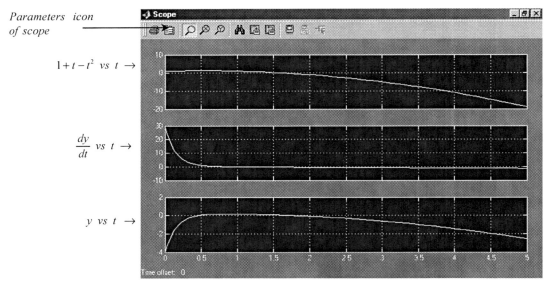

$1+t-t^2$ vs t →

$\dfrac{dy}{dt}$ vs t →

y vs t →

Figure 3.15(a) *Simulink Scope output for three input ports*

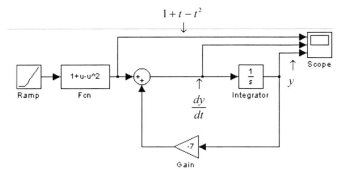

Figure 3.15(b) *Simulink model with three input Scope*

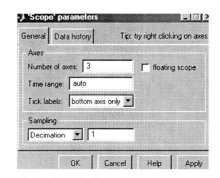

Figure 3.15(c) *Scope parameters dialog window*

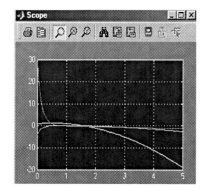

Figure 3.15(e) *Scope output of the model for figure 3.15(d)*

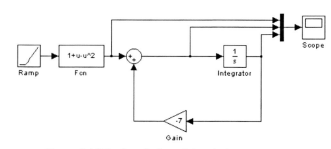

Figure 3.15(d) *Simulink model with three input Mux*

3.6 Some user wanted circumstances

Let us consider the example 1 of section 3.2 where we simulated the first order equation $\dfrac{dy}{dt}+7y=1+t-t^2$ over $0 \le t \le 5$. Suppose we want to see the variations of $1+t-t^2$, y, and $\dfrac{dy}{dt}$ in the same window as three different plots but on the common *t* variation. The model with the functional flow is presented in figure 3.15(b). Now we need a three input Scope to display just mentioned functions thereby modifying the model like figure 3.15(b). To do so, doubleclick the Scope and click the parameter's icon of the Scope (relative position of the parameter's icon is shown in figure 3.15(a)). The Scope parameter's dialog window appears as shown in figure 3.15(c) in which you

change the number of axes from 1 to 3 (it were two if we had two inputs) under the 'General → Axes' and click okay. You should see three input ports in your Scope by doing so. Connect the functional lines with the Scope input ports according to the order mentioned in figure 3.15(b). Finally, save your model and run it to see

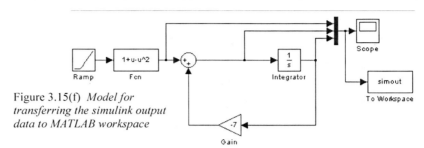

Figure 3.15(f) *Model for transferring the simulink output data to MATLAB workspace*

the figure 3.15(a) where $1+t-t^2$, y, and $\frac{dy}{dt}$ are graphed sequentially following the input functional order. Autoscale setting of the figures has to be done individually but all of them share the common time axis.

Another situation can be the displaying of $1+t-t^2$, $\frac{dy}{dt}$, and y in the same scope but not in three different plots. The model we need is presented in figure 3.15(d) in which we placed one Mux block (the block can multiplex multiple vectors and be brought in the model file following the library link *'Simulink → Signal Routing → Mux'*). When you bring the block in the model file, the number of input ports might be two but we need one more. So, doubleclick the Mux block and change the number of inputs from 2 to 3 in its parameter window. Construct the model as shown in figure 3.15(d), save it, and finally run the model file to see the Scope output (with the autoscale setting) as shown in figure 3.15(e).

Edit plot icon

Figure 3.15(g) *Upper portion of MATLAB figure window*

The shapes so displayed in the figure 3.15(e) are identical to the ones displayed in figure 3.15(a) but the Scope does not provide the ready made option for differentiating them. Visually one can differentiate various plots by first connecting one Scope to each functional flow. See the individual Scope output and then identify the individual plot in the combined scope output. But it might be necessary to print or store the figure. We can take the help of MATLAB Command Prompt to manipulate the output of the Mux. One can send the Scope input data to MATLAB Command Window by appending one more block called 'To Workspace' (which can be reached via *'Simulink → Sinks → To Workspace'*) in the model as presented in figure 3.15(f). Save and run the model, go to MATLAB Command Window, and perform the following:

MATLAB Command
>>who ↵   ← The symbol ↵ stands for Enter Key of the Keyboard

Your variables are:

simout tout

The command who displays the variables present in MATLAB workspace. Referring to figure 3.15(f), the block 'To Workspace' contains simout inside it, which means the output of the simulation is assigned to simout as vector in the sequence $1+t-t^2$, $\frac{dy}{dt}$, and y (because we multiplexed in this order in figure 3.15(d)). The x-axis or time information is stored in tout. To plot the simulink output, we take the help of command simplot as follows:

MATLAB Command
>>simplot(tout,simout) ↵

61

To differentiate various plots, one can use the command legend (the same action can be performed via *'MATLAB Figure Window → Insert → Legend'*) that results the figure 3.15(h).

>>legend('Input','Derivative','Output') ↵

We differentiate the functions $1 + t - t^2$, $\frac{dy}{dt}$, and y by the names Input, Derivative, and Output respectively. Once you have the plot, you can manipulate the plot for the title, labeling, background color, or other graphical properties from the MATLAB figure window. Since our text is written in black and white form, we present the color graphics in black and white form. Let us explain how one can change the line color or style. The upper portion of the MATLAB figure window is shown in figure 3.15(g) where you see the Edit plot icon. Click the Edit plot icon, bring the mouse pointer to any line or plot, and click the right button of the mouse. You see the options for the line width, line style, or line color wherefrom you change the line width, color, or style to any wanted form.

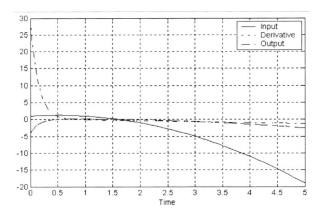

Figure 3.15(h) *Simulink output plotted in MATLAB figure window*

3.7 Some practical examples

So long we modeled and simulated various order differential equations to the context of pure mathematics. Once the simulink notion and concept of implementing the differential equations are well understood, the simulation style and technique can be applied to study the behavior of any physical system that is governed by the ordinary differential equations. Two token examples are going to be presented in the following just to mention where this kind of modeling may find its application.

$R = 0 \ to \ 100 \ \Omega \quad C = 2\mu F \quad L = 1mH$

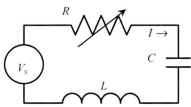

Figure 3.16(a) *A series electrical circuit*

♣ ♦ Example 1

The world of electrical engineers is composed of three basic passive elements – resistance, inductance, and capacitance. Every element in this universe possesses this three fundamental electrical properties. However, figure 3.16(a) presents an electrical circuit in which the resistance R, the inductance L, and the capacitance C are connected in series. Depending on the presence of these three basic elements, the circuit response (which is usually the current in the circuit as labeled by I) might be different. The current in the circuit may reach to its steady state either oscillatorily or nonoscillatorily. Very often oscillatory behavior of current yields instability of the electrical system represented by the series circuit but as far as quick response is concern slight oscillatory nature is wanted. How much oscillation the electrical system can tolerate determines the trade-off between the circuit parameters that correspond to the oscillatory and non oscillatory cases. For instance, the resistance may have certain limit of current withstandabilty specified by the manufacturer which must have to be maintained preventing from the damage of the electrical circuit. The voltage V_s can represent any electric voltage input such as sinusoidal, ramp, step or other with respect to time. Let us assume that the V_s is a ramp function i.e. $V_s(t) = t$ and component in the circuit can not withstand more than $3 \mu A$. The values of L and C are fixed as shown in figure 3.16(a) and their initial conditions are zero. R can change from 0 to 100 Ω.

Our objective of the simulation is to find the safest value of R which will not damage the circuit and the behavior of the current with and without the oscillation for $0 \le t \le 2\,ms$.

⎙ Simulink Solution

The voltage developed in the three elements are given by RI, $L\frac{dI}{dt}$, and $\frac{\int_0^t I dt}{C}$ for R, L, and C respectively. The units of V_s, R, L, C, and I should be in volt, ohm, henry, farad, and ampere respectively.

Kirchhoff's voltage law around the circuit yields $RI + L\dfrac{dI}{dt} + \dfrac{\int_0^t I\,dt}{C} = V_s$. Even though the equation forms an integro differential system, transforming the equation to the differential semblance is rather convenient. Differentiating both sides, one obtains $R\dfrac{dI}{dt} + L\dfrac{d^2I}{dt^2} + \dfrac{I}{C} = \dfrac{dV_s}{dt}$ whose simulink acceptable form is $\dfrac{d^2I}{dt^2} = \dfrac{1}{L}\dfrac{dV_s}{dt} - \dfrac{I}{LC} - \dfrac{R}{L}\dfrac{dI}{dt}$. The gains associated with $\dfrac{dV_s}{dt}$, I, and $\dfrac{dI}{dt}$ are $\dfrac{1}{L}$, $-\dfrac{1}{LC}$, and $-\dfrac{R}{L}$ respectively. The third gain needs to use a variable one since R changes from 0 to 100 Ω. With the given parameter values and taking the units into account, $\dfrac{1}{L}$, $-\dfrac{1}{LC}$, and $-\dfrac{R}{L}$ become 10^3, $-\dfrac{1}{2}\times10^9$, and -10^3R respectively. They can be coded as 1e3, −0.5e9, and −1e3 R

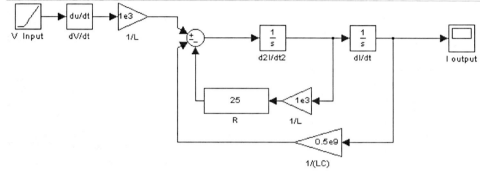

Figure 3.16(b) *Simulink model for the series electrical circuit current analysis*

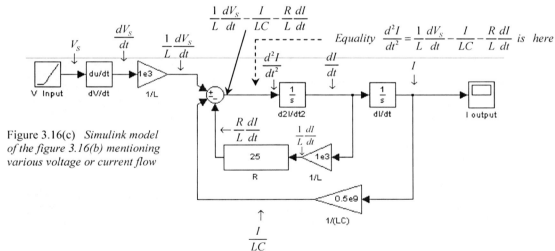

Figure 3.16(c) *Simulink model of the figure 3.16(b) mentioning various voltage or current flow*

respectively while setting the gains of the Gain blocks of the simulink model. A Slider Gain block offers the flexibility on the model so that we change the value of the R from 0 to 100 without changing the other parameters. The third gain is split in two parts −1e3 and R. Two Integrator blocks are needed to obtain I from $\dfrac{d^2I}{dt^2}$. Figures 3.16(b) and 3.16(c) show the model and its flow explanation for the problem at hand. However, the modeling procedure is mentioned as follows:

⇒ *open a new simulink model file and click the library browser icon*
⇒ *bring a Ramp block ('Simulink → Sources → Ramp') and rename it as V Input*
⇒ *bring a Derivative block ('Simulink → Continuous → Derivative') and rename it as dV/dt*
⇒ *bring a Gain block ('Simulink → Math Operations → Gain'), rename it as 1/L, set its gain to 1e3, and stretch the block to display the gain*
⇒ *bring an Integrator block ('Simulink → Continuous → Integrator') and rename it as d2I/dt2*

⇒ *bring another Integrator block and rename it as dI/dt*

⇒ *select the previous Gain block, perform the copy, paste, and flip operation on the block, and rename it as 1/L (It will display 1/L1 by the action. Delete 1 and press Enter to rename it again 1/L)*

⇒ *bring a Slider Gain block ('Simulink → Math Operations → Slider Gain'), rename it as R, select and doubleclick the block to see the block parameter window as shown in figure 3.16(d), insert 0 and 100 in the low and high slots of the window, and flip the block*

⇒ *the middle slot in the Slider Gain window corresponds to the present gain depending on the slider position*

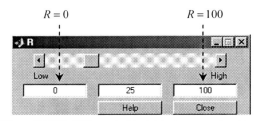

Figure 3.16(d) *Slider gain block window named as R*

⇒ *bring one more Gain block, rename it as 1/(LC), set its gain as 0.5e9, flip it, and stretch the block*

⇒ *bring one scope block ('Simulink → Sinks → Scope') and rename it as I output*

⇒ *bring one sum block ('Simulink → Math Operations → Sum'), select and doubleclick the block, make sure + − − in the List of signs to operate it as a three input Sum block*

⇒ *place all these blocks relatively and connect them according to the figure 3.16(b)*

⇒ *set the start and stop time ('Simulation → Simulation parameters → Solver') as 0 and 2e-3 to insert $0 \leq t \leq 2\,ms$*

⇒ *finally save your model, click the start simulation icon, and doubleclick I output block to see the current response*

Current in ampere

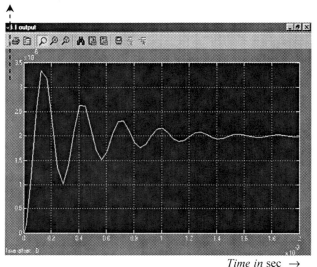

Time in sec →

Figure 3.16(e) *I output for R=5*

With the slider gain block R, you can set any value from 0 to 100 (in ohms) and see the corresponding I output response in ampere. When R=5, you see the I output response with the autoscale setting as depicted in figure 3.16(e). Observing the figure, one can remark that there are some oscillations at the beginning of the interval $0 \leq t \leq 2\,ms$, the oscillations die down with increasing time, and the current tends to settle to $2\,\mu A$. The maximum current in figure 3.16(e) is greater than $3\,\mu A$ so R=5 is not a proper choice. Start changing R from 5 to 6, 7, 8, ... etc in your model by the slider and see the corresponding I output response. Whenever you change R, click the start simulation icon to see the corresponding change.

Conclusion: By the action 'change and see', you will find that R=10 is the safest limit for the circuit which produces less than $3\,\mu A$ maximum current. Whenever we connect the circuit, we should never make $R < 10\Omega$. The value of R from 10Ω to 100Ω will not damage the circuit. So our conclusion is without the connection or the circuit board experimentation, we are able to predict the behavior of a series electrical circuit by dint of simulink that is where the success of modeling and simulation lies.

However, your instructor or boss may demand hardcopy output from the modeling and simulation. Let us say you want to compare three different I outputs for R=5, R=10 and R=50 in the same plot. Just bring one 'To Workspace' block ('*Simulink → Sinks → To Workspace*') and connect that as shown in figure 3.16(f) with the existing model of figure 3.16(b). Let us perform the following procedure:

⇒ *select the slider gain block R in your model and set the value of R to 5 by sliding the slider*

⇒ *doubleclick To Workspace block, change the variable name from simout to R5, change the save format from structure to array, and click OK (the action is displayed in figure 3.16(g))*

⇒ *from the model menu bar, click 'Simulation → Simulation parameters → Solver', change the solver option types from the variable step to fixed step with the method ode5 (Dormand-Prince), and click OK*

64

⇒ *click start simulation icon to run the model, and the action sends the output to MATLAB Command Window*

⇒ *select again the slider gain block R in your model, set the value of R to 10 by sliding the slider, change the variable name from R5 to R10 in the To Workspace block window*

⇒ *click the start simulation icon to run the model again*

⇒ *set the slider gain to 50, change the variable name to R50, and run the model finally*

Now go to the MATLAB Command Window and check what variables are present in the MATLAB Workspace by the following:

MATLAB Command

>>who ↵

Your variables are:

R10   R5   R50   tout

The execution says that the three different cases of I output are stored in the workspace variables R5, R10, and R50 for $R = 5$, $R = 10$, and $R = 50$ respectively and tout holds the time information for $0 \leq t \leq 2\,ms$. We can plot them as shown in figure 3.16(h) in the same figure from the workspace as follows:

>>plot(tout,R5,'k-',tout,R10,'k--',tout,R50,'k:') ↵

>>legend('for R=5','for R=10','for R=50') ↵

The command plot can graph y versus x data (not expressions). The argument 'k' indicates the plot line color to be black (because the text is in black and white form). The characters -, -., and -- beside k provide control on the line style such as continuous, dash dot, and broken respectively. You can also edit the graph from the MATLAB figure window once it is plotted.

❖ ❖ Example 2

Let us consider the mechanical system of figure 3.17 in which a rolling cart moves on a smooth surface. The mechanical system is composed of three basic passive elements – mass of the cart (m), damping element (D), and compliance (K). Accordingly, these three elements cause to exhibit three kinds of mechanical force – inertia, damping, and spring. The force F is an externally applied one and the passive elements of the cart system are given by $m = 0.2\,Kg$, $D = 2\,N/meter/s$, and $K = 10\,meter/N$. We wish to see the displacement (x) and velocity $\left(\dfrac{dx}{dt}\right)$ responses for the cart system subject to a sudden thrust F at $t = 0$ over the time $0 \leq t \leq 1$. The cart is having a velocity of $2\,meter/s$ in the direction of x at $t = 0$.

☞ *Simulink Solution*

The inertia, damping, and spring forces offered by the cart are given by $m\dfrac{d^2x}{dt^2}$, $D\dfrac{dx}{dt}$, and $\dfrac{x}{K}$ respectively. These three forces are also termed as the resisting forces and their sum must be equal to the external force hence one can write the dynamics of the mechanical system as $m\dfrac{d^2x}{dt^2} + D\dfrac{dx}{dt} + \dfrac{x}{K} = F$. Theoretically, the sudden thrust F can be defined as the Dirac delta function $\delta(t)$ whose definition is zero existence, infinite amplitude, and area unity. In

Figure 3.16(f) *Connecting To Workspace block with the I output block of fig. 3.16(b)*

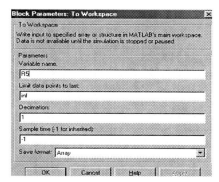

Figure 3.16(g) *Parameter window for To Workspace block*

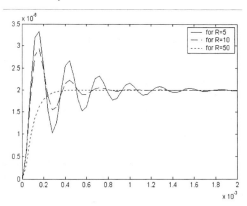

Figure 3.16(h) *I output for R=5, R=10, and R=50 for the series electrical circuit plotted in the same graph*

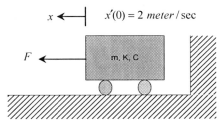

Figure 3.17 *A cart pulled by an external force F*

reality no such mechanical source exists and the practical sources always exhibit finite thrust. Again the duration of the thrust F can not be zero instead it has to be finite. The concept of absolute infinity is undefined. For the simulation purpose, we take some force which has the high finite value and exists for a short period of time. Figures 3.18(a), 3.18(b), 3.18(c), and 3.18(d) depict the ideal Dirac delta function, unit step function, shifted unit step function, and finite value-short duration function respectively.

Let us define the thrust as $F = 10\,N$ and $t_0 = 0.1\,sec$ (so that the area is unity) whose symbolic representation is $F = 10u(t) - 10u(t - 0.1)$. To model the finite force source, we bring two Step blocks (*'Simulink* →

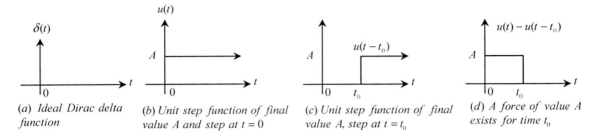

| (a) Ideal Dirac delta function | (b) Unit step function of final value A and step at t = 0 | (c) Unit step function of final value A, step at t = t₀ | (d) A force of value A exists for time t₀ |

Figure 3.18(a)-(d) *Formation of a short existent finite force from the unit step function*

Sources → *Step')* and set their step and final value as indicated in figure 3.19(a) and then add them by a Sum block to obtain the finite thrust.

Employing the suitable form for the simulation, we have $\dfrac{d^2x}{dt^2} = \dfrac{F}{m} - \dfrac{D}{m}\dfrac{dx}{dt} - \dfrac{x}{mK}$, essentially a second order system, and the equality needs to exercise two

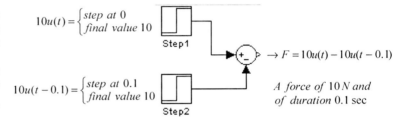

$$10u(t) = \begin{cases} step\ at\ 0 \\ final\ value\ 10 \end{cases}$$

$$10u(t - 0.1) = \begin{cases} step\ at\ 0.1 \\ final\ value\ 10 \end{cases}$$

$\to F = 10u(t) - 10u(t - 0.1)$

A force of $10\,N$ *and of duration* 0.1 *sec*

Figure 3.19(a) *Modeling of the thrust F of value 10N and of duration 0.1 sec*

Integrator blocks to obtain the displacement x from the acceleration $\dfrac{d^2x}{dt^2}$. The simulink model is presented in figure 3.19(b) whose modeling process is described as follows:

⇒ *open a new simulink model file and click the library browser icon*

⇒ *bring a Step block, rename it as Step1, doubleclick the block to set its step time and final value as 0 and 10 respectively*

⇒ *copy the Step1 and paste it to see Step2, doubleclick the Step2 block to set its step time and final value as 0.1 and 10 respectively*

⇒ *bring a Sum block and doubleclick the block to change its List of Signs to +-*

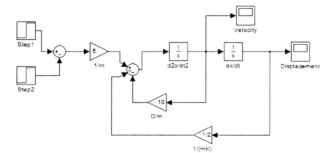

Figure 3.19(b) *Simulink model for the rolling cart system of the figure 3.17*

⇒ *bring a Gain block, rename it as 1/m, doubleclick the block to change its gain to 5* $\left(\because \dfrac{1}{m} = \dfrac{1}{0.2} = 5\right)$

⇒ *bring another Sum block and doubleclick the block to change its List of Signs to +--*

⇒ *bring one Integrator block, rename the block as d2x/dt2 and doubleclick the block to set its initial value to 2 since* $x'(0) = 2\ meter/\sec$

66

⇒ *bring another Integrator block and rename it as dx/dt, (assume that the displacement is 0 at $t = 0$)*

⇒ *bring another Gain block, rename it as D/m, doubleclick the block to change its gain to 10* $\left(\because \dfrac{D}{m} = \dfrac{2}{0.2} = 10 \right)$, *and flip the block*

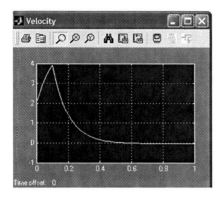

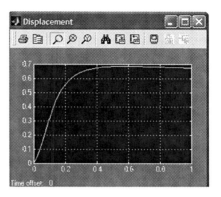

Figure 3.19(c) *Velocity response for the rolling cart system of the figure 3.17*

Figure 3.19(d) *Displacement response for the rolling cart system of the figure 3.17*

⇒ *also get one more Gain block, rename it as 1/(mK), doubleclick the block to change its gain to 1/2* $\left(\because \dfrac{1}{mK} = \dfrac{1}{0.2 \times 10} = \dfrac{1}{2} \right)$, *and flip the block*

⇒ *bring two Scope blocks and rename them as Displacement and Velocity*

⇒ *place all these blocks relatively and connect them according to the figure 3.19(b)*

⇒ *set the start and stop time as 0 and 1 to insert $0 \leq t \leq 1$*

⇒ *finally save your model, click the start simulation icon, and doubleclick the blocks Velocity and Displacement to see their responses as presented in figures 3.19(c) and 3.19(d) respectively*

Referring to just mentioned figures, the velocity and the displacement of the cart system reach to a maximum and then decrease. The thrust we applied is 10 N. We can increase the thrust and see how the velocity and displacement reach further. Consequently we can form a relationship or table between the thrust and the maximum displacement from our simulink model. Let us imagine that a rocket propulsion system is to be made. Necessarily a rocket system possesses the three basic properties { m , D , K } which can come from the measurement. Once we have these parameters, we can find the thrust needed to overcome the earth's gravitational force employing our simulink model. So simulink can help us design a rocket propulsion system and that is where simulink can be fruitful.

3.8 Handling the singularities or infinity like situations

Before we introduce the way of handling the infinity like situations in simulink, let us observe the asymptotic behavior of the curves $y = \dfrac{1}{x - 2}$ and $y = -\dfrac{1}{x^3 - 2}$ whose plots are depicted in figures 3.20(a) and 3.20(b) respectively. Depending on the type of the differential equations and initial conditions, one may obtain different sorts of solutions for the same system. There is a possibility that we may get a solution of the differential equation under some circumstance like the curves $y = \dfrac{1}{x - 2}$ or $y = -\dfrac{1}{x^3 - 2}$ in which case a switching is always associated with either from $-\infty$ to $+\infty$ or from $+\infty$ to $-\infty$. The term absolute infinity is undefined and computer can not deal with the absolute infinity. Instead we set some predetermined high value of the output (again the degree of highness is also relative) and if the output exceeds the high limit, we call that our integrator reached to a saturation.

Anyhow let us recall the example 2 of section 3.2 where we encountered this sort of situation. Now we wish to provide the remedy. Let us try to run the model of figure 3.3(b) again, you see some error message as shown in the attached figure 3.20(c) indicating a saturation point at x =0.6597513726886849. Looking into the Scope output, one can infer that the saturation is toward the $+\infty$ for the example at hand. Since the infinity like situations always happen in pair that is $+\infty$ to $-\infty$ or $-\infty$ to $+\infty$, we can state that slightly before x =0.6597513726886849 the solution is $+\infty$ (for finiteness of the graph it is 25 from the Scope) and slightly after $x = 0.6597513726886849$ the

solution should be $-\infty$ (or -25 for the graphical finiteness). Therefore the required interval $0 \le x \le 2$ is split in two parts: from $x = 0$ to $x = 0.6597513726886849$ and from $x = 0.6597513726886849$ to $x = 2$. We simulate the model part by part but in the second part the setpoint for the initial value of the Integrator is -25. The reader can set any large value other than 25 as long as the reader is satisfied with the graphical shape. Since all blocks in a simulink model share the same horizontal axis/time value or x, first we run the model for the first part and send the data to MATLAB workspace, after that do the same thing for the second part of the interval, and then combine them. However, you need to remodel the model of figure 3.3(b) as shown in figure 3.20(d) and perform the following in the model:

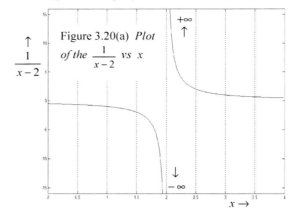

Figure 3.20(a) *Plot of the* $\dfrac{1}{x-2}$ *vs* x

⇒ *doubleclick the Integrator block to see the block parameter window, check the limit output button, set the upper saturation limit as 25, set the lower saturation limit −25, make sure that the initial condition is 10, keep the other settings unchanged, and click OK to see the icon outlook of the Integrator like figure 3.20(d) which indicates saturation*

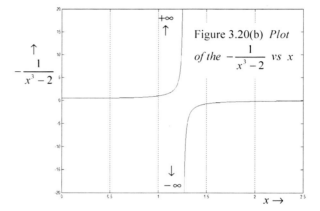

Figure 3.20(b) *Plot of the* $-\dfrac{1}{x^3-2}$ *vs* x

⇒ *bring one To Workspace block following the link 'Simulink → Sinks → To Workspace' and connect the block with the functional line to Scope as shown in figure 3.20(d)*

⇒ *set the start and stop time as 0 and 0.6597513726886849 respectively in the solver settings*

⇒ *doubleclick the To Workspace block to see the block parameter window, change the Variable name to simout1, change the Save format to Array, click OK, and run the model*

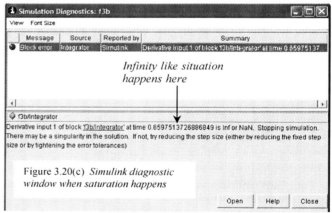

Figure 3.20(c) *Simulink diagnostic window when saturation happens*

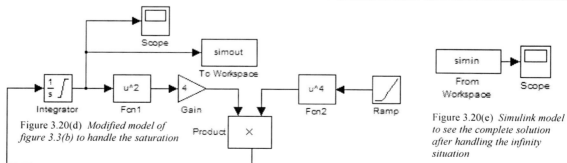

Figure 3.20(d) *Modified model of figure 3.3(b) to handle the saturation*

Figure 3.20(e) *Simulink model to see the complete solution after handling the infinity situation*

⇒ *go to MATLAB Command Window and execute who to see the variables present as follows:*
>>who ↵

Your variables are:

simout1 tout

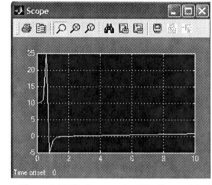

⇒ *to keep the horizontal axis or x information (carried out by tout) to some variable x1, perform the following in the Command Window:*
>>x1=tout; ↵

⇒ *again go to the simulink model, doubleclick the Integrator block to set the initial value as −25 (this serves the purpose of −∞), set the solver start and stop time as 0.6597513726886849 and 2 respectively, doubleclick the To Workspace block, change the variable name to simout2, and run the model for the second interval*

Figure 3.20(f) *Scope output for the model of figure 3.20(e)*

⇒ *go to MATLAB Command Window and use the command who to see the variables present*
>>who ↵

Your variables are:

simout1 simout2 tout   x1

⇒ *the return says the following:*
  simout1 → solution of the differential equation from x=0 to x=0.6597513726886849
  simout2 → solution of the differential equation from x=0.6597513726886849 to x=2
  x1 → corresponding values of the x at which simout1 occurs
  tout → corresponding values of the x at which simout2 occurs
  and each of the four variables is a column matrix

⇒ *now we form a rectangular matrix in MATLAB command window having two columns where the first and second columns refer to the independent variable values and dependent variable values respectively by the following command (because that is the operational format of simulink):*
  >>simin=[[x1;tout] [simout1;simout2]]; ↵

The command [x1;tout] puts the column matrix x1 on top of the column matrix tout, and so happens for the command [simout1;simout2]. The command [[x1;tout] [simout1;simout2]] places the two columns so formed side by side. Anyhow, the complete solution is assigned to the variable simin. As far as the horizontal axis or x axis is concern, one can not see the output retained in simin in the same model unless the start and stop times are changed to 0 and 2 respectively. Open a new simulink model file, bring a From Workspace block following the link '*Simulink → Sources → From Workspace*', bring a Scope, and connect the two blocks as shown in figure 3.20(e), set the start and stop time as 0 and 2 respectively, run the model, doubleclick the Scope to see the complete output as shown in figure 3.20(f) which makes sense with the analytical solution displayed in figure 3.3(a). In figure 3.3(a), the line between $+\infty$ to $-\infty$ is not possible that is why the plot is empty at the singular point. But simulink always connects line between consecutive points that is why there is a line during the switching from +25 to −25 in figure 3.20(f). You can play with another saturation limit value other than 25 to obtain a better shape of the solution.

3.9 Some factors regarding the simulation of differential equations

Nowadays technical and nontechnical problems are not isolated in nature, for instance, physics, chemistry, mathematics, engineering etc, rather, they are interdisciplinary in nature. System dynamics being represented by differential equations are very important in modeling and simulation for a large number of continuous systems. While working in SIMULINK, following factors might help reader model and simulate successfully a particular problem:

1. The first step of finding the simulink model for a differential equation is to rearrange the differential equation so that the highest order derivative is on the left side of the equation. The algebraic equality of the equation must be maintained once you model the right side of the equation.
2. Initial conditions are always necessary in an Integrator block and they start with $t = 0$.
3. The number of Integrator blocks employed in a simulink model must be equal to the order of the differential equation.

4. The differential equations utilize Integrator blocks instead of the Derivative ones so that proper algebraic equation is maintained.

5. The expression of the form $\frac{1}{t}$ is avoided in simulink model because of the appearance of singularities or infinity like situations at $t = 0$. If it is inevitable to employ such expression, choose some small value or constant such as 10^{-5}. You can even use MATLAB's epsilon for this purpose, which can be called just by eps. Write eps in MATLAB Command prompt and execute to see the zero value for your computer.

6. Modeling a differential equation in simulink must need to apply exact order, start and stop time, and initial conditions.

7. Even though the Integrator block explicitly represents Laplace s domain but the solution found by the block is in the time domain.

8. When you employ Fcn block to provide some input or excitation function for a differential equation, make sure that the function defined by the Fcn block does not generate values like negative element with fractional power {for instance, $(-5)^{0.4} = 0.5883 + j1.8105$ }. The situation when the complex numbers are being generated in the model functional line might standstill your simulation.

3.10 Block links used in this chapter

This chapter mainly discusses the differential equation simulation problems of various orders and types for which we adopted the blocks as presented in the table 3.A.

Table 3.A   Necessary blocks for the differential equation modeling as found in simulink library

Block name	Representative Symbol/Function	Icon Outlook	Block name	Representative Symbol/Function	Icon Outlook
Integrator	$\int u\,dt$ or $\frac{1}{s}$		Step	$u(t-1)$ (default)	
Link: *Simulink → Continuous → Integrator*			Link: *Simulink → Sources → Step*		
Scope	Oscilloscope or function display		Gain	Multiplication by a constant	
Link: *Simulink → Sinks → Scope*			Link: *Simulink → Math Operations → Gain*		
Fcn	Entering functional expression considering the independent variable as u		Sum	Σ or	
Link: *Simulink → User-Defined Functions → Fcn*			Link: *Simulink → Math Operations → Sum*		
Product	Multiplies two or more functions on a common independent variable		Mux	Multiplexing:	
Link: *Simulink → Math Operations → Product*			Link: *Simulink → Signal Routing → Mux*		
Ramp	Generates function like $y(t) = t$		To Workspace	Send the simulink model output data to MATLAB Command Window	
Link: *Simulink → Sources → Ramp*			Link: *Simulink → Sinks → To Workspace*		
Slider Gain	Offers provision for variable gain through a slider		From Workspace	Obtains data from MATLAB Command Window to feed the data in a simulink model	
Link: *Simulink → Math Operations → Slider Gain*			Link: *Simulink → Sources → From Workspace*		

Modeling Difference Equations

We intend to highlight the modeling and simulation of difference equations of various types and order in this chapter mostly in pure mathematics context. Differential equations describe the dynamics of a continuous system whereas the difference equations are the discrete counterpart of the differential equations and represent a discrete system. Today's world is becoming more and more digital in all regards ranging from the theoretical analysis to everyday life. Never have we been so associated with computers as we did in last one or two decades. Analytical solution and SIMULINK modeling of the difference equations are very closely aligned to the corresponding analytical solution approach and modeling of the differential equations. A differential equation considers the value of the dependent variable over the whole domain of the independent variable on the contrary a difference equation only concerns the value of the dependent variable at some specific interval of the independent variable. Hidden the analysis, the system is continuous but we analyze the system just considering the samples. To transform a differential equation in algebraic equation, very often we employ Fourier or Laplace transform whereas Z transform transforms a difference equation to an algebraic equation. We carry out SIMULINK implementation on a substantial number of difference equations in this chapter so that the reader achieves mastery in simulating the equations.

4.1 Difference equations and the unit delay operator

Differential equations apply to the continuous case where we concern about the rates of change of the dependent variables with respect to the independent variable. Both the rate of change and the independent variable are assumed to be continuous. The advent of fast speed digital computer is redirecting many of the continuous system problems to be modeled in discrete form. Instead of dealing with the whole domain of the independent variable, functions or signals can be analyzed just by taking the samples of the function that is how a discrete analysis differs from the continuous one. A difference equation is the discrete counterpart of the ordinary differential equation. Difference equations relate the dependent variables in terms of the discrete sample time difference rather than the continuous rates of changes. The equation $y_k = y_{k-1} + y_{k-2}$ is an example of the difference equation, where k is an integer. The equation says that for any k, the output y is the sum of the preceding two terms. That is, when $k = 2$, $y_2 = y_1 + y_0$,

$k = 3$, $y_3 = y_2 + y_1$,

$k = 4$, $y_4 = y_3 + y_2$, and

.... so on.

Another style of writing the difference equation $y_k = y_{k-1} + y_{k-2}$ is $y[k] = y[k-1] + y[k-2]$. The order of a difference equation is the difference between the largest and smallest indices appearing in the equation. Thus the order of the difference equation $y_k = y_{k-1} + y_{k-2}$ is $k - (k-2) = 2$. The notions of the derivatives of differential and difference equations are little different. The discrete first derivative of any sequence y_k is defined as $\dot{y}_k$ or $Dy_k = y_{k+1} - y_k$. The second derivative is given by $\ddot{y}_k$ or $D^2 y_k = D(Dy_k) = D(y_{k+1} - y_k) = y_{k+2} - y_{k+1} - (y_{k+1} - y_k) = y_{k+2} - 2y_{k+1} - y_k$. In a similar fashion the higher order derivatives can be obtained, which are as follows:

$D^3 y_k = y_{k+3} - 3y_{k+2} + 3y_{k+1} - y_k$,

$D^4 y_k = y_{k+4} - 4y_{k+3} + 6y_{k+2} - 4y_{k+1} + y_k$,

$D^5 y_k = y_{k+5} - 5y_{k+4} + 10y_{k+3} - 10y_{k+2} + 5y_{k+1} - y_k$, and

.... so on.

Let us be familiarized with some difference equations:

$y[k] = y[k-1] \Rightarrow$ no input function, first order, independent variable is k, dependent variable is $y[k]$, k is integer, and $y[k]$'s are discrete integers

$3y_k = -4y_{k+1} + 2^{-k} \Rightarrow$ input function is 2^{-k}, first order, independent variable is k, dependent variable is $y[k]$, k is integer, and $y[k]$'s are discrete fractional

$y[m+1] - 4y[m] - 8y[m-1] - m = 0 \Rightarrow$ input function is m, second order, independent variable is m, dependent variable is $y[m]$, m is integer, and $y[m]$'s are discrete

$3y_k + y_{k-3} = -4y_{k-4} + 2k + 3k^2 \Rightarrow$ input function is $2k + 3k^2$, fourth order, independent variable is k, dependent variable is $y[k]$, k is integer, and $y[k]$'s are discrete

The dependent variable $y[k]$ is also termed as a sequence. We find the solution of a difference equation in simulink by employing the Unit Delay block. One can reach to the Unit Delay block in simulink library following the link *'Simulink $\rightarrow$ Discrete $\rightarrow$ Unit Delay'*. Open a new simulink model file and bring the block in the model file whose appearance is shown in figure 4.1. The difference equation can be transformed to an algebraic equation by the Z transform.

The Unit Delay block contains $\frac{1}{z}$ inside since the unilateral Z transform (which considers $k \geq 0$) of a sequence $y[k-1]$ is $\frac{1}{z} Y(z)$, where $Y(z)$ is the Z transform of the sequence $y[k]$.

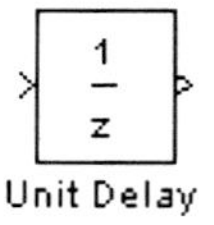

Figure 4.1 Unit Delay block

The purpose of the simulation of a difference equation is to obtain the wave shape of the sequence $y[k]$ in k domain. If we have $y[k]$ or $y[k+1]$, $y[k-1]$ or $y[k]$ can be obtained by unit delaying as follows:

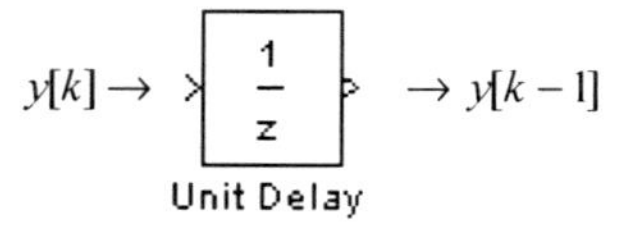

$y[k] \rightarrow$ [Unit Delay $\frac{1}{z}$] $\rightarrow y[k-1]$ $y[k+1] \rightarrow$ [Unit Delay $\frac{1}{z}$] $\rightarrow y[k]$

Unit delay of $y[k]$ to obtain $y[k-1]$ 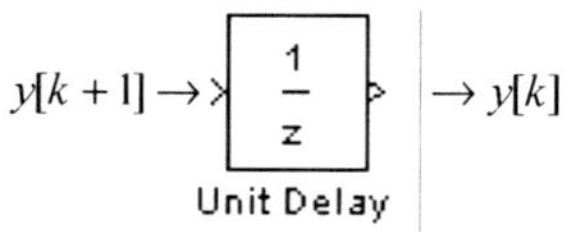   *Unit delay of $y[k+1]$ to obtain $y[k]$*

The unit delaying operation has to be carried out twice if we want to obtain $y[k-2]$ from $y[k]$ or $y[k]$ from $y[k+2]$ as presented schematically below:

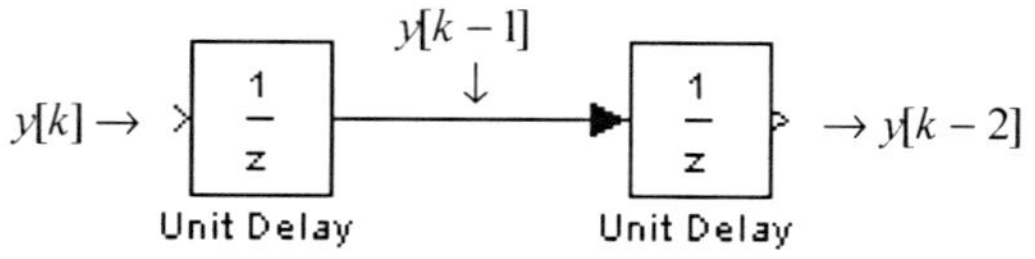

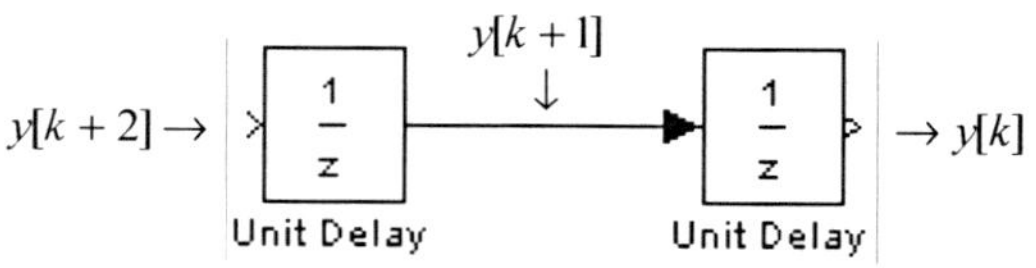

$y[k] \rightarrow$ [Unit Delay $\frac{1}{z}$] $\xrightarrow{y[k-1]}$ [Unit Delay $\frac{1}{z}$] $\rightarrow y[k-2]$ $y[k+2] \rightarrow$ [Unit Delay $\frac{1}{z}$] $\xrightarrow{y[k+1]}$ [Unit Delay $\frac{1}{z}$] $\rightarrow y[k]$

Depending on the order of the difference equation, we may need to employ the unit delay block multiple times. For instance, modeling of $y[k-3]$ from $y[k]$ can happen as follows:

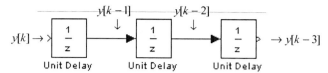

Similarly, the implementation of $y[k]$ from $y[k+3]$ can be realized as follows:

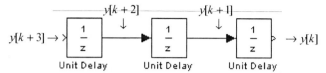

It is important to mention that the horizontal or time axis of the simulink Scope block simulates the discrete independent variable k. In the subsequent sections we discuss the modeling and simulation of various order difference equations.

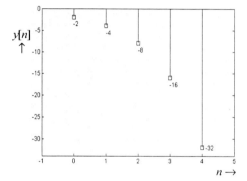

Figure 4.2(a) *Stem plot of $y[n]$ versus n*

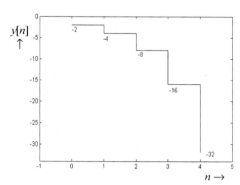

Figure 4.2(b) *Stair plot of $y[n]$ versus n*

4.2 Modeling a first order difference equation

The difference between the largest and smallest indices appearing in the equation is always 1 in a first order difference equation and we need one unit delay block and one initial condition to simulate the equation in simulink. The equation may contain the expression relating $y[k]$ and $y[k-1]$ or $y[k+1]$ and $y[k]$.

Let us begin with the simulation of the difference equation $y[n] - 2y[n-1] = 0$ without the input excitation over the domain $n \leq 4$ and with the initial condition $y[0] = -2$.

First we see the analytical solution. In order to solve the equation, the method of induction (inserting different n's) can be applied:

$$y[0] = -2, \quad y[1] = 2 \times (-2), \quad y[2] = 2 \times (-4), \quad y[3] = 2 \times (-8) \quad \ldots. \text{ etc.}$$

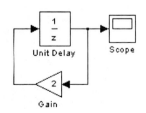

Figure 4.2(c) *Simulink model for $y[n] - 2y[n-1] = 0$*

$$y[0] = -2 \rightarrow$$

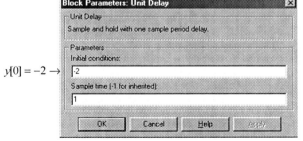

Figure 4.2(d) *Unit Delay block parameter window*

By inspection, one can write the general solution as $y[n] = -2^{n+1}$ where $n \geq 0$ and n is an integer. Within the given domain of n, we tabulate the $y[n]$ as follows:

n	$y[n]$
0	-2
1	-4
2	-8
3	-16
4	-32

If you plot the tabular $y[n]$ versus n, the plot takes the shape of stem plot of figure 4.2(a) in which both the n and $y[n]$ are discrete. The value of $y[n]$ can be kept constant within the consecutive interval of n that results the stair plot of figure 4.2(b). Our simulink Scope will provide us the stair plot. We restate our purpose as

$$\underbrace{\textit{with } y[n] - 2y[n-1] = 0 \textit{ and } y[0] = -2}_{\textit{problem space}} \quad \textit{and} \quad \underbrace{y[n] \textit{ versus } n \textit{ plot of figure } 4.2(b) \textit{ for } 0 \le n \le 4}_{\textit{solution space}}.$$

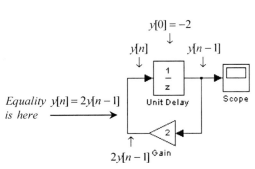

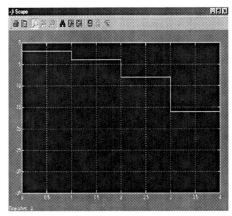

Figure 4.2(e) *Simulink model of figure 4.2(c) mentioning the sequence flow*

Figure 4.2(f) *Simulink Scope output for the model of figure 4.2(c)*

In order to model the problem in simulink, we must rearrange the equation so that the highest index is on the left side hence $y[n] = 2y[n-1]$. A unit delay can return $y[n-1]$ taking $y[n]$ as input. Modeling should be such that the equality between $y[n]$ and $2y[n-1]$ is maintained. Passing $y[n-1]$ through a Gain block of gain 2 can provide $2y[n-1]$. Figure 4.2(c) presents the model of the equation whose sequence flow is mentioned in figure 4.2(e). The procedure for the modeling is as follows:

⇒ *open a new simulink model file and click the library browser icon*

⇒ *bring a Unit Delay block in the model file following the link 'Simulink → Discrete → Unit Delay', doubleclick the block to set the initial condition $y[0] = -2$ in the dialog window as shown in figure 4.2(d)*

⇒ *bring one Scope block in the model file following the link 'Simulink → Sinks → Scope'*

⇒ *bring one Gain block in the model file following the link 'Simulink → Math Operations → Gain', flip the block, doubleclick the block, and set its gain to 2*

⇒ *place the three blocks relatively and connect them according to the figure 4.2(c)*

⇒ *in order to insert $n \le 4$, click in your model menu bar 'Simulation → Simulation parameters → Solver', set the start and stop time as 0 and 4 respectively, set the Solver options Type as Fixed-Step and discrete, and click OK*

⇒ *click the start simulation icon and doubleclick the scope to see the scope output as shown in figure 4.2(f) with autoscale setting*

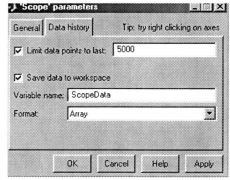

Figure 4.2(g) *Down links for the Scope parameter's window*

As you see, the analytical solution displayed in figure 4.2(b) and the simulink output displayed in figure 4.2(f) are identical. The horizontal and vertical axes of the figure 4.2(f) correspond to n and $y[n]$ respectively. One important point is that the Scope block must be connected to the lowest index sequence. Referring to figure 4.2(c), the Scope block is connected to the sequence output $y[n-1]$ not to the $y[n]$. Some enthusiastic reader might be interested to view the numerical data output from the simulation. Doubleclick the Scope block and click the parameters' icon (figure 3.15(a) is referred to for the relative position of the icon) of the Scope block from the menu bar. You see two down links in the dialog window: General and Data History as shown in figure 4.2(g). Click the Data history, check the Save data to workspace, change the format to Array, click OK, go to the simulink model file, and click the start simulation icon. The action you did is you ran your model and sent the output data to MATLAB workspace. Referring to figure 4.2(g), you see the variable name as ScopeData. In order to see the variables present in MATLAB, go to MATLAB Command Window and execute the following:

MATLAB Command

$\quad\quad\quad$ >>who ⏎

$\quad\quad\quad$ Your variables are:

$\quad\quad\quad$ ScopeData tout

As you see, there are two variables: ScopeData and tout. Just to see what the ScopeData contains:

$\quad\quad\quad$ >>ScopeData ⏎

$\quad\quad\quad$ ScopeData =

$$
\begin{array}{cc}
0 & -2 \\
1 & -4 \\
2 & -8 \\
3 & -16 \\
4 & -32 \\
\uparrow & \uparrow \\
n & y[n]
\end{array}
$$

The tabular data we presented before and the contents of ScopeData are identical. The ScopeData is basically a rectangular matrix whose first and second columns represent the independent integer n and the dependent sequence $y[n]$ values respectively. The variable tout in the workspace just contains the first column of the ScopeData or the independent integer n information. In a simplistic way, we mentioned the simulation of a first order difference equation. However, we close the section by presenting two more examples on simulating the first order difference equation.

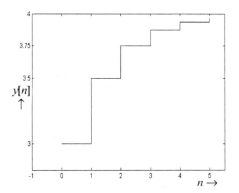

Figure 4.3(a) *Stair plot of $y[n]$ versus n for the example 1*

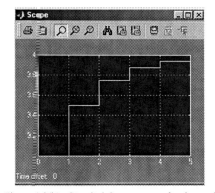

Figure 4.3(b) *Simulink Scope output for the model of figure 4.3(c)*

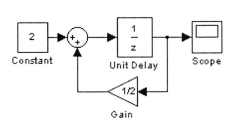

Figure 4.3(c) *Simulink model for the example 1*

Equality $y[n+1] = 2 + \dfrac{1}{2}y[n]$ is down

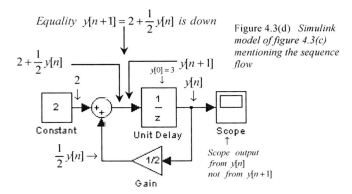

Figure 4.3(d) *Simulink model of figure 4.3(c) mentioning the sequence flow*

75

❖ ❖ Example 1

Simulate the difference equation $2y[n+1] = y[n] + 4$ over $0 \le n \le 5$ with the initial condition $y[0] = 3$.

⚏ Analytical Solution

Applying the method of induction, one would get the solution as $y[n] = 4 - \left(\dfrac{1}{2}\right)^n$ and the stair plot of which is displayed in figure 4.3(a). Our expectation is as follows:

$$\underbrace{\overbrace{}^{problem\ space}}_{with\ 2y[n+1] = y[n] + 4\ and\ y[0] = 3} \quad and \quad \underbrace{\overbrace{}^{solution\ space}}_{y[n]\ versus\ n\ plot\ of\ figure\ 4.3(a)\ for\ 0 \le n \le 5} .$$

⚏ Simulink Solution

The model that can simulate the problem is shown in figure 4.3(c) accompanying the sequence flow diagram in figure 4.3(d). First, rearrange the equation to keep the highest index on the left side hence $y[n+1] = \dfrac{1}{2} y[n] + 2$. The input function is 2, which is a constant. The gain with $y[n]$ is $\dfrac{1}{2}$, and a sum block can add $\dfrac{1}{2} y[n]$ and 2. Let us describe the model building procedure as follows:

⇒ *bring one Unit Delay, one Scope, and one Gain blocks in a new simulink model file as we brought in the beginning example*

⇒ *doubleclick the Unit Delay block and set the initial condition $y[0] = 3$ as we did before*

⇒ *doubleclick the Gain block, change its gain to 1/2, and flip the block*

⇒ *bring one Constant block in the model file following the link 'Simulink → Sources → Constant', doubleclick the block to change the constant value to 2, and click OK*

⇒ *bring one Sum block in the model file following the link 'Simulink → Math Operations → Sum'*

⇒ *place different blocks relatively and connect them according to the figure 4.3(c)*

⇒ *set the start and stop time as 0 and 5 respectively to insert $0 \le n \le 5$ and set the Solver options Type as Fixed-Step and discrete as we did before*

⇒ *click the start simulation icon and doubleclick the Scope to see the Scope output as shown in figure 4.3(b) with the autoscale setting*

One can compare that the analytical plot of figure 4.3(a) and the simulink output of figure 4.3(b) are identical.

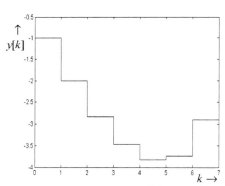

Figure 4.4(a) *Stair plot of $y[k]$ versus k for the example 2*

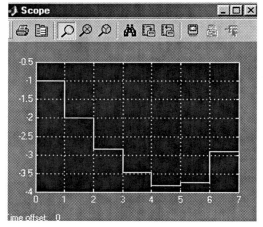

Figure 4.4(b) *Simulink Scope output for the model of figure 4.4(c)*

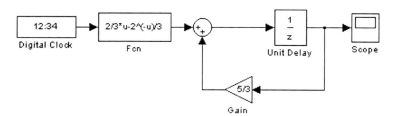

Figure 4.4(c) *Simulink model for the example 2*

❖ ❖ Example 2

Find the wave shape of $y[k]$ over $0 \leq k \leq 7$ for the difference equation $3y[k+1] - 5y[k] - 2k + 2^{-k} = 0$ with the initial condition $y[0] = -1$.

⎙ *Analytical Solution*

The reader can find asistance how to solve the difference equations analytically by going through [1] and [16]. For space reason, we provide the analytical solution of the difference equation as $y[k] =$

$\frac{3}{14}\left(\frac{5}{3}\right)^{k} + \frac{2}{7}\left(\frac{1}{2}\right)^{k} - k - \frac{3}{2}$ which when

plotted displays the shape as shown in figure 4.4(a). We refer the question statement as follows:

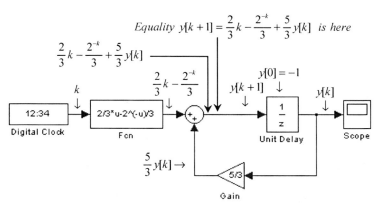

$\text{Equality } y[k+1] = \frac{2}{3}k - \frac{2^{-k}}{3} + \frac{5}{3}y[k] \text{ is here}$

Figure 4.4(d) *Simulink model of figure 4.4(c) mentioning the sequence flow*

$$\underbrace{\text{problem space}}_{\text{with } 3y[k+1] - 5y[k] - 2k + 2^{-k} = 0 \text{ and } y[0] = -1} \text{ and } \underbrace{\text{solution space}}_{y[k] \text{ versus } k \text{ plot of figure 4.4(a) for } 0 \leq k \leq 7}.$$

⎙ *Simulink Solution*

The simulation model is depicted in figure 4.4(c) followed by the sequence flow diagram in figure 4.4(d). Keeping the highest index on the left side, we have the equation $y[k+1] = \frac{5}{3}y[k] + \frac{2}{3}k - \frac{2^{-k}}{3}$. The gain we need for $y[k]$ is $\frac{5}{3}$. The input excitation $\frac{2}{3}k - \frac{2^{-k}}{3}$ is a function of the independent variable k and whose simulink code is 2/3*u-2^(-u)/3. The code contains the independent variable u instead of k because of the block input definition. The Digital Clock of figure 4.4(c) simulates k so the Clock generates 0, 1, 2, 3, 4, 5, 6 and 7 over the given domain $0 \leq k \leq 7$. If the domain were $0 \leq k \leq 3$, it would generate 0, 1, 2, and 3. The modeling procedure is mentioned as follows:

⇒ *bring one Unit Delay, one Scope, one Gain, and one Sum blocks in a new simulink model file*

⇒ *doubleclick the Unit Delay block, set the initial condition to $y[0] = -1$, doubleclick the Gain block, set its gain to 5/3, and flip the block*

⇒ *bring one Digital Clock and one Fcn blocks in the model file following the links 'Simulink → Sources → Digital Clock' and 'Simulink → User-Defined Functions → Fcn' respectively*

⇒ *doubleclick the Fcn block, type 2/3*u-2^(-u)/3 in the parameter expression of the dialog window to enter $\frac{2}{3}k - \frac{2^{-k}}{3}$, click OK, and enlarge the Fcn block to see the expression inside it*

⇒ *place different blocks relatively and connect them according to the figure 4.4(c)*

⇒ *set the start and stop time as 0 and 7 respectively with Solver options Type as Fixed-Step and discrete*

⇒ *click the start simulation icon and doubleclick the scope to see the output as shown in figure 4.4(b)*

The horizontal and vertical axes of figure 4.4(b) represent k and $y[k]$ respectively. The analytical and simulink outputs shown in figures 4.4(a) and 4.4(b) are the replica of each other. The scope is connected with the sequence flow of $y[k]$ not with the flow of $y[k+1]$.

4.3 Modeling a second order difference equation

In this section we present the modeling of the second order difference equation which can be identified by noticing the difference between the largest and smallest indices appearing in the equation as 2. In consideration of modeling the equation, two Unit Delay blocks and two initial conditions are necessary.

As a beginning example, let us consider the equation $3y[k-1] - 2y[k+1] - y[k] = 0$ with the initial conditions $y[0] = \pi$ and $y[1] = -1$. The largest and smallest indices in the equation are $k+1$ and $k-1$ thereby indicating the order 2. The equation does not contain any excitation function and we wish to see the output $y[k]$ for $0 \leq k \leq 6$.

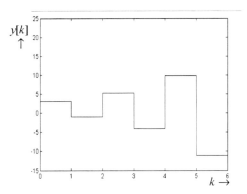

Figure 4.5(a) *Stair plot of* $y[k]$ *versus* k

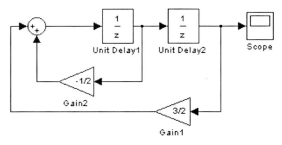

Figure 4.5(b) *Simulink model for* $3y[k-1] - 2y[k+1] - y[k] = 0$

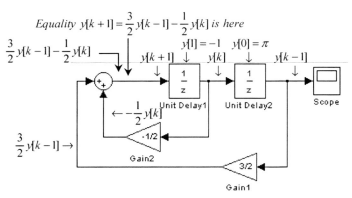

Figure 4.5(c) *Simulink model of figure 4.5(b) mentioning the sequence flow*

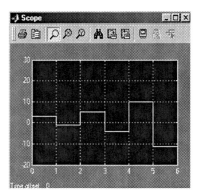

Figure 4.5(d) *Simulink Scope output for the model of figure 4.5(b)*

The analytical solution of the equation is given by $y[k] = \frac{2}{5}\left(-\frac{3}{2}\right)^k (\pi + 1) + \frac{3}{5}\pi - \frac{2}{5}$, the stair plot of which is presented in figure 4.5(a). Our objective of the simulation is to obtain the wave shape of figure 4.5(a) in simulink Scope.

To commence with simulink, the first step is to rearrange the equation as $y[k+1] = \frac{3}{2}y[k-1] - \frac{1}{2}y[k]$. Two Unit Delay blocks are necessary in order to obtain $y[k-1]$ from $y[k+1]$. Two Gain blocks associated with $y[k-1]$ and $y[k]$ require the gains as $\frac{3}{2}$ and $-\frac{1}{2}$ respectively. However, figures 4.5(b) and 4.5(c) show the simulink model and the model with the sequence flow respectively. We mention the modeling as follows:

$\Rightarrow$ *open a new simulink model file and click the library browser icon*

$\Rightarrow$ *bring one Unit Delay block in the model file, rename the block as Unit Delay1, copy the block, and paste it to see the block Unit Delay2, doubleclick the Unit Delay1, set the initial condition $y[1] = -1$ by typing -1, doubleclick the Unit Delay2, and set the initial condition $y[0] = \pi$ by typing pi*

$\Rightarrow$ *bring one Gain block, rename it as Gain1, doubleclick the block to set the gain as $\frac{3}{2}$ (code is 3/2), flip the block, copy the Gain1 block in the menu bar, paste it to see the block Gain2, doubleclick the block to set the gain as $-\frac{1}{2}$ (code is -1/2), and if necessary, enlarge the block to display the gain*

$\Rightarrow$ *bring one Scope and one Sum blocks in the model file, place various blocks relatively, and connect them according to the figure 4.5(b) (notice that the Scope is connected with the least index sequence)*

$\Rightarrow$ *set the start and stop time as 0 and 6 respectively to insert $0 \le k \le 6$ and set the Solver options Type as Fixed-Step and discrete*

⇒ *click the start simulation icon and doubleclick the Scope to see the Scope output as shown in figure 4.5(d) (identical with the analytical plot of figure 4.5(a)) with the autoscale setting*

Now we present two more examples on modeling the second order difference equation in simulink.

❖ ◆ Example 1

Simulate the equation $3y[m-1] - 4y[m+1] + y[m] - e^{-m} = m$ to obtain $y[m]$ over $0 \le m \le 9$ with the initial conditions $y[0] = -1$ and $y[1] = 2$.

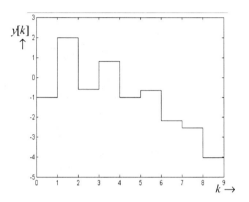

Figure 4.6(a) *Stair plot of $y[k]$ versus k*

Figure 4.6(b) *Simulink scope output for the model of figure 4.6(c)*

In horizontal axis,
$1 \Leftrightarrow k = 0$
$10 \Leftrightarrow k = 9$

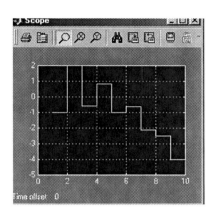

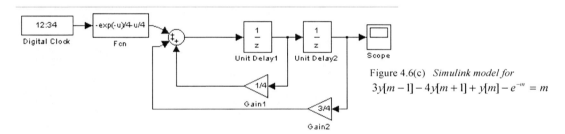

Figure 4.6(c) *Simulink model for*
$3y[m-1] - 4y[m+1] + y[m] - e^{-m} = m$

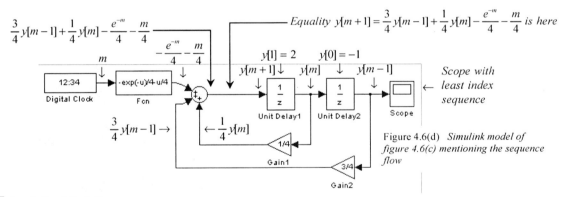

Figure 4.6(d) *Simulink model of figure 4.6(c) mentioning the sequence flow*

⬚ *Analytical Solution*

The indices m, $m-1$, or $m+1$ bear relative meanings as regards to a sequence. If we have constant coefficient difference equation, all positive index difference equation, or constant input difference equation, the lower limit of the independent variable domain such as $0 \le m \le 9$ does not affect the solution of the equation. Furthermore, simulink assumes that the value of the sequence is zero before t or $m = 0$. If the input excitation is some function of the independent variable or index, the solution becomes dependent on the index m hence the domain description and the initial conditions must be consistent. In this problem, $y[0]$ occurs for $m = 1$ but we demand the output to be at $m = 0$, which is inconsistent. The problem should have been stated as follows:

Simulate the equation $3y[m-1] - 4y[m+1] + y[m] - e^{-m} = m$ to obtain $y[k]$ over $0 \le k \le 9$ with the initial conditions $y[0] = -1$ and $y[1] = 2$, where k is a dummy variable.

79

However, carrying out the method of induction, one would get the discrete solution of the difference equation in tabular form as follows:

$y[k]$	−1	2	−0.5920	0.8182	−1.0019	−0.6414	−2.1634	−2.5225	−4.0035	−4.8929
k	0	1	2	3	4	5	6	7	8	9
m	1	2	3	4	5	6	7	8	9	10

The stair plot of the above tabular data is shown in figure 4.6(a). We view the simulation as follows:

$$\overbrace{\text{with } 3y[m-1]-4y[m+1]+y[m]-e^{-m}=m}^{\textit{problem space}} \text{ and } \overbrace{y[k] \textit{ versus } k \textit{ plot of figure 4.6(a)}}^{\textit{solution space}} .$$
$$\text{and } y[0]=-1 \text{ and } y[1]=2 \qquad\qquad \text{for } 0 \le k \le 9$$

⊟ Simulink Solution

Figures 4.6(c) and 4.6(d) present the simulink model and the model with the sequence flow description respectively. The important elements of the modeling are as follows:

⇒ *the rearranged equation, the gain for $y[m]$, and the gain for $y[m-1]$ are $y[m+1] = \dfrac{3}{4} y[m-1] +$*

$\dfrac{1}{4} y[m] - \dfrac{e^{-m}}{4} - \dfrac{m}{4}$, $\dfrac{1}{4}$, *and* $\dfrac{3}{4}$ *respectively*

⇒ *the input function* $-\dfrac{e^{-m}}{4} - \dfrac{m}{4}$ *has the simulink code -exp(-u)/4-u/4 assuming u as the independent variable*

⇒ *the least index $m-1$ is shifted by one sample from the reference $m=0$ hence we take the start and stop time of the simulation as 1 and 10 respectively*

⇒ *open a new simulink model file and click the simulink library icon*

⇒ *bring one Unit Delay block, rename the block as Unit Delay1, doubleclick the block to set the initial condition $y[1] = 2$, copy the Unit Delay1 block, paste the block to see the Unit Delay2, doubleclick the Unit Delay2 to insert $y[0] = -1$, and bring one Scope block*

⇒ *bring one Gain block, rename the block as Gain1, doubleclick the block to set its gain as 1/4, flip the block, copy the Gain1, paste the block to see the Gain2, doubleclick the Gain2 to insert its gain as 3/4*

⇒ *bring one Sum block, doubleclick the block to insert $+++$ in the List of signs in the dialog window to convert the block as a three input Sum block*

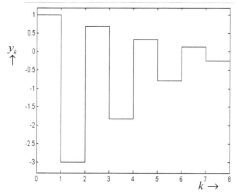

Figure 4.7(a) *Stair plot of y_k versus k*

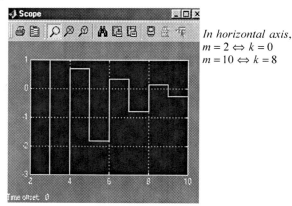

In horizontal axis,
$m = 2 \Leftrightarrow k = 0$
$m = 10 \Leftrightarrow k = 8$

Figure 4.7(b) *Simulink Scope output for the model of figure 4.7(c)*

⇒ *bring one Digital Clock and one Fcn blocks, doubleclick the Fcn block to enter the functional expression for $-\dfrac{e^{-m}}{4} - \dfrac{m}{4}$ as -exp(-u)/4-u/4 in the parameters expression of the dialog window, enlarge the block to see the inside expression*

⇒ *place various blocks relatively and connect them according to the figure 4.6(c), set the start and stop time as 1 and 10 respectively, and set the Solver options Type as Fixed-Step and discrete*

⇒ *finally click the start simulation icon and doubleclick the Scope to see the Scope output as shown in figure 4.6(b) (same as that of the analytical plot of figure 4.6(a)) with the autoscale setting*

❖ ❖ Example 2

The examples we illustrated so far did not comprise the equations with variable coefficients. This example considers the coefficient of the sequence being function of the index. Let us simulate the equation $3y_{m-2} - (m+2)y_m = 2^{-m}$ to obtain y_k over $0 \leq k \leq 8$, where k is a dummy variable with the initial conditions $y_0 = 1$ and $y_1 = -3$.

▱ Analytical Solution

Inserting from $m = 2$ to $m = 10$ in the given equation, one obtains the following tabular solution:

$y[k]$	1	−3	0.6875	−1.825	0.3333	−0.7866	0.1230	−0.2631	0.0365
k	0	1	2	3	4	5	6	7	8
m	2	3	4	5	6	7	8	9	10

We graphed above tabular data in figure 4.7(a) and list our expectations as follows:

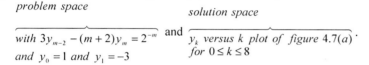

problem space

$$\overbrace{\text{with } 3y_{m-2} - (m+2)y_m = 2^{-m}}$$ and $$\overbrace{y_k \text{ versus } k \text{ plot of figure 4.7(a)}}$$.
and $y_0 = 1$ and $y_1 = -3$ for $0 \leq k \leq 8$

solution space

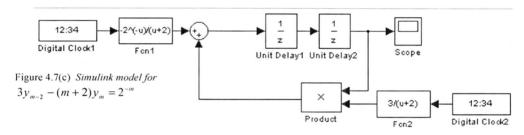

Figure 4.7(c) *Simulink model for* $3y_{m-2} - (m+2)y_m = 2^{-m}$

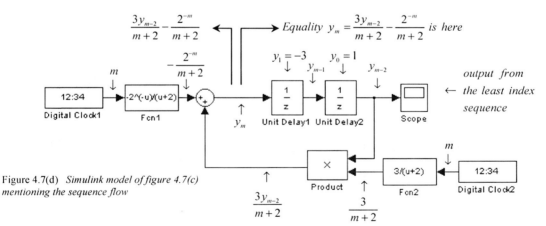

Figure 4.7(d) *Simulink model of figure 4.7(c) mentioning the sequence flow*

▱ Simulink Solution

Although the equation does not contain the intermediate sequence y_{m-1}, we still and all need two Unit Delay blocks. A simulink model might be dependent on the arrangement of the equation. For instance, the rearranged equations $y_m = \dfrac{3y_{m-2} - 2^{-m}}{m+2}$ and $y_m = \dfrac{3y_{m-2}}{m+2} - \dfrac{2^{-m}}{m+2}$ can layout two simulink models despite they return the identical solution. Choosing the second form, the modeling is delineated in figure 4.7(c) followed by the sequence descriptory representation in figure 4.7(d). As usual, the modeling procedure is explicated below:

⇒ *the rearranged equation says that the gain for* y_{m-2} *is* $\dfrac{3}{m+2}$, *the code of which is 3/(u+2) considering u as the independent variable*

81

$\Rightarrow$ *considering u as the independent index, we write the input function* $-\dfrac{2^{-m}}{m+2}$ *as –2^(-u)/(u+2)*

$\Rightarrow$ *open a new simulink model file and click the simulink library icon*

$\Rightarrow$ *bring one Unit Delay block, rename the block as Unit Delay1, doubleclick the block to set the initial condition* $y_1=-3$, *copy the Unit Delay1 block, paste the block to see the Unit Delay2, doubleclick the Unit Delay2 to insert* $y_0=1$, *and bring one Scope block*

$\Rightarrow$ *bring one Digital Clock block, rename the block as Digital Clock1, bring one Fcn block, rename the block as Fcn1, and doubleclick the block to enter the functional expression for* $-\dfrac{2^{-m}}{m+2}$ *as –2^(-u)/(u+2) in the parameters expression of the dialog window*

$\Rightarrow$ *bring one Sum and one Product blocks and flip the Product block*

$\Rightarrow$ *copy the Fcn1, paste it to see Fcn2, doubleclick the block to enter the functional expression for* $\dfrac{3}{m+2}$ *as 3/(u+2) in the parameters expression of the dialog window, and flip the block*

$\Rightarrow$ *copy the Digital Clock1, paste it to see the Digital Clock2 (both the Digital Clock1 and Digital Clock2 serve the purpose for generating the discrete integer m), and flip the Digital Clock2*

$\Rightarrow$ *place various blocks relatively and connect them according to the figure 4.7(c), set the start and stop time as 2 and 10 respectively, and set the Solver options Type as Fixed-Step and discrete*

$\Rightarrow$ *finally click the start simulation icon and doubleclick the Scope to see the Scope output as shown in figure 4.7(b) (same as that of the analytical plot of figure 4.7(a)) with the autoscale setting*

Now we switch to the next section for simulating the higher order difference equations.

4.4 Modeling a third order difference equation

We focus the modeling and simulation of the third order difference equations in consideration of the example of higher order ones in this section. As far as the definition of the order is concern, the difference between the largest and smallest indices appearing in the equation must be 3. The employment of three Unit Delay blocks and three initial conditions is obvious.

Let us project the concepts and ideas utilized in previous sections in conjunction with one more Unit Delay block and one initial condition considering the beginning example of a third order difference equation without the input function and with three initial conditions, $5y[k+3]-4y[k+2]+2y[k+1]-3y[k]=0$, $y[0]=-1$, $y[1]=1$, and $y[2]=-3$. We intend to see the output for $0\le k\le 10$.

By insertion of the value of k from 0 to 10, we found the following table on the rearranged equation $y[k+3]=\dfrac{4}{5}y[k+2]-\dfrac{2}{5}y[k+1]+\dfrac{3}{5}y[k]$:

$y[k]$	−1	1	−3	−3.4	−0.92	−1.176	−2.6128	−2.1718	−1.398	−1.8173	−2.1978
k	0	1	2	3	4	5	6	7	8	9	10

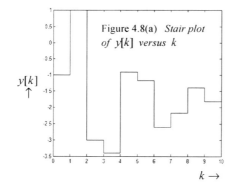

Figure 4.8(a) *Stair plot of $y[k]$ versus k*

$k \rightarrow$

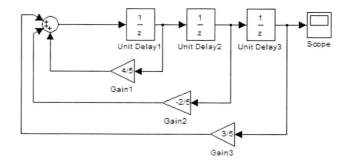

Figure 4.8(b) *Simulink model for* $5y[k+3]-4y[k+2]+2y[k+1]-3y[k]=0$

Depicted in figure 4.8(a) is the stair plot of the above tabular data. Figure 4.8(b) shows the model that can simulate the third order difference equation without the excitation. Since the sequence flow description is more meaningful for understanding rather than the model, we also included the sequence flow description in figure 4.8(c). We believe that the reader is by now able to construct the model for the difference equations of the first and second order. Open a new simulink model file and construct the model as shown in figure 4.8(b).

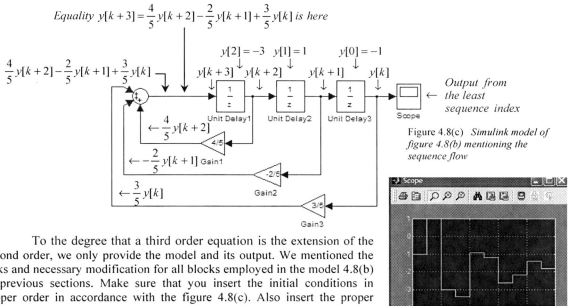

Equality $y[k+3] = \frac{4}{5}y[k+2] - \frac{2}{5}y[k+1] + \frac{3}{5}y[k]$ *is here*

Figure 4.8(c) *Simulink model of figure 4.8(b) mentioning the sequence flow*

To the degree that a third order equation is the extension of the second order, we only provide the model and its output. We mentioned the links and necessary modification for all blocks employed in the model 4.8(b) in previous sections. Make sure that you insert the initial conditions in proper order in accordance with the figure 4.8(c). Also insert the proper start-stop time and correct solver settings. Successful model construction and simulation should yield you the output of the Scope with the autoscale setting as illustrated in figure 4.8(d), which is the perfect match of the analytical stair plot of the figure 4.8(a). Very briefly we present now two more third order difference equation simulation.

Figure 4.8(d) *Simulink Scope output for the model of figure 4.8(b)*

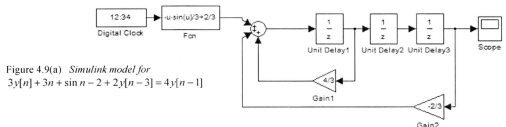

Figure 4.9(a) *Simulink model for*
$3y[n] + 3n + \sin n - 2 + 2y[n-3] = 4y[n-1]$

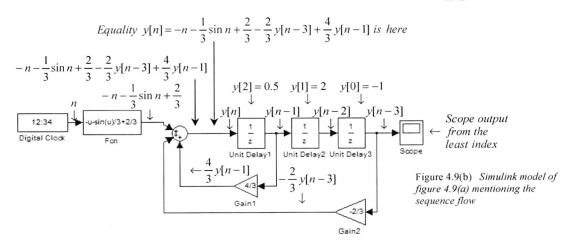

Equality $y[n] = -n - \frac{1}{3}\sin n + \frac{2}{3} - \frac{2}{3}y[n-3] + \frac{4}{3}y[n-1]$ *is here*

Figure 4.9(b) *Simulink model of figure 4.9(a) mentioning the sequence flow*

❖ ❖ Example 1

Simulate the third order difference equation $3y[n] + 3n + \sin n - 2 + 2y[n-3] = 4y[n-1]$ to obtain the sequence $y[k]$ over $0 \le k \le 6$ with the initial conditions $y[0] = -1$, $y[1] = 2$, and $y[2] = 0.5$.

The equation possesses some input function. Not only does the rearranged form $y[n] = -n - \frac{1}{3}\sin n + \frac{2}{3} -$

$\frac{2}{3}y[n-3] + \frac{4}{3}y[n-1]$ put the convenient form for simulink modeling but also it is worthwhile for finding the

analytical solution in discrete or tabular form and that is presented as follows:

$y[k]$	−1	2	0.5	−1.0470	−5.8105	−12.0943	−20.6679
k	0	1	2	3	4	5	6
n	3	4	5	6	7	8	9

Graphing the above table we see the stair plot of figure 4.9(c). The necessary simulink model and its sequence flow schematics are presented in figures 4.9(a) and 4.9(b) respectively. Construct the model and run it. The stair plot of the figure 4.9(c) and the simulink Scope output displayed in figure 4.9(d) with the autoscale setting easily make sense about their identicalness.

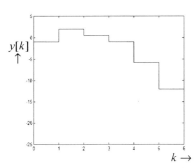

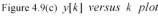

$k \rightarrow$

Figure 4.9(c) $y[k]$ *versus* k *plot*

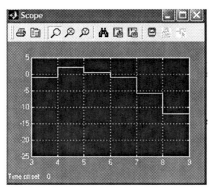

In horizontal axis of the scope

$n = 3 \Leftrightarrow k = 0$
$n = 9 \Leftrightarrow k = 6$

Figure 4.9(d) *Simulink Scope output for the model of figure 4.9(a)*

❖ ❖ Example 2

Not necessarily a difference equation holds the linear form or constant coefficient terms sometimes the largest index sequence might be the function of division, multiplication, or powered form of the intermediate

sequences such as $(n+1)^2 y[n] + 2\cos n - \left(\frac{2y[n-3] - 4y[n-1]}{y[n-2] - 2}\right)^2 = 0$. Let us obtain the solution $y[k]$ for $0 \le k \le 7$

subject to the initial values $y[0] = 2$, $y[1] = -1$, and $y[2] = 3$.

✒ Solution

In order to verify the simulink output's consistency, we recurrently provide the analytical data as tabular

form. The equation in order becomes $y[n] = \frac{1}{(n+1)^2}\left[-2\cos n + \left(\frac{2y[n-3] - 4y[n-1]}{y[n-2] - 2}\right)^2\right]$ whose thorough

computation inserting $n = 3$ to $n = 10$ is presented in the following table for different k:

$y[k]$	2	−1	3	0.5682	0.7826	0.0958	−0.0314	−0.0112
k	0	1	2	3	4	5	6	7
n	3	4	5	6	7	8	9	10

The stair plot of the above tabular data essentially yields the figure 4.10(c). We refer to the figures 4.10(a) and 4.10(b) for the simulink model and its sequence flow description respectively. A lot of mathematical manipulations are hidden in the modeling that becomes self-expressive from the functional flow indication of figure 4.10(b). We suggest you to follow the arrow indication to learn about the functional flow. Just to explain the modeling in an easy way (referring to figure 4.10(a)),

⇒ both the Digital Clock1 and Digital Clock2 generate the discrete integer n

⇒ the block Fcn3 generates $\frac{1}{(n+1)^2}$ taking n as the input but the functional expression code for Fcn1

must be set as $1/(u+1)^2$ considering u as the independent variable

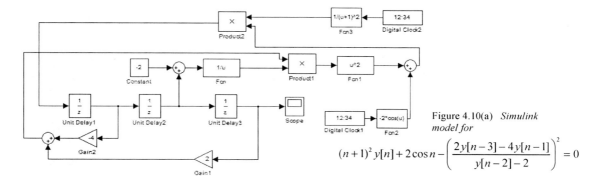

Figure 4.10(a) *Simulink model for*

$$(n+1)^2 y[n] + 2\cos n - \left(\frac{2y[n-3] - 4y[n-1]}{y[n-2] - 2} \right)^2 = 0$$

Equality $y[n] = \frac{1}{(n+1)^2} \left[-2\cos n + \left(\frac{2y[n-3] - 4y[n-1]}{y[n-2] - 2} \right)^2 \right]$ *is here*

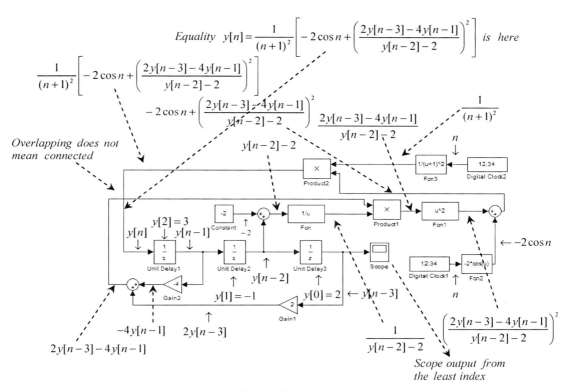

Figure 4.10(b) *Simulink model of figure 4.10(a) mentioning the sequence flow*

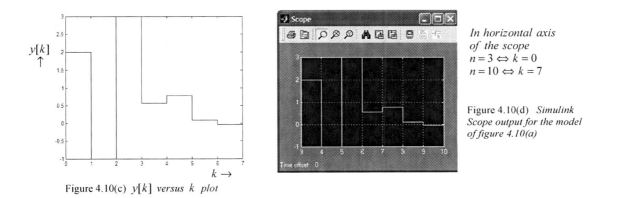

Figure 4.10(c) $y[k]$ *versus* k *plot*

In horizontal axis of the scope
$n = 3 \Leftrightarrow k = 0$
$n = 10 \Leftrightarrow k = 7$

Figure 4.10(d) *Simulink Scope output for the model of figure 4.10(a)*

$\Rightarrow$ *one can form the term* $\dfrac{1}{(n+1)^2}\left[-2\cos n+\left(\dfrac{2y[n-3]-4y[n-1]}{y[n-2]-2}\right)^2\right]$ *from the product of* $\dfrac{1}{(n+1)^2}$ *and*

$\left[-2\cos n+\left(\dfrac{2y[n-3]-4y[n-1]}{y[n-2]-2}\right)^2\right]$ *(performed by the block Product2)*

$\Rightarrow$ *the term* $\left[-2\cos n+\left(\dfrac{2y[n-3]-4y[n-1]}{y[n-2]-2}\right)^2\right]$ *is the addition of* $-2\cos n$ *and* $\left(\dfrac{2y[n-3]-4y[n-1]}{y[n-2]-2}\right)^2$

$\Rightarrow$ *the block Fcn takes the input as* $-2+y[n-2]$ *and returns the output as* $\dfrac{1}{-2+y[n-2]}$ *because the functional code is 1/u*

$\Rightarrow$ *the blocks Gain1 and Gain2 provide the gains 2 and –4 for* $y[n-3]$ *and* $y[n-1]$ *of the numerator* $2y[n-3]-4y[n-1]$

$\Rightarrow$ *the division form* $\dfrac{2y[n-3]-4y[n-1]}{y[n-2]-2}$ *is converted to multiplication form as* $\{2y[n-3]-4y[n-1]\}\times$

$\dfrac{1}{y[n-2]-2}$ *(performed by the block Product1)*

$\Rightarrow$ *the block Fcn1 takes* $\dfrac{2y[n-3]-4y[n-1]}{y[n-2]-2}$ *as input and returns* $\left(\dfrac{2y[n-3]-4y[n-1]}{y[n-2]-2}\right)^2$ *as output by holding the functional code u^2*

$\Rightarrow$ *however, open a new simulink model file and bring all necessary blocks in your model file as shown in figure 4.10(a), place and connect them relatively according to the figure, insert the initial conditions, flip or enlarge the blocks if it is necessary, set the exact functional expression for the Fcn blocks, set the start and stop time as 3 and 10 to insert* $k=0$ *and* $k=7$ *respectively, and set the Solver options Type as Fixed-Step and discrete*

$\Rightarrow$ *finally, click the start simulation icon and doubleclick the Scope to see the Scope output as shown in figure 4.10(d) (same as that of the analytical plot of figure 4.10(c)) with the autoscale setting*

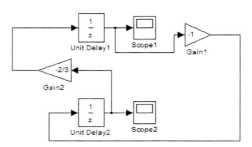

Figure 4.11(a) *Simulink model for the system*
$x[n]+y[n+1]=0$ *and* $2y[n]+3x[n+1]=0$

Comparing the two figures, we can best infer as stating that simulink is a very effective tool for simulating the higher order difference equations.

4.5 Modeling a system of difference equations

The first and second order difference equations form the framework for some of the most complex discrete time systems. A clear

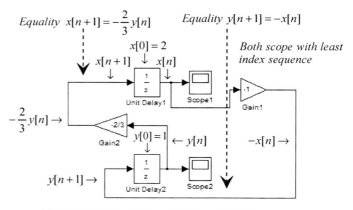

Figure 4.11(b) *Simulink model of figure 4.11(a) mentioning the sequence flow*

understanding of these elementary equations is necessary before you examine how simulink can be advantageous in simulating a system of difference equations. A practical discrete system may necessitate employing multiple outputs coupled by some mathematical relationship and sharing a common index variation that is how a system of difference equations is evolved. Our effort here is to start with the following system in which no input functions are associated:

$$x[n] + y[n+1] = 0$$
$$2y[n] + 3x[n+1] = 0$$

We wish to see the outputs $x[k]$ and $y[k]$ over $0 \le k \le 6$ covering the conditions $x[0] = 2$ and $y[0] = 1$.

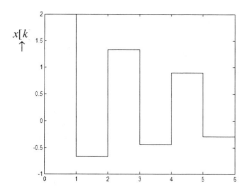

Figure 4.11(c) *Stair plot for $x[k]$ versus k of the system*

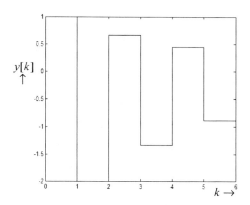

Figure 4.11(d) *Stair plot for $y[k]$ versus k of the system*

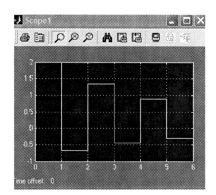

Figure 4.11(e) *Simulink Scope1 output for $x[k]$ for the model of figure 4.11(a)*

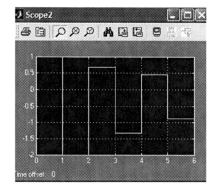

Figure 4.11(f) *Simulink Scope2 output for $y[k]$ for the model of figure 4.11(a)*

Before we rush to the simulation, let us compute the analytical outputs for the system by inserting $n = 0$ to $n = 6$ in the set:

$$n = 0 \Rightarrow x[0] + y[1] = 0 \text{ and } 2y[0] + 3x[1] = 0 \xrightarrow{\text{applying the initial conditions}} 2 + y[1] = 0 \text{ and } 2 + 3x[1] = 0 \xrightarrow{\text{solution}}$$

$$x[1] = -\frac{2}{3} \text{ and } y[1] = -2$$

$$n = 1 \Rightarrow x[1] + y[2] = 0 \text{ and } 2y[1] + 3x[2] = 0 \xrightarrow{\text{from the previous result}} -\frac{2}{3} + y[2] = 0 \text{ and } 2 \times (-2) + 3x[2] = 0$$

$$\xrightarrow{\text{solution}} x[2] = \frac{4}{3} \text{ and } y[2] = \frac{2}{3}, \dots \text{ and so on.}$$

Continuing this way, one can prepare the following tabular output for $x[k]$ and $y[k]$:

$x[k]$	2	−0.6666	1.3333	−0.4444	0.8888	−0.2963	0.5926
$y[k]$	1	−2	0.6666	−1.3333	0.4444	−0.8888	0.2963
k	0	1	2	3	4	5	6

Attached figures 4.11(c) and 4.11(d) are also the stair plots for $x[k]$ and $y[k]$ respectively and that is what we are after. To commence with simulink, both equations need to be arranged so that the largest index sequence as regards to $x[k]$ and $y[k]$ is on the left side of the equations from what cause

$$x[n+1] = -\frac{2}{3}y[n] \text{ and } y[n+1] = -x[n].$$

Despite a Unit delay block can produce $x[n]$ taking $x[n+1]$ as input, the first equality is not formed from $x[n]$ instead from $y[n]$ followed by a gain $-\frac{2}{3}$. In like manner, the second equality is not conduced from $y[n]$ but from $x[n]$ with a gain of -1. Depicted figures 4.11(a) and 4.11(b) represent the simulink model and the sequence flow version respectively. Successful model construction and simulation should return you the Scope1 and Scope2 outputs as exhibited by the figures 4.11(e) and 4.11(f) for $x[n]$ and $y[n]$ respectively.

Now we mention two extended examples of the simulation in which some input function and higher order sequence are involved.

♦ ♦ Example 1

Simulate the system
$$\begin{cases} 3y[n-1] + x[n] = 2^{-n} \\ 2 - nx[n-1] + 3y[n-1] + y[n] = 0 \end{cases}$$
to obtain the discrete wave shapes for $x[k]$ and $y[k]$ over the domain $0 \le k \le 5$ considering the initial values $x[0] = 1$ and $y[0] = -1$.

☞ Solution

The smallest and the largest indexes associated with the two equations are $(n-1, n)$ and $(n-1, n)$ for both sequences. That is why each of the sequences $x[k]$ and $y[k]$ is of order 1 and each one needs one Unit Delay block for the simulation. The suitable form for the simulink is given by
$$\begin{cases} x[n] = 2^{-n} - 3y[n-1] \\ y[n] = -2 + nx[n-1] - 3y[n-1] \end{cases}$$. As

attached, figures 4.12(a) and 4.12(b) represent the simulink model for the system and its sequence flow descriptory counterpart respectively. The Unit Delay1 and Unit delay2 correspond to $x[k]$ and $y[k]$ respectively. Equality for both difference equations must have to be maintained. Since both sequences are

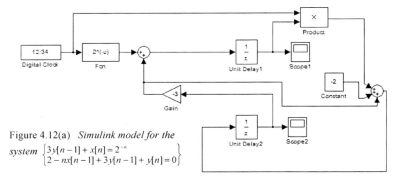

Figure 4.12(a) *Simulink model for the system* $\begin{cases} 3y[n-1] + x[n] = 2^{-n} \\ 2 - nx[n-1] + 3y[n-1] + y[n] = 0 \end{cases}$

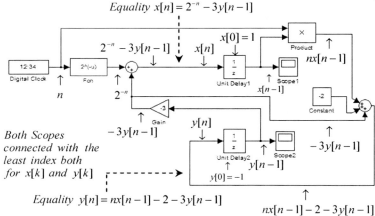

Figure 4.12(b) *Simulink model of figure 4.12(a) mentioning the sequence flow*

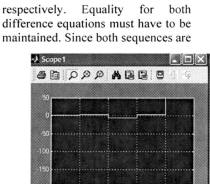

Figure 4.12(c) *Simulink Scope1 output for $x[k]$ for the model of figure 4.12(a)*

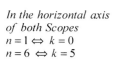

In the horizontal axis of both Scopes
$n = 1 \Leftrightarrow k = 0$
$n = 6 \Leftrightarrow k = 5$

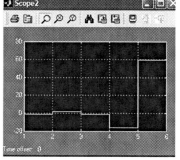

Figure 4.12(d) *Simulink Scope2 output for $y[k]$ for the model of figure 4.12(a)*

lagging by one sample, we should insert the start and stop time as 1 and 6 for $k = 0$ and $k = 5$ respectively. Having constructed and run the model successfully, simulink should turnout the Scope harvests as shown in figures 4.12(c) and 4.12(d) respectively with the autoscale setting.

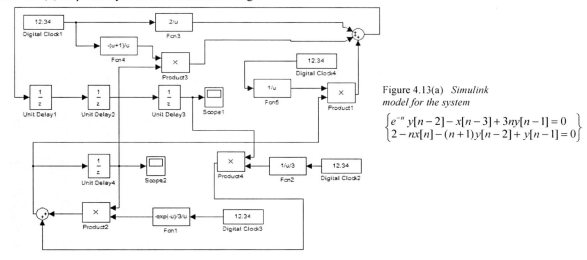

Figure 4.13(a) *Simulink model for the system*

$$\begin{cases} e^{-n}y[n-2] - x[n-3] + 3ny[n-1] = 0 \\ 2 - nx[n] - (n+1)y[n-2] + y[n-1] = 0 \end{cases}$$

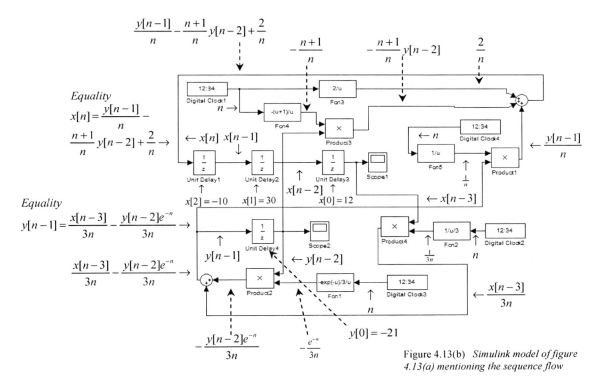

Figure 4.13(b) *Simulink model of figure 4.13(a) mentioning the sequence flow*

❖ ❖ Example 2

This example includes a system where mixed order sequences are involved. Let us find the dependent discrete waves of $x[k]$ and $y[k]$ for the system $\begin{cases} e^{-n}y[n-2] - x[n-3] + 3ny[n-1] = 0 \\ 2 - nx[n] - (n+1)y[n-2] + y[n-1] = 0 \end{cases}$ over the domain $0 \le k \le 13$ contingent to the initial conditions $x[0] = 12$, $x[1] = 30$, $x[2] = -10$, and $y[0] = -21$.

⌨ Solution

The smallest and largest indexes for $x[k]$ are $n-3$ and n, so are $n-2$ and $n-1$ for $y[k]$ respectively. One noticeable point is that $x[k]$ has the order 3 conversely $y[k]$ is of order 1. Nevertheless, their initial values do not start with the same index n: $x[k]$ with $n=3$ and $y[k]$ with $n=2$. The model referred to figure 4.13(a) and the sequence flow diagram referred to figure 4.13(b) correspond to the system of the difference equations we are

discussing. The suitable algebraic form for the simulink is $\begin{cases} y[n-1] = -\dfrac{e^{-n}}{3n}y[n-2] + \dfrac{x[n-3]}{3n} \\ x[n] = \dfrac{2}{n} - \dfrac{n+1}{n}y[n-2] + \dfrac{y[n-1]}{n} \end{cases}$ in which the second

equation holds the $x[k]$ and the first does $y[k]$ as far as the highest index is concern. The blocks Unit Deay1, Unit

Delay2, and Unit Delay3 take care of the second equation to obtain $x[n-3]$ from $x[n]$ in the set whereas the Unit Delay4 recovers $y[n-2]$ from $y[n-1]$ as placed in the first equation. All Digital Clocks in the model generate the same index n. Since the reader is familiar with bringing various

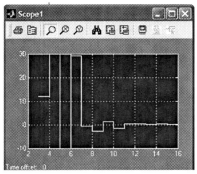

In each Scope
$n = 3 \Leftrightarrow k = 0$
$n = 16 \Leftrightarrow k = 13$

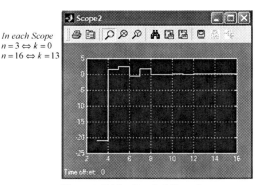

Figure 4.13(c) *Simulink Scope1 output for $x[k]$ for the model of figure 4.13(a)*

Figure 4.13(d) *Simulink Scope2 output for $y[k]$ for the model of figure 4.13(a)*

blocks and constructing the model from the beginning examples, the model construction procedure is not mentioned. Depicted figures 4.13(c) and 4.13(d) confirm the simulation outputs for $x[k]$ and $y[k]$ respectively both with the autoscale setting. As displayed by the Scopes, both horizontal axes share $n=3$ to $n=16$ or in other words $k=0$ to $k=13$ thereby calling for setting the start and stop time of the simulation as 3 and 16 respectively. Also it can be pointed out that the starting index is chosen from the largest delay among multiple sequences here for instance $n-3$ (from $x[k]$ not from $y[k]$). Initially, we see $x[n-3]=x[0]$ and $y[n-2]=y[1]$ but $x[k]$ and $y[k]$ independently attain their wave shapes so we do not correlate them at $n=3$. However, we bring to an end of the chapter by summarizing the links of various blocks employed in the preceding simulation.

4.6 Block links used in this chapter

We addressed in this chapter mainly the simulation of various types and order of difference equations toward what we employed the simulink blocks as presented in the table 4.A.

Table 4.A   Necessary blocks for modeling the difference equations as found in simulink library

Block name	Representative Symbol/Function	Icon Outlook	Block name	Representative Symbol/Function	Icon Outlook
Digital Clock	Integer m, $m \geq 0$	12:34 Digital Clock	Unit Delay	Delays $x[n]$ to $x[n-1]$ by one sample	$\frac{1}{z}$ Unit Delay
Link: *Simulink $\rightarrow$ Sources $\rightarrow$ Digital Clock*			Link: *Simulink $\rightarrow$ Discrete $\rightarrow$ Unit Delay*		
Gain	Multiplication by a constant	1 Gain	Scope	Function viewer	Scope
Link: *Simulink $\rightarrow$ Math Operations $\rightarrow$ Gain*			Link: *Simulink $\rightarrow$ Sinks $\rightarrow$ Scope*		
Constant	Provides a constant value	1 Constant	Sum	Σ or	+
Link: *Simulink $\rightarrow$ Sources $\rightarrow$ Constant*			Link: *Simulink $\rightarrow$ Math Operations $\rightarrow$ Sum*		
Fcn	Entering functional expression considering the index as u	f(u) Fcn	Product	Multiplies two or more functions on a common index	$\times$ Product
Link: *Simulink $\rightarrow$ User-Defined Functions $\rightarrow$ Fcn*			Link: *Simulink $\rightarrow$ Math Operations $\rightarrow$ Product*		

Modeling Common Problems of Control Systems

In this chapter we outline the modeling of common problems of the control system. As elementary building blocks, an electrical system is composed of resistance-inductance-capacitance, a mechanical system is composed of mass-damping-spring, whereas a control system always considers being composed of some transfer functions expressed by some function of Laplace variable s if the system is continuous. We focused mainly the continuous time system. How the step response can be visualized for the transfer function and state space oriented systems is presented at the beginning of the chapter. In later part of the chapter the overture is to address the performance analysis through steady state, overshoot, settling time computation ... etc. Not only that, error performance analysis through the wing of simulink is also discoursed in conjunction with the so learned PID controller design. Once the notion of the elementary continuous modeling is reasoned out, the extension of the modeling to the discrete counterpart and the complex control system can easily be conducted.

5.1 Control system and its model

The fundamental notion in the control system theory is that some blocks can portray the system, and the blocks have some inputs and outputs. To this context, simulink is the best package to handle the control system problems because simulink program modules also assume that all class of problems can be put through blocks. A control system which is formed from many interconnected components provides some definitive responses. The definitive responses are always user defined and can be any physical parameter connected such as displacement, current, angular speed, three-dimensional movement of a robot … etc. As far as we define the required response and supplied input to a control system in some mathematical model either by the transfer function, the state space model, or differential equation, deterministic parameter analysis of the control system is no complicity nowadays because never had we so much advancement in computational tools as we do today. The advent of the faster computer processor allows us to analyze, understand, and visualize many control problems even at our desktop PCs without accessing to some sophisticated control instrumentation lab for which analysis simulink is an appropriate tool. Many facets of today's life are utilizing the control engineering in some way or other ranging from the automatic car driving to the Martian surface exploration. We believe that the most effective method to learning is to discover the past and existing ideas and concepts through the hand-on experience that can pave the way to confront the unanswered and challenging problems. For this reason the simplistic control classroom examples are selected and the finished simulink solutions are presented.

5.2 Step response of a control system from the transfer function

Very often a continuous time control system is characterized by some Laplace transform transfer function $H(s)$ which in general takes the rational form of the polynomial of s. Let us say the transfer function is $H(s) = \dfrac{N(s)}{D(s)}$, where $N(s)$ and $D(s)$ are the numerator and the denominator polynomials in s respectively. The step response of a continuous time system means the output of the system when a step function is the input to the system. Theoretically, first we take the Laplace transform of the input function and then multiply the transform with the system transfer function. The step response is obtained by taking the inverse Laplace transform of the multiplied transfer function. Let us see how simulink simulates the step response of a control system.

♦ ♦ Example 1

Find the step response $y(t)$ of the first order control system $H(s) = \dfrac{1}{-80s + 7}$ as shown in figure 5.1(a) over $0 \leq t \leq 20$.

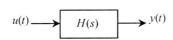

Figure 5.1(a) *Step input to a control system*

Figure 5.1(b) *Simulink model for finding the step response*

⌂ Analytical solution

The step function $u(t)$ has the Laplace transform $\dfrac{1}{s}$. So the step response should have the transfer function $\dfrac{1}{s} \times \dfrac{1}{-80s + 7} = \dfrac{1}{-80s^2 + 7s}$ and whose inverse Laplace transform is $y(t) = \dfrac{1}{7} - \dfrac{1}{7}e^{\frac{7}{80}t}$. The plot of $y(t)$ for $0 \leq t \leq 20$ is depicted in figure 5.1(c). Our objective is to find the simulink solution like the figure 5.1(c).

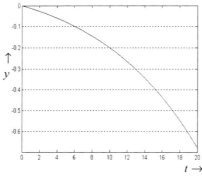

Figure 5.1(c) *Plot of* $y = \dfrac{1}{7} - \dfrac{1}{7}e^{\frac{7}{80}t}$ *vs t for* $0 \leq t \leq 20$

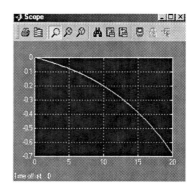

Figure 5.1(d) *Step response from simulink Scope*

⏧ Simulink solution

Figure 5.1(b) shows the simulink model for simulating the problem. The numerator and denominator of $H(s)$ are entered as the polynomial coefficients but in descending power of s. So we have the polynomial coefficients [1] and [−80   7] for the numerator and denominator respectively regarding the example at hand. The first step of finding the step response is to open a new simulink model file and perform the following procedure:

⇒ *click the library browser icon*
⇒ *bring a Step block in the model file following the link 'Simulink → Sources → Step'*
⇒ *since the step for u(t) starts at t =0, doubleclick the block to change its default step time from 1 to 0 in the block parameter's window (figure 3.7(e))*
⇒ *bring a Transfer Fcn block in the model file following the link 'Simulink → Continuous → Transfer Fcn'*
⇒ *and doubleclick the Transfer Fcn block to open the block parameter window and insert the specifications of H(s) as shown in figure 5.1(e)*

Numerator coefficients of H(s) are here

Denominator coefficients of H(s) are here

⇒ *bring a Scope block in the model file following the link 'Simulink → Sinks → Scope'*
⇒ *place the three blocks relatively and connect them according to the figure 5.1(b)*

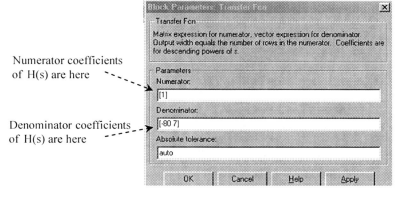

Figure 5.1(e) *Transfer Fcn block parameter window*

⇒ *to insert $0 \le t \le 20$,* click 'Simulation → Simulation parameters → Solver' from the menu bar of the model file to enter the start time as 0 and stop time as 20, keep the other settings as they are, and click OK (figure 3.1(e))*
⇒ *save the model file in your directory and click the start simulation icon*
⇒ *doubleclick the Scope block and change its axes settings to autoscale*

You would see the figure 5.1(d) as the scope output, and that is the expected step response.

♦ ♦ Example 2

Assume that we have a negative feedback control system as shown in figure 5.2(a) whose forward path transfer function, feedback function, and input function are $H(s) =$

$\begin{cases} zero: & -4 \\ poles: 0, 2, and 4, & F(s) = s, \text{ and } x(t) = 0.7u(t-0.7) \text{ respectively.} \\ gain: -5 \end{cases}$

Find the shifted step response of the control system for $0 \le t \le 1.2$.

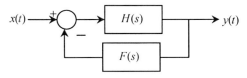

Figure 5.2(a) *A negative feedback system with shifted step function as input*

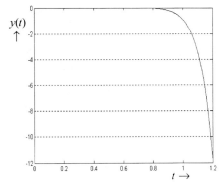

Figure 5.2(b) *Plot of y(t) vs t for the example 2*

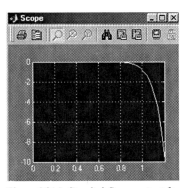

Figure 5.2(c) *Simulink Scope output for the example 2*

⬭ Analytical Solution

A negative feedback control system as shown in figure 5.2(a) has the equivalent transfer function $\frac{Y(s)}{X(s)} =$

$\frac{H(s)}{1+F(s)H(s)}$, where $Y(s)$ and $X(s)$ represent the Laplace transforms for $y(t)$ and $x(t)$ respectively. The given gain,

zero, and poles form $H(s) = \frac{-5(s+4)}{s(s-2)(s-4)}$ and Laplace transform of $x(t)$ is $\frac{7}{10s}e^{-\frac{7}{10}s}$ therefrom $Y(s) =$

$\frac{7H(s)e^{-\frac{7}{10}s}}{10s\{1+H(s)F(s)\}} = \frac{-7(s+4)e^{-\frac{7}{10}s}}{2s^2(s+1)(s-12)}$ and $y(t) = -\frac{287}{180}u(t-0.7) + \frac{21}{26}u(t-0.7)e^{-(t-0.7)} - \frac{7}{234}u(t-0.7)e^{12t-\frac{42}{5}}$

$+\frac{7}{6}tu(t-0.7)$ following the inverse Laplace transform. If one plots $y(t)$, the function takes the shape as displayed in

figure 5.2(b). Let us restate our purpose:

$$\overbrace{\text{with } H(s) = \begin{cases} gain: -5 \\ zero: \ -4 \\ poles: 0, \ 2, \ and \ 4 \end{cases}}^{problem\ space} , F(s) = s, \text{ and } x(t) = 0.7u(t-0.7) \quad \text{and} \quad \overbrace{y(t) \ plot \ of \ figure \ 5.2(b) \ for \ 0 \le t \le 1.2}^{solution\ space}.$$

⬭ Simulink Solution

The model for simulating the problem is presented in figure 5.2(d). Now the transfer function is no longer in numerator-denominator form instead it is in gain-pole-zero form. Also the step function is not the standard one instead it is shifted at $t = 0.7$ and the final value is 0.7. Neither the numerator-denominator nor the gain-pole-zero form of the transfer function can conceive the Laplace variable s. From the properties of Laplace transform, the differential operator $\frac{dy}{dt}$ is equivalent to s that is why the negative feedback path of figure 5.2(d) holds a derivative block in the model. However, to construct the simulink model of figure 5.2(d),

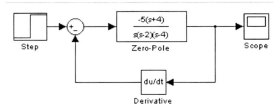

Figure 5.2(d) *Simulink model for finding the shifted step response of the example 2*

    ⇒ *open a new simulink model file and click the simulink library icon*

    ⇒ *to model $x(t) = 0.7u(t-0.7)$, bring one Step block in the model file, doubleclick the block, change its Step time from the default 1 to 0.7, and change its final value from 1 to 0.7 (figure 3.7(e))*

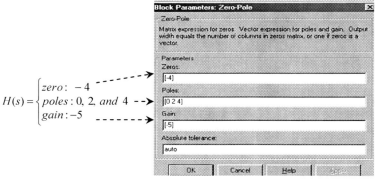

$$H(s) = \begin{cases} zero: \ -4 \ \dashrightarrow \\ poles: 0, \ 2, \ and \ 4 \ \dashrightarrow \\ gain: -5 \ \dashrightarrow \end{cases}$$

Figure 5.2(e) *Zero-Pole block parameter window*

    ⇒ *bring one Sum block in the model file following the link 'Simulink → Math Operations → Sum', doubleclick the block, change its List of signs to + −, and click OK*

    ⇒ *bring one Zero-Pole block in the model file following the link 'Simulink → Continuous → Zero-Pole', doubleclick the block, enter the zero, poles (one space gap between the poles), and gain as shown in figure 5.2(e), and click OK*

    ⇒ *bring one derivative block following the link 'Simulink → Continuous → Derivative' and flip the block*

    ⇒ *bring one Scope block, place the blocks relatively, and connect them according to the figure 5.2(d)*

    ⇒ *click 'Simulation → Simulation parameters → Solver' from the model menu bar to enter the start and stop times as 0 and 1.2 respectively for the insertion of $0 \le t \le 1.2$, keep the other settings as they are in the parameter window (figure 3.1(e))*

⇒ *click the start simulation icon and finally doubleclick the Scope block to see the output as shown in figure 5.2(c) with the autoscale setting*

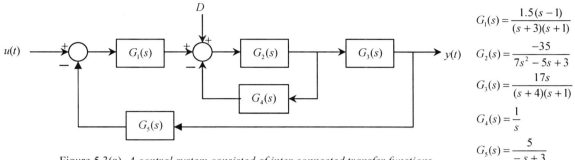

$$G_1(s) = \frac{1.5(s-1)}{(s+3)(s+1)}$$

$$G_2(s) = \frac{-35}{7s^2 - 5s + 3}$$

$$G_3(s) = \frac{17s}{(s+4)(s+1)}$$

$$G_4(s) = \frac{1}{s}$$

$$G_5(s) = \frac{5}{-s+3}$$

Figure 5.3(a) *A control system consisted of inter connected transfer functions*

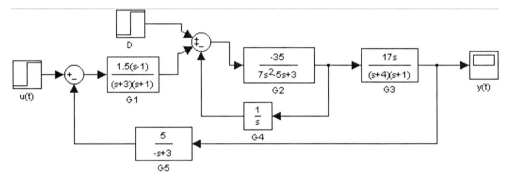

Figure 5.3(b) *Simulink model for the control system of figure 5.3(a)*

❖ ❖ Example 3

A practical continuous time control system may contain many interconnected transfer functions with the positive and negative feedback, example of which can be the control system of the figure 5.3(a). All transfer functions are attached on the right side of the figure. A constant disturbance signal D of value 0.2 appears at $t = 0.1\sec$. We wish to see the output $y(t)$ for a unit step input over $0 \le t \le 2$.

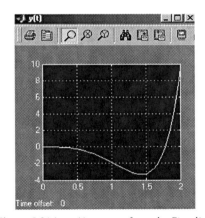

Figure 5.3(c) $y(t)$ *output from the Simulink model of figure 5.3(b)*

⬗ Simulink Solution

Figure 5.3(b) presents the model for analyzing the control system of figure 5.3(a). The disturbance D can be put in analytical form as $D(t) = 0.2u(t - 0.1)$. The transfer functions $G_1(s)$ and $G_3(s)$ can best be entered to simulink as pole-zero-gain form because the numerator and denominator are already in factored form hence their

pole-zero-gains are $\begin{cases} zero: 1 \\ poles: -3 \ and \ -1 \\ gain: 1.5 \end{cases}$ and $\begin{cases} zero: 0 \\ poles: -1 \ and \ -4 \\ gain: 17 \end{cases}$

respectively. Since $G_2(s)$ and $G_5(s)$ are not in factored form, they best can be entered in terms of the polynomial

coefficients. Their numerator/denominator coefficient codings are $\dfrac{[-35]}{[7 \ -5 \ 3]}$ and $\dfrac{[5]}{[-1 \ 3]}$ respectively. Hence, the

elements of the modeling are as follows:

⇒ *open a new simulink model file and click the simulink library icon*
⇒ *bring a Step block, doubleclick the block (figure 3.7(e)) to change its step time from 1 to 0 to model u(t), click OK, and then rename the block as u(t)*
⇒ *bring one Sum block, doubleclick the block to change its List of signs to + −, and click OK*

$\Rightarrow$ *bring one Zero-Pole block, doubleclick the block to set* $\begin{cases} zero:1 \\ poles:-3\ and -1 \\ gain:1.5 \end{cases}$, *and rename the block as*

G1

$\Rightarrow$ *bring another sum block, doubleclick the block to change its List of signs to + + −, and click OK*

$\Rightarrow$ *bring one Transfer Fcn block, doubleclick the block to set* $\dfrac{[-35]}{[7\ -5\ 3]}$, *and rename the block as G2*

$\Rightarrow$ *bring another Zero-Pole block, doubleclick the block to set* $\begin{cases} zero:0 \\ poles:-1\ and -4 \\ gain:17 \end{cases}$, *rename the block as*

G3

$\Rightarrow$ *bring one Integrator block, flip the block, and rename it as G4*

$\Rightarrow$ *bring another Transfer Fcn block, doubleclick the block to set* $\dfrac{[5]}{[-1\ 3]}$, *rename the block as G5, and flip the block*

$\Rightarrow$ *bring another Step block, doubleclick the block to change its step time from 1 to 0.1 and final value from 1 to 0.2 to model* $D(t)=0.2u(t-0.1)$, *click OK, and then rename the block as D*

$\Rightarrow$ *bring one Scope block, rename it as y(t), place various blocks relatively, and connect them according to the figure 5.3(b)*

$\Rightarrow$ *set the solver start and stop time as 0 and 2 respectively keeping the other settings unchanged*

$\Rightarrow$ *click the start simulation icon and finally see the y(t) block output by doubleclicking it with the autoscale setting as shown in figure 5.3(c).*

5.3 Step response of a control system from state space representation

Apart from the transfer function terminology, a continuous time control system can be modeled by the state space matrix form given by $\begin{cases} \dot{x}=Ax+Bu \\ y=Cx+Du \end{cases}$, where { A , B , C , D } are called the state space matrices. The state x is

hidden in the control system and is associated with the internal parameters of the system. For instance, if the system is an electrical circuit, the states are the inductor current and capacitor voltage. If the system is a mechanical one, the velocity of the damping and the displacement due to the compliance constitute the states. The matrix A carries the system parameter's information. For an electrical

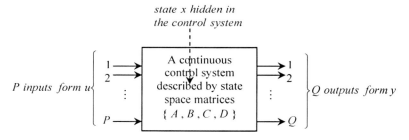

Figure 5.4(a)   *State space model of a control system*

circuit, A is purely a function of resistance, inductance, and capacitance. For a mechanical object, A is composed of inertia, damping, and compliance. In general, orders of A , B , C , and D are $N\times N$, $N\times P$, $Q\times N$, and $Q\times P$ respectively, all matrices have real constant elements, P and Q are the numbers of inputs and outputs respectively, and P , Q , and N are integers. Figure 5.4(a) shows the schematics of the representation.

Find the step response over $0\le t\le 2$ for a single input-single output control system described by the state space matrix $A=\begin{bmatrix} -2 & -1 \\ 2 & 3 \end{bmatrix}$,

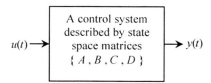

Figure 5.4(b)   *A single input - single output control system*

$B=\begin{bmatrix} -1 \\ -2 \end{bmatrix}$, $C=[-2\quad 1]$, and $D=[-1]$ as shown in figure 5.4(b). Both the analytical and simulink solutions are presented in the following.

☞ *Analytical Solution*

Once the state space matrices are known, the transfer function of the control system is readily obtained from the expression $C(sI - A)^{-1}B + D$, where s is Laplace transform variable, I is an identity matrix of the same order as that of A, and $(sI - A)^{-1}$ indicates the matrix inverse of $sI - A$. Hence, the transfer function becomes [−2

$$1]\left(s\begin{bmatrix} 1 & 0 \\ 0 & 1 \end{bmatrix} - \begin{bmatrix} -2 & -1 \\ 2 & 3 \end{bmatrix}\right)^{-1}\begin{bmatrix} -1 \\ -2 \end{bmatrix} + [-1] = \frac{-s^2 + s - 12}{s^2 - s - 4}$$. The step function has the Laplace transform $\frac{1}{s}$ therefore the

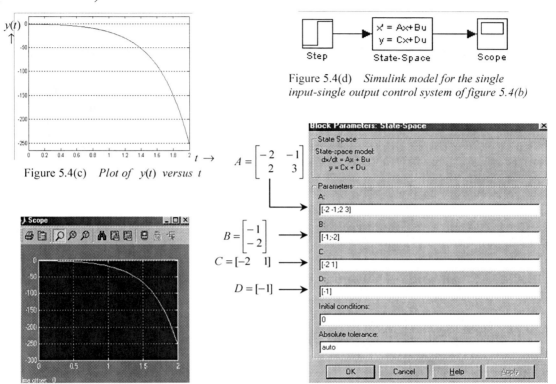

Figure 5.4(d)   *Simulink model for the single input-single output control system of figure 5.4(b)*

Figure 5.4(c)   *Plot of $y(t)$ versus t*

$$A = \begin{bmatrix} -2 & -1 \\ 2 & 3 \end{bmatrix}$$

$$B = \begin{bmatrix} -1 \\ -2 \end{bmatrix}$$

$$C = [-2 \quad 1]$$

$$D = [-1]$$

Figure 5.4(e)   *Simulink Scope output for the model of figure 5.4(d)*

Figure 5.4(f)   *Dialog window for the State-Space block*

Laplace transform and its inverse for the output are given by $Y(s) = \dfrac{-s^2 + s - 12}{s(s^2 - s - 4)}$ and $y(t) = 3 - 4e^{\frac{t}{2}}\cosh\dfrac{t\sqrt{17}}{2} +$

$\dfrac{4}{\sqrt{17}}e^{\frac{t}{2}}\sinh\dfrac{t\sqrt{17}}{2}$ respectively. When $y(t)$ is plotted, it takes the shape of figure 5.4(c) for $0 \le t \le 2$. Let us rephrase the problem as

$$\underbrace{\text{with state space matrices } \{A, B, C, D\}}_{\text{\textit{problem space}}} \text{ and } \underbrace{y(t) \text{ plot like figure 5.4(c) for } 0 \le t \le 2}_{\text{\textit{solution space}}}.$$
and u(t) input

☞ *Simulink Solution*

We presented the simulink model in figure 5.4(d). Let us perform the following to bring about the model:

⇒ *open a new simulink model file and click the simulink library icon*

⇒ *bring a Step block, doubleclick the block to change its step time from 1 to 0 to model u(t), and click OK*

⇒ *bring a State-Space block in the model file following the link 'Simulink → Continuous → State-Space', doubleclick the block to enter the state space matrix descriptions in accordance with the simulink coding of matrices as shown in figure 5.4(f) (rows are separated by semicolons, elements in a row are separated by one space, and the third brace [ ] contains the matrix), and click OK*

⇒ *bring one Scope block, place various blocks relatively, and connect them according to the figure 5.4(d)*

⇒ *set the solver start and stop time as 0 and 2 respectively keeping the other settings unchanged*

⇒ *finally click the start simulation icon and doubleclick the Scope block to see the output as shown in figure 5.4(e) with the autoscale setting*

Analytical and simulink solutions are identical as expected. We present few examples for the step responses of a control system on the state space representations in the following.

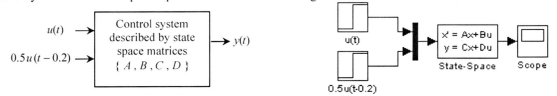

Figure 5.5(a)   *A two inputs-single output control system*　　　Figure 5.5(b)   *Simulink model for the example 1*

✦ ✦ Example 1

Let us consider the state space matrices of a control system as follows: $A = \begin{bmatrix} -1 & 1 \\ 2 & 1 \end{bmatrix}$, $B = \begin{bmatrix} -1 & 4 \\ -2 & -4 \end{bmatrix}$, C =[−1   1], and D =[−1   3]. The state space matrices' dimension determines the numbers of inputs and outputs especially the dimension of the matrix D. Here, the dimension of D is 1×2 hence the values of P and Q are 1 and 2 respectively. Referring to the figure 5.4(a), one can infer that there should be two inputs and one output. Let us assume that the two inputs are the step function $u(t)$ and the shifted step $0.5u(t-0.2)$. Figure 5.5(a) shows the control system we are discussing. Our objective of the simulation is to obtain the output wave shape of $y(t)$ for $0 \leq t \leq 1$.

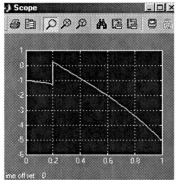

⬚ *Simulink Solution*

The two inputs first need to be multiplexed before passing through the control system modeled by the states space matrices. Presented model in figure 5.5(b) can simulate the problem. Let us mention briefly the model building process:

Figure 5.5(c)   *Simulink Scope output for the model of figure 5.5(b)*

⇒ *open a new simulink model file and click the simulink library icon*

⇒ *bring one Step block, doubleclick the block, change its step time to 0, click OK, and rename the block as u(t)*

⇒ *bring another Step block, doubleclick the block, change its step time to 0.2, change its final value to 0.5, click OK, and rename the block as 0.5u(t-0.2)*

⇒ *bring one two input Mux block following the link 'Simulink → Signal Routing → Mux'*

⇒ *bring one State-Space block, doubleclick the block, enter the matrices A, B, C, and D by the simulink codes [-1 1;2 1], [-1 4;-2 -4], [-1 1], and [-1 3] respectively in the dialog window as we did before*

⇒ *bring one Scope block, place various blocks relatively, and connect them according to the figure 5.5(b)*

⇒ *set the solver start and stop times as 0 and 1 respectively keeping the other settings unchanged*

⇒ *finally click the start simulation icon and doubleclick the Scope block to see the output as shown in figure 5.5(c) with the autoscale setting*

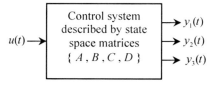

Figure 5.6(a)   *A single input - three outputs control system*

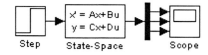

Figure 5.6(b)   *Simulink model for the example 2*

✦ ✦ Example 2

Find the step responses for a control system described by the state space matrices $A = \begin{bmatrix} -2 & -1 \\ 2 & 3 \end{bmatrix}$, $B = \begin{bmatrix} -1 \\ -2 \end{bmatrix}$,

$$C = \begin{bmatrix} -1 & 0 \\ 2 & 3 \\ 0 & 1 \end{bmatrix}, \text{ and } D = \begin{bmatrix} -2 \\ 3 \\ 1 \end{bmatrix} \text{ over } 0 \le t \le 0.5.$$

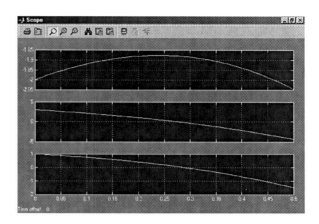

⟑ Simulink Solution

We have seen the step responses for the single input-single output and multiple inputs-single output in previous examples but this example implements the step response for a control system having single input-multiple outputs. Order of the matrix D (which is 3×1) tells us that there are three outputs (equal to the number of the rows of D) and one input (equal to the number of the columns of D). Figure 5.6(a) depicts the control system whose simulink model is presented in figure 5.6(b). The model returns the three outputs together. To separate them, we need a demultiplexing operation. Some

Figure 5.6(c)   *Simulink Scope output for the model of figure 5.6(b)*

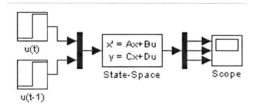

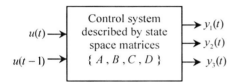

Figure 5.7(a)   *A two inputs-three outputs control system*

Figure 5.7(b)   *Simulink model for the example 3*

elements of the model building process are as follows:

⇒ *bring one Step, one State-Space, and one Scope blocks in a new simulink model file*

⇒ *change the step time to 0 to model $u(t)$, doubleclick the State-Space block and enter the matrix codes as [-2 -1;2 3], [-1;-2], [-1 0;2 3;0 1], and [-2;3;1] for A, B, C, and D respectively*

⇒ *bring one Demux block following the link 'Simulink → Signal Routing → Demux', doubleclick the block to change its number of outputs from 2 to 3, and click OK*

⇒ *doubleclick the Scope block, click the parameters icon in the dialog window (figure 3.15(a)), and change the number of axes from 1 to 3 because of three inputs*

⇒ *place various blocks relatively according to figure 5.6(b), connect them, set the solver start and stop time as 0 and 0.5 respectively, and click the start simulation icon*

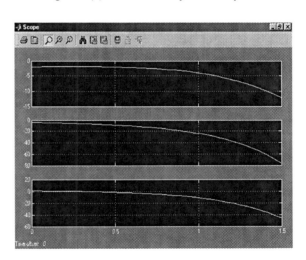

Figure 5.7(c)   *Simulink Scope output for the model of figure 5.7(b)*

⇒ *doubleclick the Scope block to view the three outputs as shown in figure 5.6(c) (bring the mouse pointer on each plot and set the autoscale for each)*

✦ ✦ Example 3

The state space matrices given by $A = \begin{bmatrix} -2 & -1 \\ 2 & 3 \end{bmatrix}$, $B = \begin{bmatrix} -1 & 2 \\ -2 & 0 \end{bmatrix}$, $C = \begin{bmatrix} -1 & 0 \\ 2 & 2 \\ 0 & 1 \end{bmatrix}$, and $D = \begin{bmatrix} -2 & 0 \\ -4 & 1 \\ 1 & 0 \end{bmatrix}$

represent multiple inputs and multiple outputs control system. Let us simulate the model by injecting two steps { $u(t)$ and $u(t-1)$ } and display the output wave shapes over $0 \le t \le 1.5$.

⑂ *Simulink Solution*

Since the order of the matrix D is 3×2, the model (figure 5.7(a)) intakes two inputs and returns three outputs. The Mux and Demux blocks we employed before apply here. The simulink model is presented in figure 5.7(b). Successful construction and simulation should provide you the Scope output of the figure 5.7(c) (autoscale setting for each).

5.4 Modeling some s domain transfer functions

In the section 5.2, we acquainted the reader how one can model the s domain transfer functions. The analyst and designer of the control system come across different kinds of the transfer functions depending on the innate dynamics of a control system. Here in this section we involve examples covering more modeling the s domain transfer functions.

✦ ✦ Example 1

Let us model the transfer function $H(s) = \dfrac{(s+4)(s+2)}{(s+j3)(s-j3)(s+5)}$. The transfer function has the complex poles, and the poles are $-j3$, $j3$, and -5. We can enter the parameter settings in the window 5.2(e) of the Zero-Pole block as $\begin{cases} \text{Zeros}:[-4 \ -2] \\ \text{Poles}:[-3i \ 3i \ -5] \\ \text{Gain}:[1] \end{cases}$ to model the transfer function.

✦ ✦ Example 2

The transfer function $H(s) = \dfrac{(s+1)^3}{(s+3)^4(s-3)^3}$ contains multiple poles (the pole -3 of multiplicity 4 and the pole 3 of multiplicity 3) and zeroes (the zero -1 of multiplicity 3). The settings (referring to the figure 5.2(e)) necessary for the parameter window of the Zero-Pole block are $\begin{cases} \text{Zeros}:[-1 \ -1 \ -1] \\ \text{Poles}:[-3 \ -3 \ -3 \ -3 \ 3 \ 3 \ 3] \\ \text{Gain}:[1] \end{cases}$.

✦ ✦ Example 3

Sometimes the terminology pertaining to the interconnection of the blocks like series or parallel becomes useful in the modeling, for example, $H(s) = \dfrac{1}{2s^4 - 3s + 7} - \dfrac{9}{(s+3)^3}$ can be modeled by considering two individual transfer functions: $\dfrac{1}{2s^4 - 3s + 7}$ and $-\dfrac{9}{(s+3)^3}$ connected in parallel. The first one can be modeled by bringing a Transfer Fcn block and entering the coefficient coding as $\begin{cases} \text{Numerator}:[1] \\ \text{Denominator}:[2 \ 0 \ 0 \ -3 \ 7] \end{cases}$ in the parameter window of the figure 5.1(e) and the second part can be modeled by the Zero-Pole block with the settings $\begin{cases} \text{Zeros}:[\ ] \\ \text{Poles}:[-3 \ -3 \ -3] \\ \text{Gain}:[-9] \end{cases}$ in the parameter window of the figure

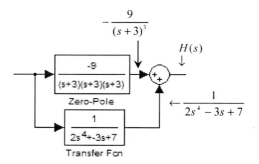

Figure 5.8(a) *Modeling a sum form transfer function in parallel sense*

5.2(e). The outputs of the two individual transfer functions are connected by a two input sum block to form the complete $H(s)$ as illustrated in the figure 5.8(a).

✦ ✦ Example 4

Similar to the example 3, the function $\dfrac{1}{2s^4 - 3s + 7} - \dfrac{9}{(s+3)^3} + \dfrac{1}{s(s+2)}$ can be modeled by three blocks connected in parallel and adding their outputs by a three input Sum block.

♣ ♦ Example 5

Not all transfer functions find the modeling in simulink, for example, $\dfrac{s^2 + s + 5}{s}$. The precondition of a transfer function to be modeled in simulink is the degree of the numerator in the transfer function must be less than or equal to that of the denominator.

♣ ♦ Example 6

One can model the function $H(s) = \dfrac{-91}{s^4}$ if the settings

$\begin{cases} \text{Zeros: } [] \\ \text{Poles: } [0\ 0\ 0\ 0] \\ \text{Gain: } [-91] \end{cases}$ occupy the parameter window of the figure 5.2(e).

♣ ♦ Example 7

The product form function can easily be modeled by placing series blocks, for instance, $H(s) = \dfrac{s + 4}{[(s + 2)^2 + 4](s^3 + 3)}$.

The expression becomes $\dfrac{1}{(s + 2)^2 + 4} \times \dfrac{s + 4}{s^3 + 3}$ following rearrangement and the figure 5.8(b) presents the modeling.

♣ ♦ Example 8

Occasionally a transfer function possesses the delay factor, for example, $H(s) = e^{-2\pi s} \dfrac{1}{s^2 + 1}$. The factor $e^{-2\pi s}$ can be modeled by the block Transport Delay located in the link *'Simulink → Continuous → Transport Delay'*, and it has a delay of 2π (whose simulink code is 2*pi). You can construct the model as shown in figure 5.8(d). Figure 5.8(c) shows the parameter window of the Transport Delay block at which we enter the Time Delay as 2*pi (but not as -2π). The Transport Delay block only intakes the $e^{-2\pi s}$ information, the other part of the Transfer function which is $\dfrac{1}{s^2 + 1}$ is taken care of by the Transfer Fcn block.

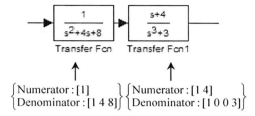

$\begin{cases} \text{Numerator: } [1] \\ \text{Denominator: } [1\ 4\ 8] \end{cases}$ $\begin{cases} \text{Numerator: } [1\ 4] \\ \text{Denominator: } [1\ 0\ 0\ 3] \end{cases}$

Figure 5.8(b) *Modeling a product form transfer function in series sense*

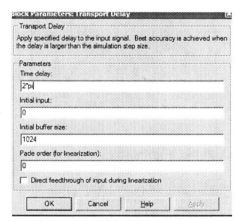

Figure 5.8(c) *Parameter window of the Transport Delay block*

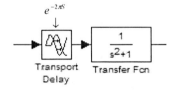

Figure 5.8(d) *Modeling the function* $e^{-2\pi s} \dfrac{1}{s^2 + 1}$

5.5 Performance analysis of a control system

Simulink model output of a prototype second order control system can further be analyzed for the steady state value, peak overshoot, rise time, peak time, and settling time. Our objective in this section is to address how one can read off these parameters from the Scope output. Also how one can readily obtain these parameters from the simulink model is mentioned.

Let us consider the negative feedback control system of figure 5.2(a) where $H(s) = \dfrac{1}{2s^2 + 0.6s + 1}$, $F(s) = 0.6$, and $x(t) = u(t)$. Open a new simulink model and bring one Step, one Transfer Fcn, one Scope, one Sum, and one Gain blocks to construct the model according to figure 5.9(a) (set the step time of the Step block to 0). Run the model for 10 seconds by setting the Solver stop time as 10 (but it is the default time, figure 3.1(e)) and doubleclick the Scope block to see the output as shown in figure 5.9(b) with the autoscale setting. The figure tells us that the oscillations are decaying but the steady state behavior of the output is not so pronounced in the figure. Let us run the model again for the Stop time as 100 seconds and that results the output of figure 5.9(c) with the autoscale setting. From this figure, the steady state behavior of the output is very transparent. Now the question is how we can read off this steady state value from the Scope. The scope menu bar retains some icons in which we can find the X-axis, Y-axis, and both X-Y axes zoom icons as indicated in figure 5.9(c). The Y axis values of the figure 5.9(c) says that the steady state value is somewhere between 0.6 and 0.7 but we are not sure about the fraction because the Y-

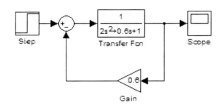

Figure 5.9(a)   *A second order control system with negative feedback and step function input*

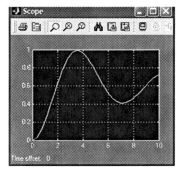

Figure 5.9(b)   *Scope output of the model 5.9(a) for $0 \leq t \leq 10$*

Zoom this area for reading off the peak overshoot

X axis zoom   *Y axis zoom*

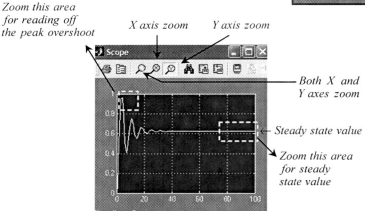

Both X and Y axes zoom

Steady state value

Zoom this area for steady state value

Figure 5.9(c)   *Scope output of the model 5.9(a) for $0 \leq t \leq 100$*

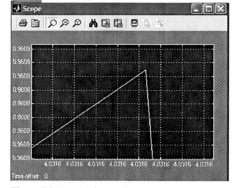

Figure 5.9(d)   *Steady state value of the output after five times zooming*

axis values are one digit separated. Click the X-Y axes zooming icon, bring the mouse pointer to any area on the plot, press the left button of the mouse, move the mouse pointer to the right, left, up, or down keeping the left button of the mouse pressed to see the target area within the plot, and then release the left button to see the effect. This action is called the zooming of the Scope output. If you zoom the area once, you may see the accuracy increased by one more digit. If you zoom it again, you append one more digit accuracy. So multiple zooming can provide more digit accurate output value. For the example at hand we performed five times zooming employing both axes near the area as shown in figure 5.9(c) so that we ended up with the figure 5.9(d). From the last figure one can easily read off the value of the steady state output as 0.625. In a similar fashion the peak

Figure 5.9(e)   *Peak overshoot value of the output after six times zooming*

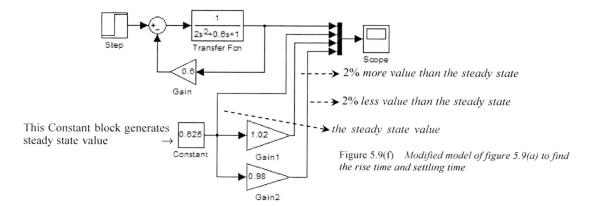

This Constant block generates steady state value

2% more value than the steady state

2% less value than the steady state

the steady state value

Figure 5.9(f)   *Modified model of figure 5.9(a) to find the rise time and settling time*

102

overshoot of the output of the figure 5.9(c) can be read off by zooming the indicated area. We exercised six times zooming with both axes icon to display the output as shown in figure 5.9(e) fromwhere the peak overshoot of the output is 0.9608 and the peak time (the time when the peak overshoot occurs) is 4.0316 seconds. During the zooming process, the reader might need to go back to the last zooming action. You can go back to the last action by rightclicking the mouse then zooming out. If you want to start over again, click autoscale following the rightclick.

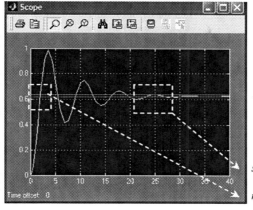

Figure 5.9(g) *Scope output of the modified model of figure 5.9(f) to find the rise time and settling time*

{
→ *upper line 2% more of steady state*
→ *middle line steady state*
→ *lower line 2% less of steady state*
}

settling time in this area

rise time in this area

Once we know the steady state value, the rise time (first time to reach the steady state value) and the settling time (time required for the system to settle within a certain percentage of the steady state value) can easily be found with little modification in modeling. For that reason we bring a Constant block (following the *link 'Simulink → Sources → Constant'*) that can generate the steady state value over the simulation period. Let us define the settling time associating ±2% of the steady state. So the settling time will be found when the output first reaches to 0.98 and 1.02 times of the steady state output. We superimpose all these signals by multiplexing with the output found in the model of the figure 5.9(a) for

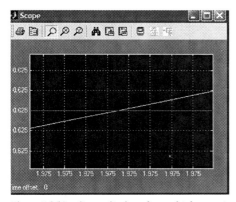

Figure 5.9(h) *Scope display after multiple zooming for rise time*

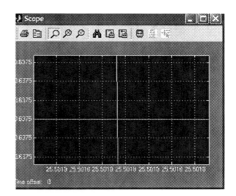

Figure 5.9(i) *Scope display after multiple zooming for settling time*

which the necessary model is shown in figure 5.9(f). To display the rise time and settling time more distinctively, we can set the stop time of the solver to 40 seconds instead of 100 seconds. Run the modified model of figure 5.9(f) to see the Scope output with the autoscale setting as depicted in figure 5.9(g). In that figure you find the targeted areas to search for the rise time and the settling time of the control system. Following several times zooming, we ended up with the Scope outputs as depicted in figures 5.9(h) and 5.9(i) therefrom we read off the two times as 1.975 seconds and 25.5018 seconds for the rise time and the settling time respectively.

Depending on the problem you handle, you may need different convenience for the same problem. We mentioned how you can find the steady state value of a control system output employing the zooming facilities. A slight modification in your model

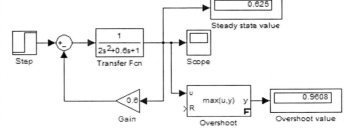

Figure 5.9(j) *Modified model of figure 5.9(a) to show the steady state value and the peak overshoot value*

of figure 5.9(f) can show you the steady state value and overshoot of your control system. The Display block shows the instantaneous value of the concern functional line at various times set by the Solver. The stop time of the Solver or the last time (here it is 100 seconds) can correspond to the steady state value without the loss of generosity. So we bring a Display block, rename the block as Steady state value, and connect the block to the output functional line as indicated in figure 5.9(j). For the maximum value of the output or peak overshoot, we need to bring a MinMax Running Resettable block following the link *'Fixed-Point Blockset → Math → MinMax Running Resettable'* located in the simulink library. Bring the block, rename it as Overshoot, doubleclick the block to see the block parameter window, set the function to max from the default min, bring another Display block, rename the block as Overshoot value, and connect them in the output functional line as shown in figure 5.9(j). Run the model to see the steady state value and the overshoot as presented in the same figure. As you notice, they are identical with those we obtained from the zooming facilities. If you go to MATLAB Command Window, you see some warning messages. The reason is we did not connect the resetting port of the MinMax Running Resettable block (labeled by R) to any input. As far as the objective of the problem is concern, we can skip that for the time being. That is what we promised to reveal.

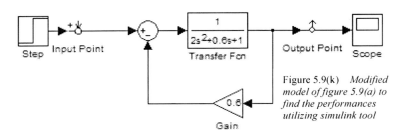

Figure 5.9(k)   *Modified model of figure 5.9(a) to find the performances utilizing simulink tool*

Figure 5.9(l)   *Pull down menu of the Tool in the simulink menu bar*

Figure 5.9(n)   *Pull down menu of simulink under LTI Viewer Window*

Figure 5.9(o)   *Plot type options offered by the LTI Viewer Window*

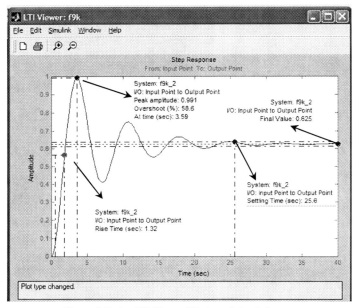

Figure 5.9(m)   *LTI Viewer Window of simulink*

✦ ✦ *An easy way of performance analysis*

The procedure we discussed so far includes the programming approach for any type of interconnected control system. But simulink affiliated tools can also provide you the ready made finding for the steady state value, peak overshoot, rise time, peak time, and settling time. To elucidate that, let us consider the model of the figure 5.9(a). The simulink tool applies for specific input and output ports in the model and that must be indicated in the model before we analyze the model. We move the Step block of the model 5.9(a) to the left so that the functional line between the Step and the Sum blocks is long enough to accommodate the block Input Point as shown in figure 5.9(k) (the link is *'Control System Toolbox → Input Point'*). All you have to do is bring the block and drop it on the functional line, the action will get the Input Point connected automatically. Let us also obtain the Output Point (whose link is *'Control System Toolbox → Output Point'*) connected with the functional line going to the Scope taking similar action. You find the menu Tool in the simulink menu bar. Click the Tool, you see Linear analysis in the pull down menu as shown in figure 5.9(l). Click the Linear analysis and simulink responds by prompting the LTI

Viewer Window (LTI for Linear Time Invariant) like the figure 5.9(m). In the LTI Viewer Window, another menu bar called simulink is there like the figure 5.9(n) and click simulink to see the pull down menu under it and click Get Linearized Model. The action activates a plot as shown in figure 5.9(m). Bring the mouse pointer in the plot area, rightclick the mouse button to see the plot type options offered by the LTI Viewer as shown in figure 5.9(o), click there Step, and again rightclick in the plot area to see the Characteristics under the menu like figure 5.9(p). You find Peak Response, Settling Time, Rise Time, and Steady State under characteristics. Rightclick four times and check all of them like the figure 5.9(p). Each time you click, you see the corresponding change in the plot area by a bold dot bullet. Bring mouse pointer in the bullet and you see the required characteristics but of coarse one at a time. We provided all four together indicated by the arrows in the figure 5.9(m) for space reason.

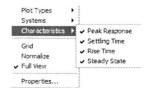

Figure 5.9(p) *Step response parameters offered by the LTI Viewer Window*

Now the option is readers. You can employ the programming approach or you can take simulink LTI Viewer Tool to find the performance of a control system.

5.6 Error indices of a feedback control system

Quantitative analysis of the error signal of a feedback control system is necessary to evaluate the operation of an adaptive control system, the optimization of parameters automatically, and the design of optimum control system. Now we discuss how one can analyze the error signals for different control systems.

Referring to the example 2 of section 5.2, we had the input and the output signals $x(t) = 0.7u(t-0.7)$ and

$$y(t) = -\frac{287}{180}u(t-0.7) + \frac{21}{26}u(t-0.7)e^{-(t-0.7)} - \frac{7}{234}u(t-0.7)e^{12t-\frac{42}{5}} + \frac{7}{6}tu(t-0.7)$$ respectively thereby forming the error

signal $e(t)$ =input−output= $x(t) - y(t) = \frac{413}{180}u(t-0.7) - \frac{21}{26}u(t-0.7)e^{-(t-0.7)} + \frac{7}{234}u(t-0.7)e^{12t-\frac{42}{5}} - \frac{7}{6}tu(t-0.7)$. Let us

plot this error signal for the same interval, and it is shown in figure 5.10(a). Our aim here is to obtain this error wave

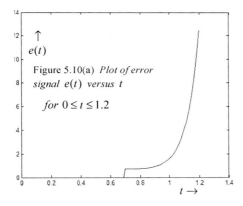

Figure 5.10(a) *Plot of error signal $e(t)$ versus t for $0 \le t \le 1.2$*

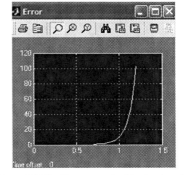

Figure 5.10(b) *Error signal $e(t)$ displayed by simulink for the model of figure 5.10(c)*

shape from the simulink. Since the error is an increasing function of time, for sure the feedback control system is performing badly. However, let us see how simulink can display the error signal. The simulink model of figure 5.2(d) is modified by adding one Scope block and then renaming the block as Error (depicted in figure 5.10(c)). Run the model and doubleclick the Error block to view the error signal displayed by simulink as shown in figure 5.10(b). The error variation $e(t)$ delineated in figures 5.10(a) and 5.10(b) ranges from 0 to somewhat 13 and from 0 to somewhat 105 respectively. What can we conclude about the error characteristics from those figures? Certainly, they are not identical but appear to have similar shape. The reason underlies in the discrepancies is the

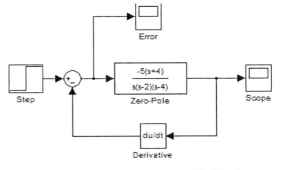

Figure 5.10(c) *Appending one Scope block by the name Error in the model of figure 5.2(d)*

reference input $x(t)$ and the output $y(t)$ may not necessarily represent the identical quantity to be measured in our hypothetical model of figure 5.10(c). For instance, $x(t)$ and $y(t)$ could represent displacement and voltage obviously they concern different units and parameters and can not be compared until some transformation from voltage to displacement or displacement to voltage is carried out. This kind of transformation obliges to engage some gain blocks in the functional or signal flow paths of the model making the input and output consistent. If you scale down the error of the figure 5.10(b) by a suitable factor for sure that will be identical with that of the figure 5.10(a). Anyhow, we would like to include two more examples on error performance investigation in the following.

♣ ♦ Example 1

Once we define the error signal $e(t)$ in our feedback model, various error performance indices are defined by the following:

Integral of the absolute magnitude of the error, $IAE = \int_0^T |e(t)| \, dt$,

Integral of the square of the error, $ISE = \int_0^T e^2(t) dt$,

Integral of the time multiplied by the absolute error, $ITAE = \int_0^T t \, |e(t)| \, dt$, and

Integral of the time multiplied by the squared error, $ITSE = \int_0^T t e^2(t) dt$.

The upper limit T of each integral is some finite time that we choose for the error analysis. Computing steady state error always needs to employ the infinite time but that is impractical for implementation. We select some finite T and analyze a control system on that.

Let us choose the model of figure 5.10(d) which is a prototype second order system often considered as the building

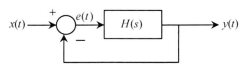

Figure 5.10(d) *A negative unity feedback system with step function as input*

block of complex control system. We wish to evaluate the error performance indices of $H(s) = \dfrac{\omega_n^2}{s^2 + 2\zeta s \omega_n + \omega_n^2}$ for

$x(t) = u(t)$ in simulink on a time interval of 0 to 15 sec, where the damping ratio is $\zeta = 0.3$ and the natural frequency is $\omega_n = 1$ radian/sec.

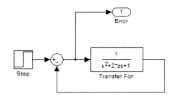

Figure 5.10(e) *Sub model only to generate error signal* $e(t)$

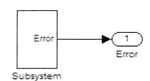

Figure 5.10(f) *Forming a simulink subsystem that can return only the error* $e(t)$

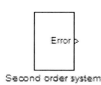

Figure 5.10(g) *Renaming the subsystem of 5.10(f) as Second order system*

☞ Simulink solution

The design problem needs to address how one can build a sub model. Here we devise a sub model that can return only the error signal $e(t)$ for 15 sec from the model 5.10(d). The given ω_n lets us write the transfer function

$H(s)$ as $\dfrac{1}{s^2 + 2zs + 1}$ where ζ is replaced by z (because the symbol ζ is not defined in the text form of simulink).

In terms of the numerator and denominator coefficients, we can write the transfer function as $\dfrac{[1]}{[1 \quad 2*z \quad 1]}$ and to

which z is fed from the MATLAB Command Window. The reason z is being fed from the command window is this problem will be extended as example 2. However, let us proceed with the modeling:

⇒ *open a new simulink model, click the library browser icon, bring a Transfer Fcn block, and doubleclick*

the block to enter the numerator and denominator codings as $\dfrac{[1]}{[1 \quad 2*z \quad 1]}$

⇒ *bring one Step block, doubleclick the block to change its step time to 0, bring one Sum block, and doubleclick the block to change its list of signs to +-*

⇒ *bring one Out1 block following the link 'Simulink → Sinks → Out1, and rename the block as Error*

⇒ *place various blocks relatively according to figure 5.10(e), connect them, and set the solver start and stop time as 0 and 15 respectively keeping other settings as default*

⇒ *go to the menu bar of the model file, click Edit, then Select all, again go to the Edit menu, click Create Subsystem to see the subsystem of figure 5.10(f), rename the subsystem as Second order system, and delete the functional line and Error node to have the outlook of the subsystem as shown in figure 5.10(g)*

Designing the sub model for the error signal $e(t)$ is worked out. Next what we need is to implement the error indices' expression for *IAE*, *ISE*, *ITAE*, and *ITSE* for which the model 5.10(h) is dedicated. How the signals are sweeping to compute different error performances is presented in figure 5.10(i). The simulink procedure adopted to build the model is as follows:

⇒ *The sub model Second order system just built is residing in the untitled model file and you can save the model by the name test in your working directory*

⇒ *bring one Abs ('Simulink → Math Operations → Abs'), one Ramp ('Simulink → Sources → Ramp'), and one Fcn ('Simulink → User-Defined Functions → Fcn') blocks in the model file test*

⇒ *doubleclick the block Fcn to enter parameter expressions as u^2 for squaring the error (considering u as the input variable of the block)*

⇒ *bring one Product ('Simulink → Math Operations → Product') block, copy the block, and paste it to see the block Product1*

⇒ *bring one Integrator block ('Simulink → Continuous → Integrator'), rename the block as Int for IAE, copy the block, and paste it, rename the block as Int for ITAE, thus obtain the other two blocks Int for ITSE and Int for ISE in the model*

⇒ *bring one Display block ('Simulink → Sinks → Display'), rename the block as IAE, copy the block, and paste it, rename the block as ITAE, thus obtain the other two blocks ITSE and ISE in the model*

⇒ *place various blocks relatively according to figure 5.10(h), connect them, and set the solver start and stop time of the main model test as 0 and 15 respectively keeping other settings as default (solver start and stop times simulate the upper and lower limits of the integration for each of the errors)*

⇒ *let us not forget that we set the denominator coefficients of the transfer function $H(s)$ during the sub model construction as [1 2*z 1], we go to the MATLAB Command Window, and execute the command >>z=0.3 to enter $\zeta = 0.3$*

⇒ *finally, click the start simulation icon and see various error performance index values as shown in figure 5.10(i)*

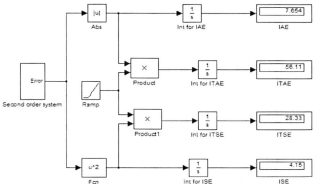

Figure 5.10(h) *Simulink model for computing the performance indices utilizing the subsystem of figure 5.10(g)*

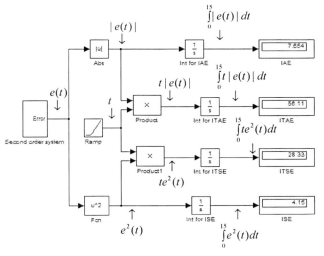

Figure 5.10(i) *Mentioning the functional flow of the model of figure 5.10(h)*

107

❖ ❖ Example 2

Example 1 considers only one case of the damping ratio ($\zeta = 0.3$). For design reasons we may need to see the error performances over the same specified time but for different damping ratio, for instance, $\zeta = 0$ to $\zeta = 2$ with step 0.05. Our objective in this problem is to obtain the example 1 mentioned errors IAE, ITAE, ITSE, and ISE for the damping ratio 0 to 2 with step 0.05 and to plot them for comparison.

✍ *Simulink solution*

The simulink model is very much sensitive to the states of the variables involved in a model or in other words the start-inbetween-stop time of the simulation. Being in a model, all blocks present in the model share the same start and stop times. So if we want to pass some value or parameter to some block that has to be conveyed out of the model. As far as the objective of the example 2 is concern, the value of the damping ratio ζ requires be passed to the sub model Second order system and we carry out that from the MATLAB Command prompt (>>). From the model of figure 5.10(h), we displayed

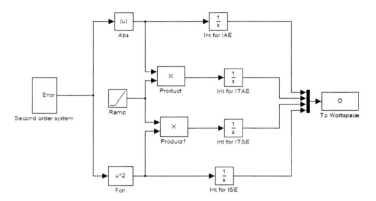

Figure 5.10(j) *Modified model of figure 5.10(h) in which IAE, ITAE, ITSE, and ISE are multiplexed*

one error on each of the indices for a certain damping ratio. We send different ζ to the sub model Second order system and obtain various performance error indices calculated and multiplexed as a four element row matrix from the MATLAB Command prompt. However to commence with simulink, we delete the display blocks IAE, ITAE, ITSE, and ISE, bring one Mux block, doubleclick the block to change its number of inputs to four for feeding all four errors, bring one To Workspace block to export the data from simulink to MATLAB workspace, and connect the model as depicted in figure 5.10(j). Doubleclick the block To Workspace and change its variable name from simout to O so that we can deal with a short variable name and also change the save format to array. In the block parameter window of To Workspace, you see the default sample time as –1. Let us save the model of figure 5.10(j) by the name diffz in the working path. Go to MATLAB Command Window and perform the following:

MATLAB Command

```
>>z=0.3; ↵          ← Setting some value of ζ
>>sim('diffz') ↵    ← We can run our simulink model diffz from the MATLAB Command prompt
>>O ↵               ← Just to see what O contains

O =

      0        0        0        0          ← This row corresponds to the start time of the solver which is 0
 0.0000   0.0000   0.0000   0.0000
 0.0002   0.0000   0.0000   0.0002
            ⋮
 7.5801  55.0027  27.7816   4.1128
 7.6540  56.1058  28.3332   4.1498          ← This row corresponds to the stop time of the solver which is 15
    ↑        ↑        ↑        ↑
   IAE     ITAE     ITSE      ISE
```

As you see, the To Workspace block variable O retains a four column matrix, the first, second, third, and fourth columns of which correspond to the indices IAE, ITAE, ITSE, and ISE respectively. Recall that we selected the time interval for the error analysis as 15. As a matter of fact, the simulink solver adaptively finds the variable time points and returns the error indices in each time. Since the default sample time is –1 of the To Workspace block, we see the error indices at all times. What if we make the sample time as 15 in the block parameter window of the To Workspace block and run the model again from the command prompt:

```
>>sim('diffz') ↵
>>O ↵

O =

      0        0        0        0          ← This row corresponds to the start time of the solver which is 0
 7.6540  56.1058  28.3332   4.1498          ← This row corresponds to the stop time of the solver which is 15
    ↑        ↑        ↑        ↑
   IAE     ITAE     ITSE      ISE
```

Due to the sample time change, we just get the indices for the start time 0 and the stop time 15 in above displayed two rows respectively. Now what we need is the second row of the matrix O for our ζ =0.3 and that can happen by the MATLAB Command O(2,:). So over the domain of ζ we pass some specific ζ to the sub model Second order system, obtain the O(2,:) for various indices, and store the output to some MATLAB variable E. These all can occur if we accomplish the following MATLAB Command:

MATLAB Command

>>E=[]; ↵

>>for z=0:.05:2 set_param('diffz/Second order system/Transfer Fcn','Denominator','[1 2*z 1]'),sim('diffz'),E=[E;O(2,:)];end ↵

The command 'E=[];' assigns an empty matrix to E so that we can keep in E the row after row outputs picked up

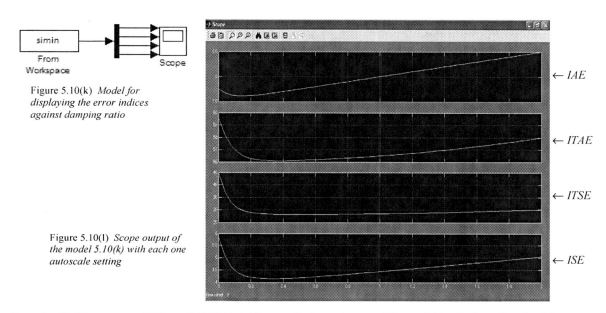

Figure 5.10(k) *Model for displaying the error indices against damping ratio*

Figure 5.10(l) *Scope output of the model 5.10(k) with each one autoscale setting*

← IAE

← ITAE

← ITSE

← ISE

from the O. The command 'for z=0:0.05:2' initiates a for loop that provides variable ζ from 0 to 2 with a step of 0.05 and which is completed by the command 'end'. The command 'set_param' (abbreviation of set parameter) outside the simulink model can pass any block parameters from the MATLAB Command Window to simulink. The first argument of the command 'set_param' refers the exact location of the block in which the parameters need to be passed. For instance, the diffz is the file name of the simulink model, Second order system is the sub model name in the file diffz, and Transfer Fcn is the specific block name in the Second order system. The file/model/block name must have to be identical, so must the space gap and the character case. However, the second argument of the 'set_param' describes the type of the parameter to be altered in the block associated with. For example, doubleclicking the block Transfer Fcn can display that one parameter has the name 'Denominator' and we want to change that according to dfferent ζ. The third argument '[1 2*z 1]' mentions what the Denominator should be. So altogether we exercised three arguments in the 'set_param'. The next command sim('diffz') is just for running the simulink model diffz from the MATLAB prompt and the command 'E=[E;O(2,:)];' is for piling up the outputs one after another. Also note that there is one space gap among 'for', 'z=0:0.05:2', and 'set_param'. The above command may not function if the simulink model is closed. Anyhow, type E in the command prompt and press enter to see the following:

E =

7.7481	59.5434	44.8011	5.8287	← This row corresponds to z or ζ =0
7.6518	57.8302	35.0755	4.8689	
⋮				
8.4749	57.8994	30.0203	5.0007	
8.4998	57.9974	30.1302	5.0311	← This row corresponds to z or ζ =2
↑	↑	↑	↑	
IAE	ITAE	ITSE	ISE	

So our purpose is accomplished, and next we need to see their plot. You can make it happen in two ways: from the menu bar of the MATLAB Command Window, click view and then workspace. You see the variable E present in the workspace. Rightclick on E, click Graph, and the click plot to see all four plots together. But the problem is the horizontal axis is not consistent with the z. To do so, perform the following:

>>z=[0:.05:2]'; ↵

>>plot(z,E) ↵

Once the graph is plotted, you can manipulate the graph from the figure window for the line color or the line style. The legend can also be inserted from the figure window by clicking *Insert → Legend* from the menu bar of the figure window of MATLAB. If you insert legend, you see the identifier as data1, data2, data3, and data4 for sure they accord the order IAE, ITAE, ITSE, and ISE respectively as we multiplexed. Another option could be first you form a variable simin in the MATLAB workspace that contains all these information by the following:

>>simin=[z E]; ↵

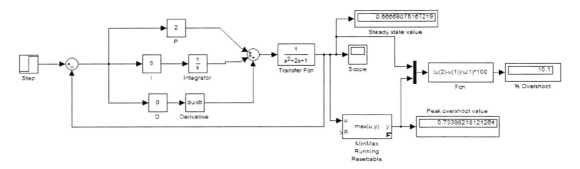

Figure 5.11(a) *The control system plant* $H(s)$ *has cascade connected PID controller*

Then, construct a new simulink model like the figure 5.10(k) where we engaged one From Workspace (*'Simulink → Sources → From Workspace'*), one Demux (*'Simulink → Signal Routing → Demux'*), and one Scope blocks. Bring the blocks in the model, doubleclick the Demux block, change its number of outputs to 4, doubleclick the Scope block, click its parameters icon, click general, and change its number of axes to 4. Run the model and doubleclick the Scope block of figure 5.10(k) to see different error indices as presented in figure 5.10(l). The horizontal axes of the Scope of the figure now simulates the ζ not the time.

Figure 5.11(b) *Block parameter window of the Slider Gain block*

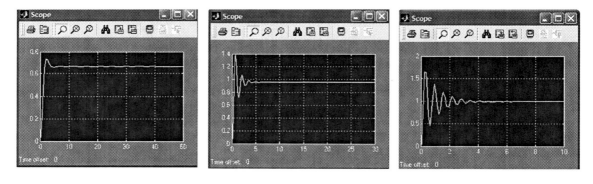

(a) for P =2 and up to 50sec   (b) for P =20 and up to 30 sec   (c) for P =100 and up to 10 sec

Figure 5.12 *Outputs of the control plant for different proportional gains keeping the integral and derivative gains inactive*

5.7 PID controller design

The PID controller is the short for the Proportional, Integral, and Derivative controller and has the transfer function $G(s) = P + \dfrac{I}{s} + Ds$. The constants P, I, and D are called the proportional, integral, and derivative gains.

Since they are functionally simple and easy to implement, the controller finds a wide range of applications. The design of the controller means finding the value of the unknown constants P, I, and D to meet some specific requirements. The first term P in the controller is just a variable number and can be implemented by a Slider gain found in '*Simulink* → *Math Operations* → *Slider Gain*'. Bring the block in a new simulink model file and doubleclick the block to see the parameter window as shown in figure 5.11(b). The figure shows the default gain as 0 to 2. But you can set any gain or number of your choice for the controller implementation. The second term $\frac{I}{s}$ is the product of constant I and the integrator operator $\frac{1}{s}$. Similarly, the third term is the product of the constant D and the Derivative operator s. Both the constants I and D can also be simulated through the Slider Gain block as far as the trial-and-error process is concern. Some interesting facts of the control engineering are being focused in the following three examples.

❖ ❖ Example 1

It is a well-known fact in the control system that only the proportional gain can not provide less settling time and with the increasing gains overshoot or oscillation phenomenon becomes vital that might turn the system unstable. We wish to visualize that phenomenon here in this example.

Let us consider that our control plant has the transfer function $H(s) = \frac{1}{s^2 + 2s + 1}$ in addition negative unity feedback, step input, and a cascaded PID controller are the interior of the system. We want to examine the output subject to different proportional gains keeping the other two gains Derivative and Integral inactive (that means $I = 0$ and $D = 0$). The systematic procedure for simulink model building of figure 5.11(a) is as follows:

⇒ *open a new simulink model file, click the library browser icon, bring one Step block, doubleclick the block to change its step time from 1 to 0, bring one Sum block, and doubleclick the block to change its list of signs to +-*

⇒ *bring one Slider Gain block, rename the block as P, set the low and high gains of the block P to 0 and 1000 respectively, slide the slider to make the proportional gain 0, copy the block, paste it two times, rename the copied blocks as I and D respectively*

⇒ *bring one Derivative block to simulate s and one Integrator block to simulate $\frac{1}{s}$*

⇒ *copy the existing Sum block in the model, paste it to the model, doubleclick the block, and change its List of signs to +++ for the summation of the PID controller signals*

⇒ *bring one Transfer Fcn block and doubleclick the block to enter its numerator and denominator polynomial coefficients as $\frac{[1]}{[1 \quad 2 \quad 1]}$*

⇒ *bring one Scope block to perceive the effect of the proportional gain, bring one Display block, rename the block as Steady state value, doubleclick the block to set its numeric format to long so that more digit accurate output can be seen, enlarge the block, bring another Display block, rename the block as Peak overshoot value, doubleclick the block to set its numeric format to long, enlarge the block, bring one MinMax Running Resettable block, and doubleclick the block to set the max mode for finding the peak*

⇒ *bring one Fcn block, doubleclick the block to enter the parameter expression as (u(2)-u(1))/u(1)*100 for computing the percentage overshoot, and enlarge the block to see the expression in it*

⇒ *bring one Mux and one more Display blocks and rename the Display block as % Overshoot*

⇒ *place various blocks relatively according to figure 5.11(a) and connect them*

We discussed about the modeling of steady state value and peak overshoot in previous section. One might question why the Fcn block is necessary. As you know, the percentage overshoot is given by $\%Overshoot = \frac{Peak\ overshoot - Steady\ state\ value}{Steady\ state\ value} \times 100\%$. The Mux block has two inputs, the upper of which is the steady state value and detected by u(1) and the lower of which is the peak overshoot and detected by u(2). With this one can easily presume that the Fcn bock just computes the percentage overshoot. However, it is the high time to experiment with the model we devised. Doubleclick the proportional gain P, set the value to 2, click simulation from the model menu, set the stop time of the solver as 50, and run the model, doubleclick the Scope to see the output with the autoscale setting as shown in figure 5.12(a).

You can read off the steady state value, the peak overshoot, and the percentage overshoot from the model output. If you go to MATLAB Common Window, you see some warning message regarding Fcn block. Let us not forget that simulink model is very much time (defined in the solver) sensitive and Display block shows the instantaneous signal entering to its input port. At the beginning of the simulation, the output is 0 so the percentage overshoot at the time $t = 0$ becomes undefined. As the time progresses, output becomes available and we do not see more warning. For the sake of the steady state value (obtained from the last value of the simulation), the warning can be ignored. Now doubleclick the proportional gain block P, set the suitable stop time and investigate the effect of proportional gain on the output by just few click operations. We brought about several simulations and tabulate the outputs in table 5.A.

Table 5.A   Solely the effect of proportional controller on the control system

P	Stop time	Steady state value	Peak overshoot	% Overshoot	Steady state error
2	50	0.6667	0.7340	10.1	0.3333
12	50	0.9233	1.2636	36.86	0.0767
20	30	0.9524	1.3890	45.84	0.0476
40	30	0.9747	1.5216	56.11	0.0253
50	20	0.9814	1.5530	58.24	0.0186
100	10	0.9900	1.6352	65.17	0.0100
500	10	0.9945	1.8607	87.1	0.0055

Supportive table 5.A filled from the simulink output discovers many momentous information about the proportional gain change. As the proportional gain increases, steady state output increases thereby reducing the steady state error and peak and percentage overshoot increase. We can see the effects in the figures 5.12(a), 5.12(b), and 5.12(c) corresponding to the proportional gains 2, 20, and 100 respectively. Now the selection of P is yours, it relies on how much overshoot your system can tolerate.

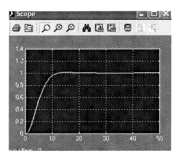 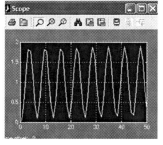 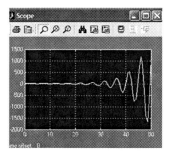

5.13(a) for I =0.2 and up to 50sec   5.13(b) for I =2 and up to 50 sec   5.13(c) for I =4 and up to 50 sec

Figure 5.13 *Outputs of the control plant for different integral gains keeping the proportional and derivative gains inactive*

❖ ❖ Example 2

In example 1, we held the integral and derivative gains completely idle. Let us now make the proportional and derivative gains inactive by setting P =0 and D = 0 respectively and investigate the performance of the integral gain controller.

⌛ Simulink solution

The model we constructed in figure 5.11(a) is hundred percent operational for the example. All we need is doubleclick the P and D blocks and set their values as 0. Next doubleclick the block I and set the test value of the integral gain you wish to see the effect on and the suitable solver time. We conducted several experiments on the model, the outcomes are presented in table 5.B.

Table 5.B   Solely the effect of integral controller on the control system

I	Stop time	Steady state value	Peak overshoot	% Overshoot	Steady state error
0.1	50	0.9981	No	No	0.0019
0.2	50	1.0000	1.0117	1.166	in the range of 10^{-7}
0.3	50	1.0000	1.0885	8.85	in the range of 10^{-7}
1	50	0.9989	1.5429	54.29	0.0011
2	50	Shows some constant oscillation with some specific frequency			
4	50	Unstable output because of increasing value			

Based on the table 5.B presented, little change in integral gain can turn our control system from a stable one to unstable unlike the proportional controller. An integral gain of I =0.2 can provide almost the steady state output with the least overshoot whereas the gain I =2 is seemingly responsible for the constant frequency oscillatory output. Just a gain I =4 is turning the control system as an unstable one. These all are delineated in figures 5.13(a), 5.13(b), and 5.13(c) respectively.

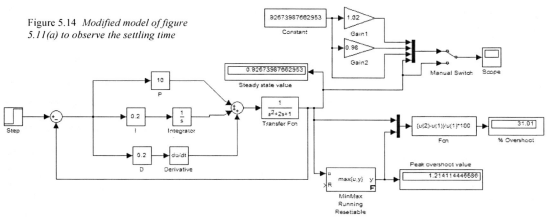

Figure 5.14 *Modified model of figure 5.11(a) to observe the settling time*

❖ ❖ Example 3

The settling time of a control system plays an important role in the design process of the PID controller. Finding the settling time from a control system output can be conducted by employing the modeling procedure of section 5.5. If you are particular about the settling time, you can modify the PID controller model of figure 5.11(a) in accordance with the figure 5.14 based on the 2% criterion or the other. The same model applies as a means of finding the proportional, integral, and derivative gains subject to two constraints – percentage overshoot and settling time.

⏚ Simulink solution

The necessary modifications carried out are we flipped the Steady state value block, brought one Constant block, two Gain blocks to be fed for 2% criterion, brought another Mux block but the number of inputs changed to four, and fed the output of the Mux to Scope via a Manual Switch (the link is *'Simulink → Signal Routing → Manual Switch'*). You could inquire why the Steady state value is not directly connected to the Scope. We are considering here the last value of the simulation as the Steady state output. Until the model is run, we do not know the Steady state value, which is to be passed to the Constant block. During the run time of the model the intermediate values within the start and stop time of the solver will also be exhibited in the Display block. Based on the instantaneous value we can not decide the settling criterion. Hence the Constant, Gain1, Gain2, and four input Mux blocks do not play any role during the finding of the Steady state value from the simulation and they need to be isolated during the Steady state output observation otherwise previous values stored in the Constant block will be superimposed in the Scope which is a complete unwanted value. This is why we provided the Manual Switch in which the down and up positions correspond to the Steady state and settling time simulation respectively. In order to change the switching all we need is just doubleclick the block. As a procedural step, one can choose the P, I, and D constants by doubleclicking the slider gain blocks, doubleclick the Manual Switch to put the connector in the down position if it is necessary, run the model, read off the Steady state value from the display block, doubleclick the Constant block to pass the Steady state value, doubleclick the Manual Switch, run the model again, and then utilize the zooming facility to examine the settling time from the Scope output. Of coarse, the process is completely a trial-and-error one. This is left as an exercise for the reader. Table 5.C is provided to present some data on the modeling so that the reader feels convenient with the simulation.

Table 5.C   Output performance of the control system due to the PID controller

P	I	D	Steady state value	Peak overshoot	% Overshoot	Settling time
0.1	0.1	0.1	1	No overshoot		35.5321 sec
1	0.2	0.1	1	No overshoot		28.6102 sec
24	0.6	1.7	1	1.2248	22.48	25.8242 sec
4	1	0.2	1	1.0616	6.16	8.7036 sec

PID Controller

Figure 5.15 *Constant value PID controller*

Once you are done with the design, you can replace the whole PID controller section (which consists of P, I, D, Derivative, Integrator, and three input sum blocks in figure 5.14) by the block PID Controller (found in *'Simulink Extras → Additional Linear → PID Controller'*) as shown in figure 5.15. You can enter the designed P, I, and D values in the parameter window of the block and that can simplify your model.

5.8 Block links used in this chapter

This chapter mainly discusses the classroom control system analysis problems for which we used the blocks as presented in the table 5.D.

Table 5.D   Necessary blocks for control system analysis as found in simulink library (not in alphabetical order)

Block name	Representative Symbol/Function	Icon Outlook	Block name	Representative Symbol/Function	Icon Outlook
Step	$u(t-1)$ (default)	Step	Derivative	$\dfrac{du}{dt}$ or s	Derivative
	Link: *Simulink → Sources → Step*			Link: *Simulink → Continuous → Derivative*	
Transfer Fcn	Transfer function when given in the polynomial coefficient form	$\dfrac{1}{s+1}$ Transfer Fcn	Integrator	$\int u\,dt$ or $\dfrac{1}{s}$	Integrator
	Link: *Simulink → Continuous → Transfer Fcn*			Link: *Simulink → Continuous → Integrator*	
State-Space	$\begin{cases}\dot{x}=Ax+Bu\\ y=Cx+Du\end{cases}$	x' = Ax+Bu y = Cx+Du State-Space	Sum	$\sum$ or	
	Link: *Simulink → Continuous → State-Space*			Link: *Simulink → Math Operations → Sum*	
Mux	Multiplexing:		Demux	Demultiplexing:	
	Link: *Simulink → Signal Routing → Mux*			Link: *Simulink → Signal Routing → Demux*	
Scope	Oscilloscope or function display	Scope	Zero-Pole	Transfer function when given in the factored form	$\dfrac{(s\text{-}1)}{s(s+1)}$ Zero-Pole
	Link: *Simulink → Sinks → Scope*			Link: *Simulink → Continuous → Zero-Pole*	
Gain	Multiplication by a constant	1 Gain	Out1	Provides output for a sub model	1 Out1
	Link: *Simulink → Math Operations → Gain*			Link: *Simulink → Sinks → Out1*	
Abs	Taking the absolute value of the input	|u| Abs	Display	Display instantaneous value of simulation	0 Display
	Link: *Simulink → Math Operations → Abs*			Link: *Simulink → Sinks → Display*	
Ramp	Generates function like $y(t)=t$	Ramp	Fcn	Entering functional expression considering the independent variable as u	f(u) Fcn
	Link: *Simulink → Sources → Ramp*			Link: *Simulink → User-Defined Functions → Fcn*	
Product	Multiplies two or more functions on a common independent variable	× Product	To Workspace	Send the simulink model output data to MATLAB Command Window	simout To Workspace
	Link: *Simulink → Math Operations → Product*			Link: *Simulink → Sinks → To Workspace*	
Constant	Generates constant value that can serve the purpose of having the steady state value	1 Constant	MinMax Running Resettable	Finds minimum or maximum within the start and stop time of the solver	u min(u,y) y R MinMax Running Resettable
	Link: *Simulink → Sources → Constant*			Link: *Fixed-Point Blockset → Math → MinMax Running Resettable*	
From Workspace	Imports data from the MATLAB workspace to simulink	simin From Workspace	Slider Gain	Provides variable gain through the graphical user interface	1 Slider Gain
	Link: *Simulink → Sources → From Workspace*			Link: *Simulink → Math Operations → Slider Gain*	
Manual Switch	Connects to one of the two inputs through the doubleclick	Manual Switch	Transport Delay	For modeling the delay factor e^{-as}	Transport Delay
	Link: *Simulink → Signal Routing → Manual Switch*			Link: *Simulink → Continuous → Transport Delay*	
Input Point and Output Point	Consider which two points in a model are to be analyzed linearly	Input Point Output Point	PID Controller	Models constant value PID controller	PID PID Controller
	Link: *Control System Toolbox → Input Point or Output Point*			Link: *Simulink Extras → Additional Linear → PID Controller*	

Modeling Some Signal Processing Problems

Signal processing has a very significant contribution in the development of today's hi-tech world ranging from everyday life to the cutting edge technologies. Electrical signals convey information at an extremely high speed that is the veiled reason why signal processing is acquiring so much consolidation at the heart of today's technology. Speedy computers and processors enhanced one more dimension to proliferate the field of signal processing. Signal processing is remarkably a vast and expansive subject. Monotonic increase in the field requires to analyze the problems of the signal processing in a smarter way, and in this regard simulink is the best tool to employ. However, signal processing is broadly divided in two categories – analog and digital. But as far as the short context of the text is concern, we did not put any demarcation between the two categories. Our aim here in this chapter is to introduce the simulink approach of handling the common signal processing problems mostly on the digital perspective.

6.1 Introduction

Now a days we live in a world of science and engineering and our surrounding world is filled with the signals. Signal emerging does not just happen in the electrical systems, its territory is becoming increasingly broader. It goes without saying that no cutting-edge development is without the influence of signal processing whether it is biomedical problems, space exploration, oceanographic analysis, meteorological prediction, sonar systems, or imaging systems. Not to be mentioned that everyday communication ranging from mobile to internet is the home of signal processing. However, all the while in previous chapters we proceeded in an exemplary approach to simulate the deterministic problems. We maintain similar approach also in this chapter to encode and implement the classroom problems relating to the signal transforms, signal operations, signal filtering, .. etc in simulink.

6.2 Analog signal, system, and sampling

An analog or continuous signal is the one that is defined or known at all instant of time. An analog system involves the signals that are continuous. The system very often is characterized by the Laplace transform transfer functions. We handled the Laplace transform transfer functions in chapter 5 for modeling the control system problems. The reader is referred to chapter 5 if the modeling of the continuous system is needed.

The terminologies continuous to discrete and analog to digital are identical. Figures 6.1(a) and 6.1(b) present their schematics. Let us say we have a function $f(t) = t$. The value of t includes the set of real numbers which can be fractional or integer. So over the whole t domain there are infinite number of points. Our finite memory computer can not hold the infinite number. What we do is we take sample at different times and manipulate the samples. The instant we take samples, we take the discrete values. Let us say we take the function at every after $t = 0.5$ that means our $f(t)$ becomes –0.5, 0, 0.5, 1.5 etc. In this case our sampling period or step size is 0.5. Physically t can represent time, displacement, or any other quantity. The simulink block Clock or Ramp can represent our independent variable t. Even though we say continuous generation can be carried out by the Clock but deep inside the system the generation is again the discrete.

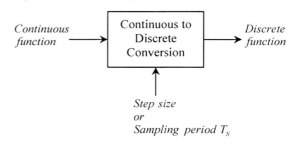

Figure 6.1(a) *Mathematical schematic for continuous function to discrete function conversion*

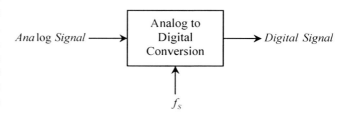

Figure 6.1(b) *Signal processing schematic for analog signal to digital signal conversion*

Simulink solver window gives you the provision for selecting the step size or in other words the T_s of the figure 6.1(a). Referring to the figure 3.1(e) that shows the simulink solver window, you find there the start and stop times that is the beginning and ending of our independent variable t respectively. For example, if the function $f(t)$ is to be generated for $0 \le t \le 3.5$, then the start time is 0 and the stop time is 3.5 respectively. In the same figure of the chapter 3, you also find the solver option type as variable and fixed step. The T_s we mentioned is the Fixed step but simulink has the ability to select the adaptive sampling period T_s indicated by Variable step under solver option type in the same

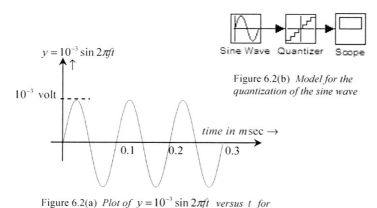

Figure 6.2(a) *Plot of* $y = 10^{-3} \sin 2\pi f t$ *versus* t *for* $0 \le t \le 0.3 m \sec$

Figure 6.2(b) *Model for the quantization of the sine wave*

Figure 6.2(c) *Quantized sine wave*

window. When you model something in simulink file, the same Clock or in other words the independent variable information is automatically shared to all blocks related to the model.

In Analog to Digital Conversion as shown in figure 6.1(b), the sampler takes sample from a continuous signal with a specific frequency f_s. The relationship between the T_s and f_s is just $T_s = \dfrac{1}{f_s}$.

6.3 Quantization of an analog signal

Quantization basically maps a continuous time signal to finite levels depending on the capacity of a digital system (usually bits) and is necessary for the analog to digital conversion (ADC). For instance, a displacement change is happening from –7 cm to 2 cm. We want to feed the information through two binary digits (00 01 10 11) of a digital system. Only four possibilities are there and accordingly we can divide our displacement signal into four levels (called quantization level). However, to quantize a signal in simulink, the range of the signal and quantization interval must be known beforehand.

Let us say we want to quantize the sinusoidal signal $y = 10^{-3} \sin 2\pi f t$ volt (plot is in figure 6.2(a)) for the time interval $0 \leq t \leq 0.3 m \sec$ with a quantization interval 0.125×10^{-3} volt, where $f = 10 KHz$. Let us bring a Sine Wave block in a new simulink model file (link: '*Simulink $\rightarrow$ Sources $\rightarrow$ Sine Wave*') to model $y = 10^{-3} \sin 2\pi f t$, doubleclick the block to enter the settings as $\begin{Bmatrix} \text{Amplitude}: 1e-3 \\ \text{Frequency (rad/sec)}: 2*pi*10e3 \end{Bmatrix}$ keeping the others as default, bring one Quantizer block in the model file following the link '*DSP Blockset $\rightarrow$ Quantizers $\rightarrow$ Quantizer*', doubleclick the block to enter the Quantization interval as 0.125e-3 in the parameters window, bring one Scope block and connect the blocks as shown in figure 6.2(b). To feed the information $0 \leq t \leq 0.3 m \sec$, change the solver stop time to 0.3e-3 ('*Simulation $\rightarrow$ Simulation parameters $\rightarrow$ Solver*'). Run the model and doubleclick the Scope to see the quantized sine wave with the autoscale setting as presented in figure 6.2(c).

Quantization error might be of interest of analysis for which the model of the figure 6.2(d) can be employed. The error is defined as the actual signal minus the quantized signal. We need one Sum block and doubleclick the block to change its list of signs to +- in the parameter window for the model of the figure 6.2(d). The error signal is displayed in the figure 6.2(e) with the autoscale setting as found from the Scope.

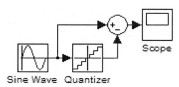

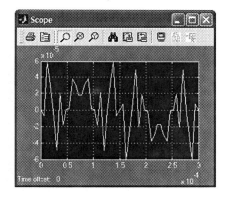

Figure 6.2(d) *Model for finding the quantization error*

Figure 6.2(e) *Quantization error of the sine wave*

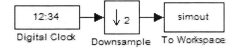

Figure 6.3(a) *Down sampling the sequence* $x[n] = n$ *by a factor 2*

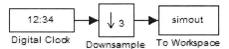

Figure 6.3(b) *Down sampling the sequence* $x[n] = n$ *by a factor 3*

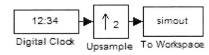

Figure 6.3(c) *Upsampling the sequence* $x[n] = n$ *by a factor 2*

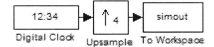

Figure 6.3(d) *Upsampling the sequence* $x[n] = n$ *by a factor 4*

6.4 Down and up sampling and padding of a digital signal

The concept of down and up samplings are solely with the discrete samples. Let us say a digital signal $x[n]$ envelope follows the ramp function ($x[n] = n$, the straight-line with a slope 45^{0}). If we choose specific sample time let us say 0.5, we should have the samples $x[n] =$[0   0.5   1   1.5   2   2.5   3] for a time interval $0 \leq n \leq 3$. When we down sample the sequence $x[n]$ by a factor of 2 (means we take the sample every after one sample starting from the beginning one), we are supposed to have [0   1   2]. Again if we down sample by a factor of 3 (means we take the sample every after two samples starting from the beginning one), we have [0   1.5]. The discrete signals can be

117

generated by the Digital Clock (link: '*Simulink* → *Sources* → *Digital Clock*'). One can assume that the output envelope from the block is a ramp function. Bring the block in a new simulink model file and doubleclick the block to enter the Sample time as 0.5. For the insertion of the discrete interval $0 \le n \le 3$, we click '*Simulation* → *Simulation parameters* → *Solver*', change the stop time to 3, change the Solver option type from Variable to Fixed step, and change the Fixed step size to 0.5. Now we bring one Downsample (link: '*DSP Blockset* → *Signal Operations* → *Downsample*') and one To Workspace (link: '*Simulink* → *Sinks* → *To Workspace*') blocks in the model file. The default setting for the Downsample is 2 and the other block helps us see the output in MATLAB workspace. Doubleclick the block To Workspace and change its Save format to Array in the parameter window of the block. Connect the blocks as shown in the figure 6.3(a) and run the model. Let us go to MATLAB command window and execute simout (inside variable of To Workspace) as follows:

MATLAB Command for	MATLAB Command for
downsampling by a factor of 2,	downsampling by a factor of 3,
>>simout ⏎	>>simout ⏎
simout =	simout =
0	0
0	0
1	1.5000
2	

As you see for the factor 2, the output is consistent with our expectation but at the beginning there is a value 0. This is due to the fact that the block Downsample has a slot for the initial value 0. Doubleclick the block Downsample and change its Downsample factor to 3 (the effect is the figure 6.3(b)), run the model, and execute simout from MATLAB workspace. The downsampled return is shown above.

The aforementioned sequence $x[n]$ can be upsampled as well. The upsampling by a factor M basically inserts $M-1$ zeroes inbetween the samples. One should bear in mind that the sample time in the Digital Clock and the sample time in the solver must be consistent with the upsampling factor. Let us say the sample time in the Digital Clock is 0.5 and we want to upsample the sequence by a factor of 2, then the solver fixed step size should be $\frac{0.5}{2} = 0.25$. If the upsample factor were 4, the solver fixed step size would be $\frac{0.5}{4} = 0.125$. The block Upsample located in the same link as the Downsample is can perform all these. So we replace the Downsample block of the figure 6.3(a) by the Upsample one, change the solver fixed step size to 0.25 (figures 6.3(c) and 6.3(d) show the model for the upsampling factor 2 and 4 respectively), and run the model. Then we execute simout in MATLAB Command Window, and the result is as follows:

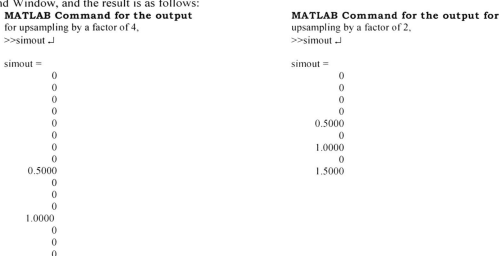

MATLAB Command for the output	MATLAB Command for the output for
for upsampling by a factor of 4,	upsampling by a factor of 2,
>>simout ⏎	>>simout ⏎
simout =	simout =
0	0
0	0
0	0
0	0
0	0.5000
0	0
0	1.0000
0	0
0.5000	1.5000
0	
0	
0	
1.0000	
0	
0	
0	
1.5000	

Padding of a sequence means adding some zeroes or other values to the sequence so that its length increases. The fast Fourier transform (FFT) becomes effective when the sequence length is the power of two. Not necessarily all sequence will have a length of the power of the two. We perform padding and then send the sequence for FFT implementation. Let us say that we have the sequence $x[n] = [1 \quad 2 \quad 7 \quad -5 \quad -6]$. The next power of two means the length of the sequence should be 8. We need to insert three zeroes at the end so that we have the new sequence $x[n] = [1 \quad 2 \quad 7 \quad -5 \quad -6 \quad 0 \quad 0 \quad 0]$. Bring one Constant (link: '*Simulink* → *Sources* → Constant*'), one Zero Pad (link: '*DSP Blockset* → *Signal Operations* → *Zero Pad*'), and one Display blocks in a new simulink model file. Enter the code of $x[n]$ as [1 2 7 –5 –6] in the Constant parameter window on doubleclicking, enlarge the block to see the contents, change the solver option type to Fixed step and discrete from the model menu bar (*Simulation → Simulation parameters → Solver)*, connect the blocks as placed in the figure 6.3(e), doubleclick the block Zero Pad and change the Number of output rows to Next power two in the parameter window of the block, and run the model. We expect the reader to see the output as shown in the Display block following the enlargement of the block. We padded the sequence $x[n]$ towards the end, padding towards the beginning of the sequence option is also found in the parameter widow of the Zero Pad. The parameter window also gives the option for the user defined padding. The padding by an element other than zero is implementable by means of the block Pad found in the same source link whose icon outlook is presented at the end of the chapter.

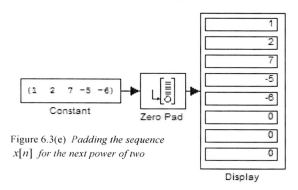

Figure 6.3(e) *Padding the sequence $x[n]$ for the next power of two*

6.5 Analog filter design

An analog filter is characterized by Laplace transfer function, for example, $H(s) = \dfrac{1}{1+2s}$. The filter has four types: low pass, high pass, band pass, and band stop. When the variable s is replaced by $j\omega$ in the expression of $H(s)$, we obtain the frequency function of the filter which is purely a complex quantity. Each one of the filters has its own kind of transfer function. Depending on the design parameter and accuracy, one can obtain different filter transfer functions for the same problem solving. In a word we can say that from the required specification, we have to find a transfer function $H(s)$. In simulink you can design the analog filters of the type $\begin{cases} \text{Butterworth} \\ \text{Chebyshev Type I} \\ \text{Chebyshev Type II} \\ \text{Bessel} \\ \text{Elliptic} \end{cases}$. While working in simulink, the transfer function of the filter is kept hidden in the model file. Let us explore the following examples on the Butterworth subclass of the analog filter design in simulink.

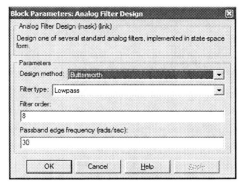

Figure 6.4(a) *Block parameter window of the Analog Filter Design*

✦ ✦ Example 1

Design a low pass continuous time Butterworth filter in simulink whose order is 4 and the pass band edge frequency (cutoff frequency) is 120 Hz.

⌘ Solution

Open a new simulink model file, bring the block Analog Filter Design following the link '*DSP Blockset → Filtering → Filter Design → Analog Filter Design*' in your model file whose icon outlook is shown in the figure 6.4(b), and doubleclick the block to see the block parameter window as shown in figure 6.4(a). In

Figure 6.4(b) *Icon outlook of the Analog Filter Design block*

that window you find the slot for the parameters $\begin{cases} \text{Design Method} \\ \text{Filter type} \\ \text{Filter order} \\ \text{Passband edge frequency} \end{cases}$. As you find in the window, the default design method and Filter type are for the Butterworth. Let us change the Filter order from the default 8 to 4. The parameter window intakes the passband edge frequency in terms of the radians per second whence the frequency code should be 2*pi*120 for the problem at hand.

✦ ✦ Example 2

Design a high pass continuous time Butterworth filter in simulink whose order is 5 and passband edge or cutoff frequency is $1KHz$.

Mohammad Nuruzzaman

⊟ Solution

The settings in the parameter window (appears as the popup menu) of the figure 6.4(a) you need are

$\left\{\begin{array}{l}\text{Design Method : Buttrworth} \\ \text{Filter type : Highpass} \\ \text{Filter order : 5} \\ \text{Passband edge frequency : } 2*pi*1e3\end{array}\right\}$. As soon as you change the settings, the icon for the highpass filter appears

(see table 6.A for the icon outlook).

✦ ✦ Example 3

Design a band pass Butterworth filter of the order 7 in simulink whose lower cutoff frequency is $10KHz$ and upper cutoff frequency is $40KHz$.

Table 6.A   Various analog filters found in simulink library (not in the alphabetical order and all of them is located in the link *DSP Blockset → Filtering → Filter Design → Analog Filter Design*)

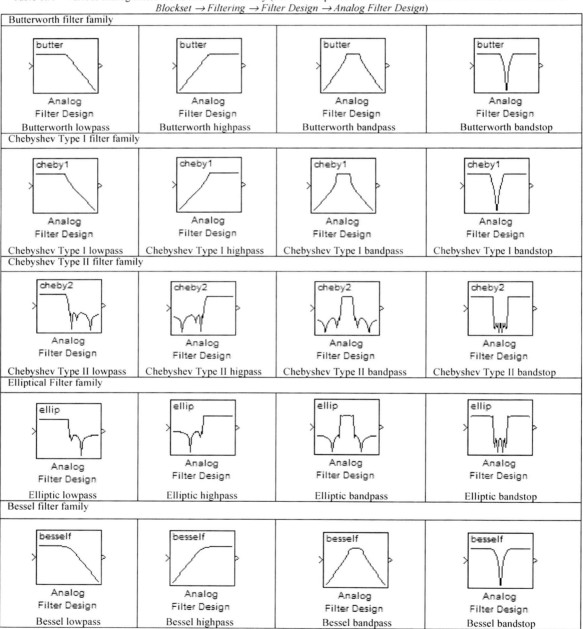

⧉ **Solution**

One needs to enter $\left\{ \begin{array}{l} \text{Design Method : Buttrworth} \\ \text{Filter type : Bandpass} \\ \text{Filter order : 7} \\ \text{Lower passband edge frequency (rads/sec): } 2*\text{pi}*10\text{e}3 \\ \text{Upper passband edge frequency (rads/sec): } 2*\text{pi}*40\text{e}3 \end{array} \right\}$ in the parameter window of the

figure 6.4(a) for the design. The icon outlook of the filter can be seen in the table 6.A.

♦ ♦ **Example 4**

Design a band stop Butterworth filter of the order 10 in simulink whose lower cutoff frequency is $1KHz$ and upper cutoff frequency is $6KHz$.

⧉ **Solution**

We enter $\left\{ \begin{array}{l} \text{Design Method : Buttrworth} \\ \text{Filter type : Bandstop} \\ \text{Filter order : 10} \\ \text{Lower passband edge frequency (rads/sec): } 2*\text{pi}*1\text{e}3 \\ \text{Upper passband edge frequency (rads/sec): } 2*\text{pi}*6\text{e}3 \end{array} \right\}$ as the settings in the parameter window

of the figure 6.4(a).

The other families of the Analog Filter Design block as presented in table 6.A can be modeled in a similar fashion.

6.6 Discrete filter modeling in z transform domain

A discrete filter can be described by the z transform transfer function in terms of the division of the numerator and denominator polynomials in z^{-1}. As an example, let us consider the filter transfer function $H(z) = \dfrac{45z^{-1} + z^{-2}}{1 - 4.5z^{-1} + 3z^{-2} - z^{-3}}$. The discrete filter function can be put in the polynomial coefficient form as the

ascending power in z^{-1} both for the numerator and denominator by $H(z) = \dfrac{[0 \quad 45 \quad 1]}{[1 \quad -4.5 \quad 3 \quad -1]}$. Because the

Figure 6.5 *Icon outlooks of the Discrete Filter block under different cases and the dialog window*

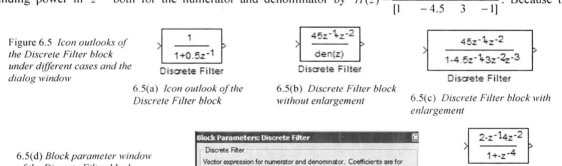

6.5(a) *Icon outlook of the Discrete Filter block*

6.5(b) *Discrete Filter block without enlargement*

6.5(c) *Discrete Filter block with enlargement*

6.5(d) *Block parameter window of the Discrete Filter block*

6.5(e) *Block outlook for* $F(z)$

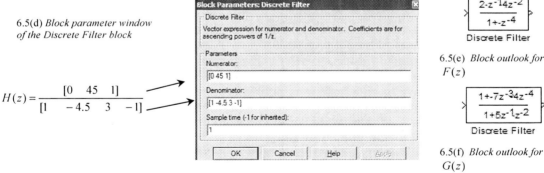

$H(z) = \dfrac{[0 \quad 45 \quad 1]}{[1 \quad -4.5 \quad 3 \quad -1]}$

6.5(f) *Block outlook for* $G(z)$

numerator does not have the constant term, we assign 0 coefficient for that. When we model the filter transfer function in simulink, we enter the numerator and denominator polynomial coefficients in the dialog window of simulink. Open a new simulink model file and click the library icon. Bring one Discrete Filter block in the model file following the link *'Simulink → Discrete → Discrete Filter'* which appears as shown in figure 6.5(a). Doubleclick the block to see the block parameter window, enter the numerator and denominator polynomial coefficients of $H(z)$ as indicated in figure 6.5(d) (one space gap between the coefficients and enclosed by the third brace), and click OK. You see the block outlook as shown in figure 6.5(b) which appears as 6.5(c) following the enlargement that is what our objective is.

Just to provide more examples, the discrete filter functions $F(z) = \dfrac{2 - z^{-1} - 4z^{-2}}{1 - z^{-4}}$ and $G(z) =$

$\dfrac{1 - 7z^{-3} - 4z^{-4}}{1 + 5z^{-1} - z^{-2}}$ can be modeled by entering the numerator and denominator coefficients as $F(z) =$

$\dfrac{[2 \quad -1 \quad -4]}{[1 \quad 0 \quad 0 \quad 0 \quad -1]}$ and $G(z) = \dfrac{[1 \quad 0 \quad 0 \quad -7 \quad -4]}{[1 \quad 5 \quad -1]}$ respectively. Their block outlooks are presented in

the figures 6.5(e) and 6.5(f) respectively.

6.7 Modeling z transform transfer function

A z transform transfer function is seen commonly in two forms either the transfer function expressed in terms of the numerator-denominator polynomials in z or expressed in terms of the gain-pole-zero form of the polynomials. Let us consider the transfer function $H(z) = \dfrac{z^2 + 45z}{-z^3 + 3z^2 - 4.5z + 1}$, the form can best be entered to simulink as the numerator and denominator polynomial coefficients form as descending powers in z. Hence, the polynomial coefficients are [1 45 0] and [−1 3 −4.5 1] for the numerator and denominator respectively. Zero is to be assumed for any missing term in the polynomials. To model the transfer function in simulink, open a new model and bring a Discrete Transfer Fcn block in the model file following the link *'Simulink → Discrete → Discrete Transfer Fcn'* whose appearance is presented figure 6.6(a). Doubleclick the block to see the dialog window of the figure 6.6(d). In that window, you type the numerator and denominator polynomial coefficients of $H(z)$ (one space gap between the coefficients and enclosed by the third brace) as indicated in the figure and click OK. The block of figure 6.6(b) can be seen due to the action you performed. Enlarge the block to display all in it like the figure 6.6(c).

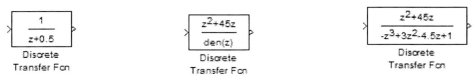

6.6(a) *Icon outlook of the Discrete Transfer Fcn block*

6.6(b) *Discrete Transfer Fcn block without enlargement*

6.6(c) *Discrete Transfer Fcn block with enlargement*

Figure 6.6 *Icon outlooks of the Discrete Transfer Fcn block under different cases and the dialog window*

$$H(z) = \frac{[1 \quad 45 \quad 0]}{[-1 \quad 3 \quad -4.5 \quad 1]}$$

6.6(d) *Block parameter window of the Discrete Transfer Fcn block*

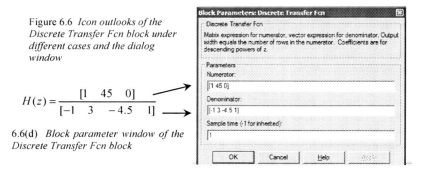

As a matter of different representation, we may have the z transfer function in the gain-pole-zero form like $G(z) = \dfrac{-34(z - 4)(z + 5)}{z(z - 2)(z + 7)(z + 0.7)}$. From the numerator and denominator factors, we see the gain, poles, and zeroes of $G(z)$ are −34, {4,−5}, and {0,2,−7,−0.7} respectively. Bring a Discrete Zero-Pole block in a new model file following the link *'Simulink → Discrete → Discrete Zero-Pole'*, the view of which is shown in figure 6.7(a). Now doubleclick the block to conduct the session with simulink. Type the gain, poles, and zeroes of $G(z)$ with proper spacing and bracing as indicated in figure 6.7(d) and click OK. The number of poles of $G(z)$ fits out of the block that is why you see the block of figure 6.7(b). Enlarge the block to display the complete discrete transfer function as shown in figure 6.7(c).

Sometimes the transfer function might have multiple poles, for instance, $H(z) = \dfrac{1}{z^3(z + 3)}$. Looking into the denominator of $H(z)$, one can say that there are three poles at $z = 0$ and the numerator is unity. If you use the

Discrete Zero-Pole block for modeling the transfer function, you can enter the poles by typing [0 0 0 –3] and the zero by typing [ ] in the dialog window. The symbol [ ] indicates empty matrix. In some other situation, you may have to deal with the single $\dfrac{1}{z}$ expression, which can be implemented by the Unit Delay block following the link *'Simulink* $\rightarrow$ *Discrete* $\rightarrow$ *Unit Delay'*. Or, mathematical manipulation might be necessary for example $H(z) = \dfrac{1}{z^3(z+3)}$ can be rearranged in two parts like $H(z) = \dfrac{1}{z^3} \times \dfrac{1}{z+3}$. The expression $\dfrac{1}{z^3}$ can be modeled by placing three Unit Delay blocks side by side or by entering the coefficients as $\dfrac{[1]}{[1\ \ 0\ \ 0\ \ 0]}$ to the Discrete Transfer Fcn block. Anyhow we present few more examples on modeling the discrete transfer functions in the following:

Discrete
Zero-Pole

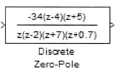

Discrete
Zero-Pole

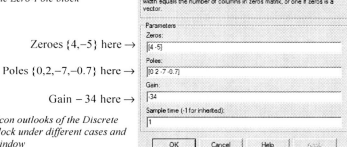

Discrete
Zero-Pole

6.7(a) *Icon outlook of the Discrete Zero-Pole block*

6.7(b) *Discrete Zero-Pole block without enlargement*

6.7(c) *Discrete Zero-Pole block with enlargement*

6.7(d) *Block parameter window of the Discrete Zero-Pole block*

Zeroes {4,–5} here →

Poles {0,2,–7,–0.7} here →

Gain – 34 here →

Figure 6.7 *Icon outlooks of the Discrete Zero-Pole block under different cases and the dialog window*

✦ ✦ Example 1

For the transfer function $H(z) = \dfrac{0.4z^2(z-4)(z+5)}{(z+7)^4(z-3)}$, we can use the Discrete Zero-Pole block setting the parameters as $\begin{Bmatrix} gain:\ 0.4 \\ zeroes:\ [0\ 0\ 4\ -5] \\ poles:\ [-7\ -7\ -7\ -7\ 3] \end{Bmatrix}$ in the dialog window.

✦ ✦ Example 2

The transfer function $H(z) = \dfrac{0.4z^3 + 3z - 6}{(z-2)^5(z+3)}$ has the higher order poles, and the numerator is in polynomial form clearly a mixing up. We better expand the denominator taking MATLAB's help. Go to MATLAB Command Window and perform the following to expand the denominator:

MATLAB Command

```
>>syms z ↵              ← Declaring the variable z as symbolic
>>y=(z-2)^5*(z+3); ↵    ← Defining the denominator (z − 2)⁵(z + 3)
>>sym2poly(expand(y)) ↵

ans =
        1   -7   10   40  -160   208  -96
```

The MATLAB Command 'expand' expands the symbolic polynomial expressions as the descending powers of z and the Command 'sym2poly' (abbreviation of symbolic to polynomial coefficients) extracts the polynomial coefficients that is how one obtains the denominator polynomial coefficients as [1 –7 10 40 –160 208 –96]. So now our $H(z)$ can be modeled by entering the numerator and denominator polynomial coefficients in the dialog window of the Discrete Transfer Fcn block as $H(z) = \dfrac{[0.4 \quad 0 \quad 3 \quad -6]}{[1 \quad -7 \quad 10 \quad 40 \quad -160 \quad 208 \quad -96]}$.

✦ ✦ Example 3

Can we model the discrete transfer function like $-2z^2 - z + \dfrac{1}{z} + \dfrac{1}{z^2}$? The answer is no because it is assumed that the z transform transfer function is unilateral not the bilateral one. Positive powers of the transform variable z happens for the negative time or index which simulink does not comply with.

✦ ✦ Example 4

We can not model the transfer function $H(z) = \dfrac{0.4z^3 + 3z - 6}{(z-2)(z+3)}$ by the Discrete Transfer Fcn block because the degree of the numerator (3) is more than that of the denominator (2). The condition we need is degree of numerator $\leq$ degree of denominator. The same condition also applies for the Discrete Zero-Pole block. For example, we can not also model $H(z) = \dfrac{(z-2)(z+3)^3}{(z-2)(z-3)}$ in simulink.

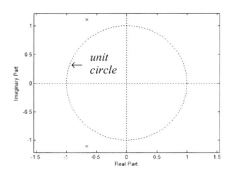

Figure 6.8(a) *Roots for the polynomial* $3z^2 + 4z + 5$ *plotted in the z plane*

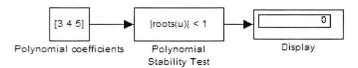

Figure 6.8(b) *Model for the stability test of the polynomial* $3z^2 + 4z + 5$

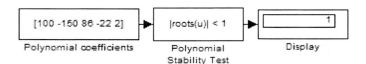

Figure 6.8(c) *Model for the stability test of the polynomial* $100z^4 - 150z^3 + 86z^2 - 22z + 2$

6.8 Stability test of a z polynomial

When we take the roots of a polynomial, the roots can be real or complex. Both the real and complex roots can have the magnitude other than unity. From the viewpoint of the stability analysis, residing the transfer function poles inside the unity circle in the complex plane is to be compelled. Our objective here is to mention the process how we can test the roots of a polynomial being inside the unit circle. Let us take the example of the polynomial $3z^2 + 4z + 5$, the polynomial has the roots $-0.6667 + j\,1.1055$ and $-0.6667 - j\,1.1055$. Each of the roots has the magnitude 1.2910 and is more than unity. The plot of the roots is shown in figure 6.8(a) in which they are placed by the mark $\times$. The simulink block Polynomial Stability Test placed in the location '*DSP Blockset $\rightarrow$ Math Functions $\rightarrow$ Polynomial Functions $\rightarrow$ Polynomial Stability Test*' can test whether the roots of a polynomial inside or outside the unit circle. For the example at hand, let us bring one Constant block in a new simulink model file, doubleclick the block to enter the coefficient code of the polynomial as [3 4 5], rename the block as Polynomial coefficients, bring one Polynomial Stability Test block in the model, bring one Display block, connect them as shown in the figure 6.8(b), run the model, and see the output 0 in the Display block. The 0 indicates that the roots are outside the unit circle.

As another example, the polynomial $100z^4 - 150z^3 + 86z^2 - 22z + 2$ has the roots $\begin{Bmatrix} 0.5 \\ 0.4 + j0.2 \\ 0.4 - j0.2 \\ 0.2 \end{Bmatrix}$, all of them

are inside the unit circle hence we should expect the output as 1. The model and its outcome are shown in the figure 6.8(c). If any root of a polynomial is having the magnitude exactly 1, the Display output should be 0.

6.9 Convolution of signals

The convolution of two discrete signals is basically the multiplication of two polynomials related to the signals. Let us consider that we have two sequences $x[n]$ =[7 8 9 −8] and $y[n]$ =[−2 1 3 4]. The polynomials (considering these are the coefficients of the polynomial in descending power of x) related with the signals are $7x^3 + 8x^2 + 9x - 8$ and $-2x^3 + x^2 + 3x + 4$ respectively. Following multiplication, one would get the resultant polynomial as $-14x^6 - 9x^5 + 11x^4 + 77x^3 + 51x^2 + 12x - 32$ or in coefficient form it becomes [−14   −9   11   77

51   12   −32]. We can expect that simulink will return these polynomial coefficients following the simulation. Let us bring

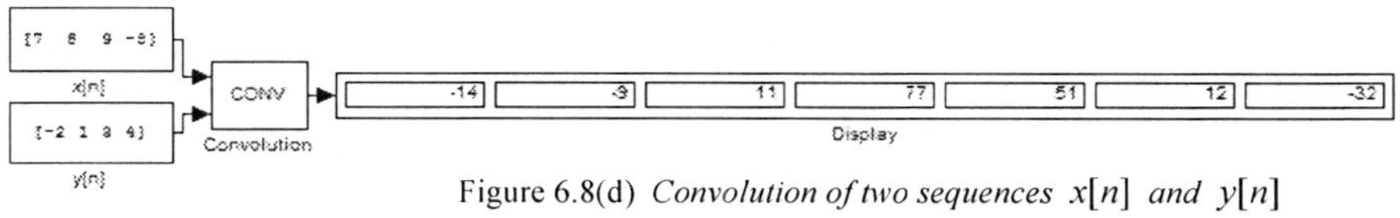

Figure 6.8(d) *Convolution of two sequences* $x[n]$ *and* $y[n]$

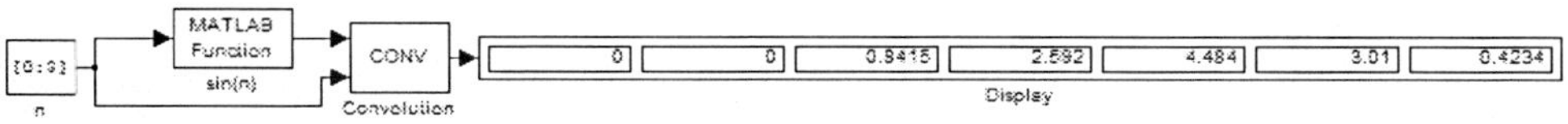

Figure 6.8(e) *Convolution of the functional sequences* $\sin n$ *and* n *for* $0 \le n \le 3$

one Constant block in a new simulink model file, doubleclick the block to enter the code of $x[n]$ as [7 8 9 −8] in the parameter window, uncheck the Interpret vector parameters as 1-D in the same window, rename the block as x[n], bring another Constant block in the model file, doubleclick the block to enter the code of $y[n]$ as [−2 1 3 4] in the parameter window, uncheck the Interpret vector parameters as 1-D in the same window, rename the block as y[n], resize each block to see the contents, bring one Convolution block in the model file following the link *'DSP Blockset* $\rightarrow$ *Signal Operations* $\rightarrow$ *Convolution'*, bring one Display block, change the solver options from the Variable Step to Fixed Step by clicking through *'Simulation* $\rightarrow$ *Simulation parameters* $\rightarrow$ *Solver* $\rightarrow$ *Solver Options* $\rightarrow$ *Type'* from the menu bar of the model, connect the blocks as shown in figure 6.8(d), run the model, and see the convolved output as presented in the same figure. The Display block needs to be resized.

Sometimes our sequence or discrete signal can follow some functional variation, for instance, $x[n] = \sin n$ and $y[n] = n$ for the interval $0 \le n \le 3$ (n integer). Within the given interval, their sample values and the convolved sequence are given by $x[n]$ =[0   0.8415   0.9093   0.1411] and $y[n]$ =[0 1 2 3] and [0   0   0.8415   2.5923   4.4842   3.0101   0.4233] respectively. Let us bring one Constant block in a new simulink model file, doubleclick the block and enter the code for generating the integers [0 1 2 3] by typing [0:3] in the parameter window, uncheck the Interpret vector parameters as 1-D in the same window, rename the block as n, bring one MATLAB Fcn block and make sure that its function is set to sin (by doubleclicking the block), rename the block as sin(n), connect the blocks along with one Convolution and one Display blocks like the figure 6.8(e), change the solver options from the Variable Step to Fixed Step as we did before, run the model, and see the conlvolved polynomial coefficients in the inside of the resized Display block.

6.10 Transform related problems

We intend to highlight the common transform implementation procedure in simulink practiced in many signal-processing problems in this section. Among them, the discrete Fourier transform (DFT), fast Fourier transform (FFT), and discrete cosine transform (DCT) can be noted.

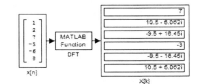

Figure 6.8(f) *Discrete Fourier transform of a sequence x[n]*

6.10.1 Discrete Fourier transform and FFT

The discrete Fourier transform (DFT) of a signal can be forward and inverse. Primarily the transform finds the information content of a signal in terms of the sinusoidal harmonics and the discrete signals take the form of a row or column matrix for one-dimensional case. The forward discrete Fourier transform $X[k]$ of a finite length sequence $x[n]$ (can be real or complex) is defined as $X[k] = \sum_{n=1}^{N} x[n] e^{-j2\pi(k-1)\frac{(n-1)}{N}}$, where k can vary from 1 to N and N is the length of the sequence $x[n]$. Both $x[n]$ and $X[k]$ are discrete in nature and having the same length and the transform is in general complex. Considering the sequence $\begin{bmatrix} n & x[n] \\ 1 & 1 \\ 2 & 2 \\ 3 & 7 \\ 4 & -5 \\ 5 & -6 \\ 6 & 8 \end{bmatrix}$, one can compute the discrete Fourier

transform of $x[n]$ employing just mentioned formula and which is given by $\begin{Bmatrix} k & X[k] \\ 1 & 7 \\ 2 & 10.5 - j6.0622 \\ 3 & -9.5 + j16.4545 \\ 4 & -3 \\ 5 & -9.5 - j16.4545 \\ 6 & 10.5 + j6.0622 \end{Bmatrix}$.

Our objective is to obtain the transform $X[k]$ in simulink and the model for implementation is presented in figure 6.8(f). To implement the model, bring one Constant block in a new simulink model file, doubleclick the block to enter the code of the $x[n]$ as [1 2 7 –5 –6 8]' (the operator ' is used so that we have a column matrix for better display) under the slot Constant value in the parameter window, uncheck the Interpret the vector parameter as 1-D in the parameter window, rename the block as x[n], resize the block, bring one MATLAB Fcn block in the model file, doubleclick the block to enter the function as fft (the function can compute the DFT employing the aforementioned formula), set the output signal type as complex in the parameter window because the values of $X[k]$ are complex, rename the block as DFT, bring one Display block, rename the block as X[k], connect the blocks as shown in figure 6.8(f), run the model, and resize the block X[k] to see the DFT output as shown in the same figure.

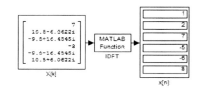

Figure 6.8(g) *FFT block computes the fast Fourier transform of the sequence $x[n]$*

The sequence $x[n]$ we chose has the length 6. When the length of the sequence is the power of 2 (for example 2, 4, 8, 16, 32, etc), the computation of the transform can happen under the algorithm of the fast Fourier transform (elaboration of fft) and for which we have the dedicated block FFT found in the link *'DSP Blockset → Transforms → FFT'*. A four element sequence example

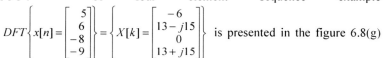

$DFT\left\{ x[n] = \begin{bmatrix} 5 \\ 6 \\ -8 \\ -9 \end{bmatrix} \right\} = \left\{ X[k] = \begin{bmatrix} -6 \\ 13 - j15 \\ 0 \\ 13 + j15 \end{bmatrix} \right\}$ is presented in the figure 6.8(g)

in which we utilized the block FFT. But you need to change the solver type from the Variable step to the Fixed step under the *Simulation → Simulation parameter → Solver* from the model file menu bar.

Inverse discrete Fourier transform (IDFT) is the recovery of the original sequence $x[n]$ from the forward transform sequence $X[k]$ whose computational formula is

given by the expression $x[n] = \frac{1}{N} \sum_{k=1}^{N} X[k] e^{+j2\pi(k-1)\frac{(n-1)}{N}}$

where n can vary from 1 to N and N is the length of the transform sequence $X[k]$. Let us recover the

sequence $x[n]$ from $\begin{Bmatrix} k & X[k] \\ 1 & 7 \\ 2 & 10.5 - j6.0622 \\ 3 & -9.5 + j16.4545 \\ 4 & -3 \\ 5 & -9.5 - j16.4545 \\ 6 & 10.5 + j6.0622 \end{Bmatrix}$. The model

of the figure 6.8(h) shows the implementation for which

Figure 6.8(h) *Inverse discrete Fourier transform of $X[k]$*

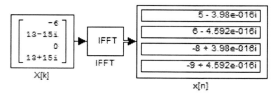

Figure 6.8(i) *IFFT block operated on the power of 2 length sequence*

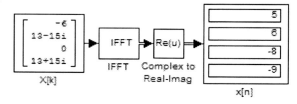

Figure 6.8(j) *Exact recovery of the original sequence x[n] by neglecting very small imaginary parts*

bring one Constant, one MATLAB Fcn, and one Display blocks in a new simulink model file, rename them as X[k], IDFT, and x[n] respectively, doubleclick the X[k] block and enter the code for $X[k]$ as [7 10.5-6.0622i -

9.5+16.4545i -3 -9.5-16.4545i 10.5+6.0622i].' (the operator .' indicates the transpose without the complex conjugate, and the transposition just places the elements as column matrix for display convenience), resize the block, uncheck the Interpret the vector parameter as 1-D in the parameter window, doubleclick the IDFT block, enter the function ifft (the function computes the inverse discrete Fourier transform intaking $X[k]$) in the parameter window, set the output signal type as real, connect the blocks according to the figure 6.8(h), run the model, and finally resize the x[n] block to see the output of simulink as shown in the figure.

When the transform sequence length is the power of two, one can use the IFFT bock located in the link '*DSP Blockset → Transforms → IFFT*', for example, the aforementioned 4-element transform sequence that is

$$IDFT\left\{ X[k]=\begin{bmatrix} -6 \\ 13-j15 \\ 0 \\ 13+j15 \end{bmatrix} \right\}=\left\{ x[n]=\begin{bmatrix} 5 \\ 6 \\ -8 \\ -9 \end{bmatrix} \right\}$$ is modeled in the figure 6.8(i). It is particularly important to mention that

the solver setting must be Fixed Step. Referring to the outcome of the figure 6.8(i), we see that the first element is 5−3.98e−016i=5− j 3.98×10^{-16}. The imaginary part appears because of the round off error due to the machine accuracy and can easily be ignored by taking the real part of the output. We just inserted one Complex to Real-Imag block (reached via '*Simulink → Math Operations → Complex to Real-Imag*') between the IFFT and x[n] blocks. Also we changed the output type of the inserted block by doubleclicking it and setting to the Real in the popup menu of the parameter window. The action is presented in figure 6.8(j). This kind of real or imaginary part rejection might also be necessary in the model of the figure 6.8(h) depending on the transform sequence $X[k]$.

6.10.2 Discrete cosine transform

The kernel of the discrete Fourier transform is complex exponential whereas the discrete cosine transform (DCT) employs the real cosine kernel. One-dimensional forward discrete cosine transform of a finite length sequence

$x[n]$ is defined as $X[k]=\sum_{n=1}^{N} w[k]x[n]\cos\dfrac{\pi(2n-1)(k-1)}{2N}$ where $w[k]=\begin{cases} \sqrt{\dfrac{1}{N}} & when \quad k=1 \\ \sqrt{\dfrac{2}{N}} & when \quad 2\le k\le N \end{cases}$, k can vary from 1 to

N, and N is the length of the sequence. Let us consider the

sequence $\begin{cases} n & x[n] \\ 1 & 1 \\ 2 & 0 \\ 3 & -2 \\ 4 & 4 \\ 5 & -6 \\ 6 & -1 \\ 7 & 5 \end{cases}$ whose transform sequence computed by

the formula of $X[k]$ is given by $\begin{cases} k & X[k] \\ 1 & 0.3780 \\ 2 & -0.7389 \\ 3 & 3.2987 \\ 4 & -3.9880 \\ 5 & 5.5708 \\ 6 & 0.2228 \\ 7 & -4.9439 \end{cases}$. The

modeling is very similar to the model of the figure 6.8(f). All we need here is to enter the MATLAB Function dct in the parameter window of MATLAB Fcn. However, the figure 6.8(k) shows the model and its simulated output.

Along with the forward discrete cosine transform, the inverse discrete cosine transform is also there for the recovery of x[n] from $X[k]$. The inverse discrete cosine transform of

$X[k]$ is given by $x[n]=\sum_{k=1}^{N} w[k]X[k]\cos\dfrac{\pi(2n-1)(k-1)}{2N}$, where

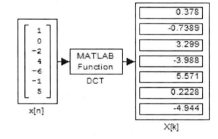

Figure 6.8(k) *Discrete cosine transform of the sequence x[n]*

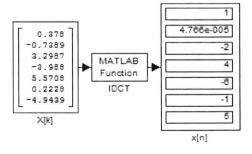

Figure 6.8(l) *Inverse discrete cosine transform of the transform sequence X[k]*

$$w[k] = \begin{cases} \dfrac{1}{\sqrt{N}} & when \quad k = 1 \\ \sqrt{\dfrac{2}{N}} & when \quad 2 \leq k \leq N \end{cases}$$, n can vary from 1 to N and N is the length of $X[k]$. Applying the inverse formula

on just obtained transform sequence of $X[k]$, one would obtain the sequence $x[n]$ we started with. The corresponding simulink model (similar to the model construction of the figure 6.8(h)) is presented in the figure 6.8(1) but the MATLAB Function should be idct in the parameter window that utilizes the inverse computational formula. Looking into the model, the second element is returned as 4.766e-005 or 4.766×10^{-5} for practical consideration that can be assumed as zero. However, the dedicated blocks both for the forward discrete cosine transform and the inverse discrete cosine transform are there in the link '*DSP Blockset $\rightarrow$ Transforms*', but one needs the length of the sequence as the power of two to make them operational also the solver settings must be the fixed step as we did for the FFT and IFFT. The icon outlooks of the DCT and IDCT blocks can be found in the table attached at the end of this chapter.

6.11 Digital filter analysis

Given a digital filter transfer function in Z domain, studying and analyzing the characteristics of the digital filter might be necessary. Among the important characteristics of a filter, we can mention pole-zero plot, stability, step response, unit impulse response, magnitude response, phase response, and group delay. Our objective of this section is to focus how simulink can be effective in finding these characteristics.

Starting with the transfer function $H(z) = \dfrac{6 - 6z^{-1}}{6 + 5z^{-1} + z^{-2}}$, we first find just mentioned characteristics of the digital filter analytically.

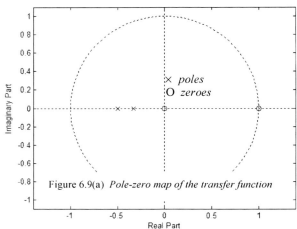

Figure 6.9(a) *Pole-zero map of the transfer function*

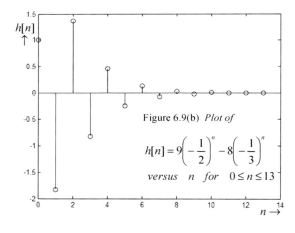

Figure 6.9(b) *Plot of*

$$h[n] = 9\left(-\frac{1}{2}\right)^n - 8\left(-\frac{1}{3}\right)^n$$

versus n *for* $0 \leq n \leq 13$

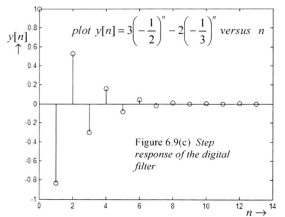

plot $y[n] = 3\left(-\dfrac{1}{2}\right)^n - 2\left(-\dfrac{1}{3}\right)^n$ versus n

Figure 6.9(c) *Step response of the digital filter*

⊟ Pole-zero map of the filter

Rearranging the transfer function provides us $H(z) = \dfrac{6z^2 - 6z}{6z^2 + 5z + 1} = \dfrac{6z(z-1)}{(3z+1)(2z+1)}$ indicating

$\left\{ \begin{array}{l} poles : -\dfrac{1}{3} \ and \ -\dfrac{1}{2} \\ zeroes : 0 \ and \ 1 \end{array} \right\}$ whose plots are shown in the figure 6.9(a).

⊟ Stability of the filter

Regarding to the poles as mapped in the figure 6.9(a), one can inspect that all poles are inside the unit circle hence the filter is stable.

☐ Unit sample response or impulse response of the filter

When the input to the filter is $\delta[n]$ (unit sample), the output of the filter is termed as the impulse response which is nothing but the inverse Z transform of the given Z transform transfer function. For the example at hand, the impulse response of the filter is given by $h[n] = 9\left(-\dfrac{1}{2}\right)^n - 8\left(-\dfrac{1}{3}\right)^n$ (taking unilateral inverse Z transform) which when plotted over $0 \le n \le 13$ takes the shape as presented in the figure 6.9(b).

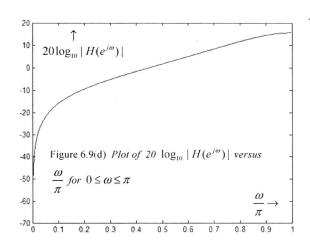

Figure 6.9(d) *Plot of* $20 \log_{10} |H(e^{j\omega})|$ *versus* $\dfrac{\omega}{\pi}$ *for* $0 \le \omega \le \pi$

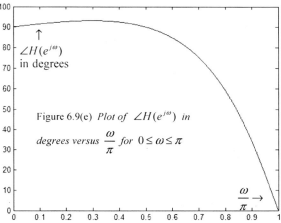

Figure 6.9(e) *Plot of* $\angle H(e^{j\omega})$ *in degrees versus* $\dfrac{\omega}{\pi}$ *for* $0 \le \omega \le \pi$

☐ Step response or unity sequence response of the filter

The step response of a filter means the response or output of the filter when the unit sequence is the filter's input. The unit sequence has the Z transform $\dfrac{z}{z-1}$. One can interpret the step response as the inverse Z transform of the $\dfrac{z}{z-1} H(z)$. For the given filter, the step response should be the unilateral inverse Z transform of $\dfrac{z}{z-1} \times \dfrac{6 - 6z^{-1}}{6 + 5z^{-1} + z^{-2}}$ which is $y[n] = 3\left(-\dfrac{1}{2}\right)^n - 2\left(-\dfrac{1}{3}\right)^n$ and the plot of the $y[n]$ versus n over $0 \le n \le 13$ is depicted in the figure 6.9(c).

Figure 6.9(f) *Plot of* $-\dfrac{d\angle H(e^{j\omega})}{d\omega}$ *versus* $\dfrac{\omega}{\pi}$ *for* $0 \le \omega \le \pi$

☐ Magnitude and phase responses of the filter

The magnitude and phase responses of $H(z)$ are obtained by inserting $z = e^{j\omega}$ in the given filter function on that account $H(e^{j\omega}) = \dfrac{6 - 6e^{-j\omega}}{6 + 5e^{-j\omega} + e^{-j2\omega}}$. Carrying out a clumsy manipulation, one would separate the magnitude and phase parts of $H(e^{j\omega})$ as $|H(e^{j\omega})| = 6\sqrt{\dfrac{2 - 2\cos\omega}{(6 + 5\cos\omega + \cos 2\omega)^2 + (5\sin\omega + \sin 2\omega)^2}}$ and $\angle H(e^{j\omega}) = \tan^{-1}\dfrac{\sin\omega}{1 - \cos\omega} + \tan^{-1}\dfrac{5\sin\omega + \sin 2\omega}{6 + 5\cos\omega + \cos 2\omega}$ respectively. Having found $|H(e^{j\omega})|$ and $\angle H(e^{j\omega})$, they can easily be plotted for different ω. For the digital domain, the angular frequency ω has a periodicity of 2π. The plots of $|H(e^{j\omega})|$ and $\angle H(e^{j\omega})$ versus ω are presented in figures 6.9(d) and 6.9(e) respectively. Jut to be consistent with simulink, we plot $20 \log_{10} |H(e^{j\omega})|$ versus $\dfrac{\omega}{\pi}$ and $\angle H(e^{j\omega})$ in degrees versus $\dfrac{\omega}{\pi}$ both for $0 \le \omega \le \pi$.

⊡ **Group delay of the filter**

Once we have the phase component $\angle H(e^{j\omega})$ of the function $H(e^{j\omega})$, the group delay is defined as $-\dfrac{d\angle H(e^{j\omega})}{d\omega}$. Differentiating with respect to ω and simplifying the expression, one would obtain $-\dfrac{d\angle H(e^{j\omega})}{d\omega} = $

$\dfrac{1}{2} - \dfrac{3}{2} \times \dfrac{15\cos\omega + 8\cos^2\omega + 5}{25 + 35\cos\omega + 12\cos^2\omega}$, plot of which can be seen in the figure 6.9(f) for the same ω domain.

With all these computations and plottings, we define our expectations from simulink as follows:

From the given Z transfer function $H(z) = \dfrac{6 - 6z^{-1}}{6 + 5z^{-1} + z^{-2}}$,

simulink should return us

$\left.\begin{array}{l} \textit{pole zero map of the figure 6.9(a)} \\ \textit{indicate the stability of the filter} \\ \textit{impulse response of the figure 6.9(b)} \\ \textit{step response of the figure 6.9(c)} \\ \textit{magnitude response of the filter like the figure 6.9(d)} \\ \textit{phase response of the filter like the figure 6.9(e)} \\ \textit{group delay behavior of the filter like the figure 6.9(f)} \end{array}\right\}$

Now let us turn our attention to simulink. To commence with the analysis, let us bring the Digital Filter Design (link: *'DSP Blockset → Filtering → Filter Designs → Digital Filter Design'*) block in a new simulink model file whose icon outlook is presented in the figure 6.9(g). On doubleclicking the block, the reader should see the Digital Filter Design window as shown in the figure 6.9(h). Our next action is to enter the given digital filter specifications into the window. From the window menu bar, let us click the

Figure 6.9(g) *Digital Filter Design icon*

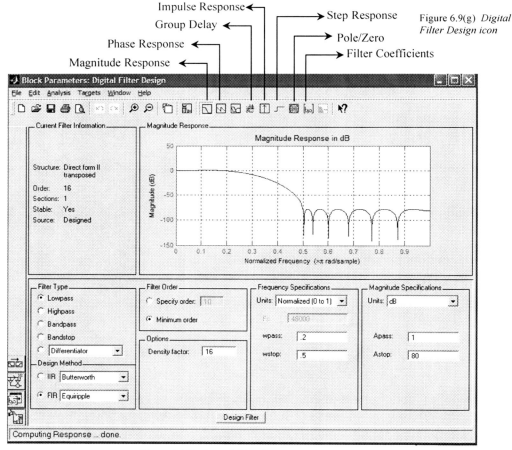

Figure 6.9(h) *Digital Filter Design window*

130

menu File prompting the pull down menu. We click the option Import Filter from the pull down menu of the File. The window responds with some changes in the lower portion keeping the upper portion almost unchanged. In the lower portion of the responded window, you find the Filter Coefficients and under which the slots for Numerator, Denominator, Sampling frequency, and Filter structure exist. The given filter $H(z)$ considering the ascending power of z^{-1} can be coded in terms of the numerator and denominator coefficients as $\begin{cases} \text{Numerator} : [6-6] \\ \text{Denominator} : [6 \ 5 \ 1] \end{cases}$. We enter the numerator and denominator coefficients in the slots of the last window keeping the other settings as default. Of coarse, one needs to clear the old coefficients before entering the coefficients for the given $H(z)$'s. Once you are done with entering the coefficients, click the Import Filter located in the lower portion of the Digital Filter Design window. You see the Magnitude Response in dB due to the action, which is exactly identical with the one we presented in figure 6.9(d). Now let us click the Analysis located in the menu bar of the Digital Filter Design window. We see a number of options related to the problem under the Analysis. Let us click from the Analysis submenu one by one and verify simulink's output consistency:

Click *Analysis* → *Magnitude Response*,   simulink responds with the plot shown in figure 6.9(d)
Click *Analysis* → *Phase Response*,   simulink responds with the plot shown in figure 6.9(e)
Click *Analysis* → *Pole/Zero Plot*,   simulink responds with the plot shown in figure 6.9(a)
Click *Analysis* → *Group Delay Response*,   simulink responds with the plot shown in figure 6.9(f)
Click *Analysis* → *Step Response*,   simulink responds with the plot shown in figure 6.9(c)
Click *Analysis* → *Impulse Response*,   simulink responds with the plot shown in figure 6.9(b)

The upper left of the window contains the slot for Current Filter Information under which you find that the filter we are analyzing is stable. The menu bar of the Digital Filter Design is furthermore furnished with the icons of the Pull down menu functions. Figure 6.9(h) is referred to for some of the icons that can perform the same operations as those conducted by the Analysis pull down functions.

Anyhow, we provided some guidelines for the digital filter analysis in this section. The reader can explore other deign tools available in the window.

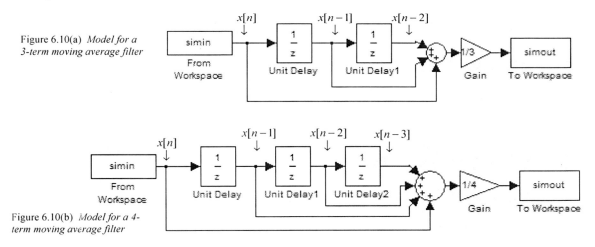

Figure 6.10(a) *Model for a 3-term moving average filter*

Figure 6.10(b) *Model for a 4-term moving average filter*

6.12 Moving average filter and phase difference of two complex signals

If a moving average filter has M terms, the difference equation for the filter is given by $y[n] = \frac{1}{M} \sum_{k=0}^{k=M-1} x[n-k] = \frac{1}{M} (x[n] + x[n-1] + \ldots + x[n-M])$. Let us consider $M = 3$ then we have $y[n] = \frac{1}{3} (x[n] + x[n-1] + x[n-2])$. Implementation of the moving average filter means the implementation of the difference equation. The equation is basically the successive delaying of the sequence $x[n]$ and then adding followed by the division of the number of terms.

Let us say we have the sequence $\begin{bmatrix} n & x[n] \\ 0 & 9 \\ 1 & 5 \\ 2 & 6 \\ 3 & 8 \end{bmatrix}$. If we consider 3-term moving average filter, what should be

the output of the filter? The output sequence length should be equal to the length of the input one. At the beginning

edge of the sequence we assume 0 hence the output of the filter should be $\begin{bmatrix} \dfrac{0+0+9}{3} & \dfrac{0+9+5}{3} & \dfrac{9+5+6}{3} \end{bmatrix}$

$\dfrac{5+6+8}{3}$$]$=[3   4.6667   6.6667   6.3333]. First of all we need to feed the data in simulink. Go to MATLAB

prompt and perform the following:

MATLAB Command

```
>>n=[0:3]';  x=[9 5 6 8]';  ↵
>>simin=[n x];  ↵
```

The first line of the command is the assignment of the index n and sequence $x[n]$ both as a column matrix to the variables n and x respectively. The second line in the command is the formation of the simulink acceptable data that is a rectangular matrix, the first and second columns of which are the index n and sequence $x[n]$ respectively. Figure 6.10(a) depicts the modeling (block links are in the table 6.B). Bring all blocks associated with the model in a new simulink file. To turn a default two input Sum block to three input one, you need to change its list of signs to +++ in the parameter window of the block. Also the gain setting of the Gain block is to be changed to 1/3. From the simulink menu bar, change the solver option type from variable to fixed and enter the fixed step size as 1 because n has the increment 1, change the stop time to 3, doubleclick the block To Workspace, change its save format to Array, and run the model. Now execute the following in the command prompt:

```
>>simout  ↵
```

```
simout =
        3.0000
        4.6667
        6.6667
        6.3333
```

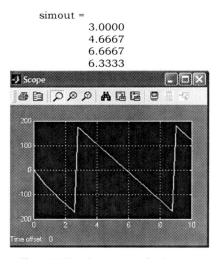

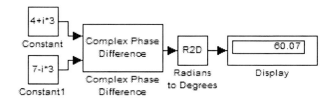

Figure 6.10(c) *Model for finding the phase difference of two complex numbers*

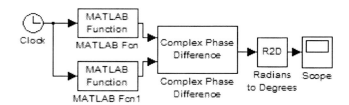

Figure 6.10(e) *Scope output for the phase difference of the two complex signals*

Figure 6.10(d) *Model for finding the phase difference of two complex signals*

That is what is expected. For the M-term moving average filter, one needs to employ M Unit Delay blocks in the model. Figure 6.10(b) illustrates the model for a 4-term moving average filter.

Now we concentrate on finding the phase difference of two complex numbers and signals. Suppose we have two complex numbers $4+j\,3$ and $7-j\,3$. When they are expressed in polar or complex exponential form, they become $5\angle 36.8799^0$ and $7.6158\angle -23.1986^0$ respectively hence the phase difference is 60.0785^0. The model of the figure 6.10(c) implements the phase difference in degrees. The code of the complex numbers fed in the Constant and Constant1 are 4+i*3 and 7-i*3 respectively. Since the output is in radians we connected the Radians to Degrees block for the conversion.

To include the example of the complex functional phase difference, let us say we have two complex functions $\sin(2+ix)$ and e^{-1+ix}. We want to see their phase difference as the function of x for $0 \le x \le 10$. The model you need is presented in the figure 6.10(d). The Clock in the model generates the independent variable x and the blocks MATLAB Fcn and MATLAB Fcn1 contain the codes of $\sin(2+ix)$ and e^{-1+ix} as sin(2+i*u) and exp(-1+i*u) considering u as the independent variable respectively. Both Fcn blocks' output must be set to complex in the parameter window of the blocks. The reader can find the block links at the end of the chapter. On running the model, the phase difference of the two complex signals is displayed by the Scope as in the figure 6.10(e) with the autoscale setting.

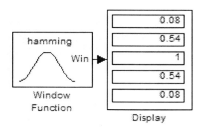

Figure 6.11(a) *Generating and displaying a Hamming window of length 5*

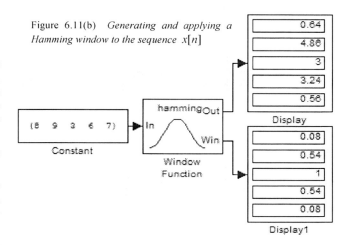

Figure 6.11(c) *Applying the Hamming window to a discrete sine wave*

6.13 Window functions

Window functions are employed in the design of the FIR filters. Simulink provides you the option for generating and applying the window functions. Let us start with the definition of the Hamming window, $w[n] =$

$$\begin{cases} 0.54 - 0.46\cos\dfrac{2\pi n}{N} & for\ 0 \le n \le N-1 \\ 0 & otherwise \end{cases}$$. The

integer N is called the window length and the generation is discrete. Let us generate a Hamming window of the length 5 hence our output sequence should be $w[n] = [\ w[0] \qquad w[1]$ $w[2] \qquad w[3] \qquad w[4]\ ] = [0.08 \quad 0.54 \quad 1 \quad 0.54$

Figure 6.11(b) *Generating and applying a Hamming window to the sequence $x[n]$*

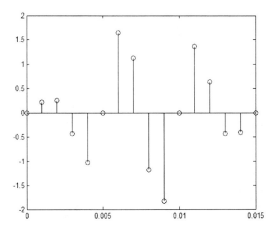

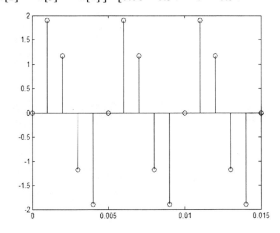

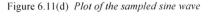

Figure 6.11(d) *Plot of the sampled sine wave*

Figure 6.11(e) *The sampled sine wave followed by the windowing*

0.08] and that is what we expect from simulink. Bring a Window Function block in a new simulink model file (link: *'DSP Blockset $\rightarrow$ Signal Operations $\rightarrow$ Window Function'*) and doubleclick the block. There are three options under parameters found in the parameter window of the block namely $\begin{Bmatrix} Operation \\ Window\ type \\ Sampling \end{Bmatrix}$. Under the popup menu of the Operation, you find the option Generate window and click that. Because the action you performed, the slot for Window length appears in the parameter window and enter there 5. The default icon is for the Hamming window.

Now bring a Display block in the model file, connect the block with the Window Function as shown in figure 6.11(a), and run the model. The Display block contents are the results of the simulation following enlargement of the block.

The next relevant query is how we can apply the window to some discrete functions. Let us say we have a sequence $x[n] = [8 \quad 9 \quad 3 \quad 6 \quad 7]$ and wish to apply aforementioned Hamming window on the sequence. The resultant output should be the multiplication of the sequence and the window function. Hence the output should be $x[n] \, w[n] = [0.64 \quad 4.86 \quad 3 \quad 3.24 \quad 0.56]$ employing the index to index multiplication. Doubleclick the block Window Function, you find the option Generate and apply window within the popup menu of Operation. Because of the option change, the block has now one input (indicated by In) and two output ports (indicated by Out and Win). The model of the figure 6.11(b) presents the implementation. The Display and Display1 connected to the Out and Win return the output and window functions respectively. If the reader wishes to apply the window without displaying, the reader should select Apply window to input under Operation in the parameter window of the block.

Assume that now we have a periodic sine wave sequence of amplitude 2 and frequency 200 *Hz* sampled at a frequency $f_s = 1000 \, Hz$ for the time interval $0 \le t \le 0.015 \sec$. With this data the sampling period or the step size should be $T_s = \dfrac{1}{1000} = 0.001 \sec$ and the simulink code of the discrete wave is 2*sin(2*pi*200*[0:0.001:0.015]). The plot of the discrete function is shown in the figure 6.11(d). There are 16 points or samples within the given interval. So one can generate a Hamming window of length 16 as we did before. Multiplying the discrete sine wave with the Hamming window results the plot of the figure 6.11(e). To implement the problem in simulink, we construct the model of the figure 6.11(c). We set the Operation mode as the Apply window to the input in the parameter window of the block Window Function on doubleclicking, set the Save format of the block To Workspace as Array on doubleclicking, and change the solver stop time to 0 from the model menu bar because of constant generation (otherwise we would see repetition of the output many times within the default 0 to 10 sec). Run the model and execute the following in the MATLAB command prompt:

MATLAB Command
```
>>t=[0:0.001:0.015]; ↵
>>stem(t,simout) ↵
```

The steps you performed should result the figure 6.11(e) that is what we demand from simulink. In the block parameter of the Window Function you find two options under the slot of sampling – periodic and symmetric. The aforementioned expression for the Hamming window is basically the periodic one. The second option symmetric just produces an even function. The block automatically assumes the window length (here it is 16) according to the input sequence length.

However, the Window Function block also possesses the option for other windows. You can click the Window type in the parameter window and find there Bartlett, Blackman, Boxcar, Chebyshev, Hann, Hanning, Kaiser, Triang, and User defined options. To be acquainted about their functional description, the reader is referred to [25] and [26].

6.14 Normalization of signals

Sometimes it is necessary to normalize a discrete signal. Let us consider the sequence $x[n] = [7 \quad 8 \quad 6 \quad -9 \quad 0]$. One way of normalizing the signal is take the magnitude which is here $\sqrt{49 + 64 + 36 + 81 + 0} = \sqrt{230}$ and divide each element in the $x[n]$ so that we have the normalized signal as [0.4616 \quad 0.5275 \quad 0.3956 \quad -0.5934 \quad 0]. Bring one Constant, one Normalization, and one Display blocks in a new simulink model file. Doubleclick the block Constant to enter the code of $x[n]$ as [7 8 6 –9 0] in the parameter window. Doubleclick the block Normalize and change its Norm to 2-norm in the parameter window of the block leaving the other as default. Connect the blocks as shown in the figure 6.11(f) and run the model. The implementation is obvious from the enlarged Display block output.

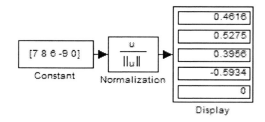

Figure 6.11(f) *Normalization of the discrete signal* $x[n]$

Another way of normalizing the signal is instead of dividing by the square root of the magnitude, divide by the square of the signal magnitude (which is here 230). Just doubleclick the block Normalization and change it norm to Squared 2-norm. Due to the norm change, the reader should see the change in the icon outlook making sense with the modeling.

Again diversified problems force us to think in different ways. For some instances such as filter responses or communication problems we are very much interested about the dB levels with respect to the maximum magnitude instead of the total magnitude of a vector. For the example of $x[n]$ at hand, the maximum of the absolute value is 9. So if we divide each element in $x[n]$ by 9 and then perform the operation $20\log_{10} x$, we end up with the sequence $[-2.1829 \quad -1.0231 \quad -3.5218 \quad 0 \quad -\infty]$. It is easy to think this way but to implement the problem we need the algorithm suitable to simulink. Anyhow the model is presented in the figure 6.11(g) whose block links are

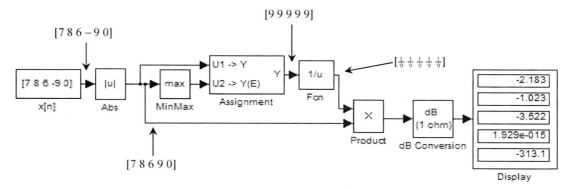

Figure 6.11(g) *Normalizing the sequence $x[n]$ with respect to maximum and then converting to dB*

tabulated in the table 6.B. Construct the model bringing all necessary blocks in a new simulink model. We renamed the Constant block as x[n]. The block MinMax finds the maximum or minimum in a constant generation. Since the default is minimum, doubleclick the block and set it for maximum. The block Assignment has two input ports labeled by U1 and U2. The U1 can take any sequence as input here it is $|x[n]|$. The U2 is the element (s) that is (are) to be assigned in the U1 (here it is the maximum of $|x[n]|$). So the output Y of the Assignment block is exactly having the same length as $|x[n]|$ is. Unavailability of the divider block makes us use the reciprocal via Fcn. Looking into the Display block, the 0 and $-\infty$ are displayed as 1.929×10^{-15} (very small) and -313.1dB (high negative dB value) respectively because approach of simulink is numerical. For simplicity we connected just five-element sequence but it can be a sequence containing hundreds of elements. The Display can then be replaced by a Scope block to view the functional variation.

6.15 IIR and FIR filters specifics

Infinite impulse response (IIR) and finite impulse response (FIR) filter names appear specifically in the digital signal processing. Although we addressed generalized filter design and analysis tool before, the objective of this section is to present the implementation in the appellation of IIR and FIR filters. Both the IIR and FIR filters are applicable to the discrete signals.

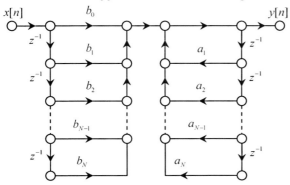

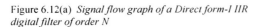

Figure 6.12(a) *Signal flow graph of a Direct form-I IIR digital filter of order N*

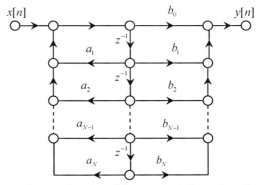

Figure 6.12(b) *Signal flow graph of a Direct form-II IIR digital filter of order N*

First we discuss the IIR structure implementation. The equation relating to the input $x[n]$ and output $y[n]$ of an IIR filter for the kind Direct form I or Direct form II is given by

discrete time equation: $y[n] - \sum_{k=1}^{N} a_k y[n-k] = \sum_{k=0}^{M} b_k x[n-k]$ and

135

the Z transform system function: $H(z) = \dfrac{\sum\limits_{k=0}^{M} b_k z^{-k}}{1 - \sum\limits_{k=1}^{N} a_k z^{-k}}$.

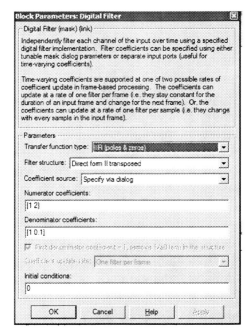

The delay element $\dfrac{1}{z}$ corresponds to a computer memory register. The multiplication usually consumes time compared to the summation in the digital hardware. Analysis of different filter structure is important in the sense that fewest constant multipliers and fewest delay elements are desired to make the digital filter system time-hardware space-cost effective. However, the figures 6.12(a) and 6.12(b) present the signal flow diagram of the IIR filter of order N for the class Direct form-I and Direct form-II respectively.

As an example we can consider the IIR digital filter with the transfer function $H(z) = \dfrac{0.1 - 0.4z^{-1} - 3z^{-2}}{1 + 0.9z^{-1} + 0.6z^{-2} - 7z^{-3}}$. Taking into account the negative sign in the denominator of the given Z transform transfer function, the filter coefficients are $\begin{cases} b_0 = 0.1, \, b_1 = -0.4, \, b_2 = -3 \\ a_1 = -0.9, \, a_2 = -0.6, \, a_3 = 7 \end{cases}$. The figures 6.12(c) and 6.12(d) realize the given IIR digital filter for the Direct forms I and II respectively. Now in terms of simulink cuisine, let us bring one Digital Filter block following the link *'DSP Blockset $\rightarrow$ Filtering $\rightarrow$ Filter Designs $\rightarrow$ Digital Filter'* in a new simulink model file

Figure 6.12(h) *Block parameter window of the Digital Filter*

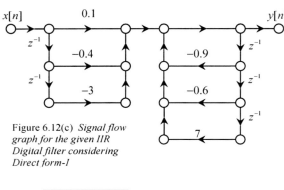

Figure 6.12(c) *Signal flow graph for the given IIR Digital filter considering Direct form-I*

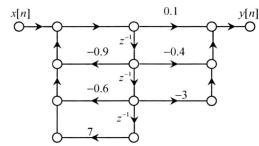

Figure 6.12(d) *Signal flow graph for the given IIR Digital filter considering Direct form-II*

Figure 6.12(e) *Icon outlook for the default Digital Filter block*

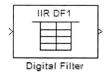

Figure 6.12(f) *Icon outlook for the IIR filter of Direct Form I*

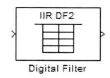

Figure 6.12(g) *Icon outlook for the IIR filter of Direct Form II*

whose default icon outlook is presented in the figure 6.12(e). Attached figure 6.12(h) is the parameter window for

the Digital Filter block in which you find the parameters $\begin{cases} \text{Transfer function type} \\ \text{Filter structure} \\ \text{Coeffient source} \\ \text{Numerator coefficients} \\ \text{Denominator coefficients} \end{cases}$. Our filter is IIR one so we do

not need to change the Transfer function type in the parameter window. In the Filter structure we need to set Direct form I from the popup menu for the realized figure 6.12(c). We enter [0.1 –0.4 –3] and [1 0.9 0.6 –7] in the slot of the Numerator and Denominator coefficients respectively. With these settings the icon outlook appears as in the figure 6.12(f). Inside the icon you find IIR DF1 which means IIR Direct form I. Again with the same filter coefficients if you select Direct Form II under Filter structure in the parameter window of the figure 6.12(h), you see the icon outlook as shown in figure 6.12(g) containing the consistent filter name IIR DF2 inside the icon.

Only the Direct forms I and II do we not see in the block Digital Filter. On clicking the popup menu of the Filter structure in the window of the figure 6.12(h), you find the available filter sub structures are

$$\left\{\begin{array}{l} \text{Direct form I} \\ \text{Direct form II} \\ \text{Direct form I transposed} \\ \text{Direct form II transposed} \\ \text{Biquad direct form II transposed (SOS)} \end{array}\right\}$$. The transposed forms can be obtained by reversing the direction of the

signal flows including the delay elements and interchanging the positions of the filter coefficients while keeping the time domain equation or the Z transform system function unchanged. In performing so, we have the filter structures of the Nth order transposed forms for the Direct forms I and II as depicted in the figures 6.13(a) and 6.13(b) respectively. Applying the example transfer function, one can easily obtain the filter structure for the transposed Direct forms I and II portrayed in the figures 6.13(c) and 6.13(d) respectively. Now in simulink implementation the step we need is click the popup menu of the Filter structure in the parameter window of the figure 6.12(h) and select the appropriate structure from the popup menu with the same filter coefficients that results the icon outlooks of the figures 6.13(e) and 6.12(e) respectively.

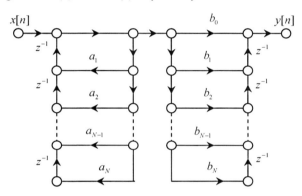

Figure 6.13(a) *Signal flow graph of a transposed Direct form-I IIR digital filter of order N*

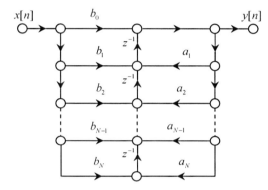

Figure 6.13(b) *Signal flow graph of a transposed Direct form-II IIR digital filter of order N*

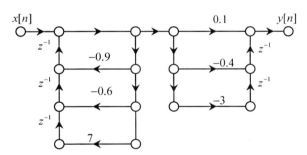

Figure 6.13(c) *Signal flow graph of the transposed Direct form-1 for the given transfer function*

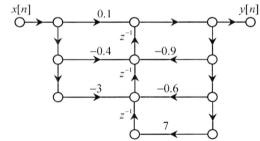

Figure 6.13(d) *Signal flow graph of the transposed Direct form-II for the given transfer function*

When the given transfer function is having only the denominator polynomials in z^{-1}, the reader can employ the IIR (all poles) option under the Transfer function type popup menu of the figure 6.12(h) for which simulink prompts with the icon of the figure 6.13(f).

Now we would like to introduce the FIR filter terminologies. The FIR filters are basically the all-zero filters except the poles at $z = 0$. The discrete

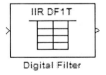

Figure 6.13(e) *Icon outlook for the transposed Direct Form 1*

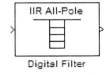

Figure 6.13(f) *Icon outlook for the All Pole IIR filter*

time domain and the Z transform domain expressions for the filters of order N are given by $y[n] = \sum\limits_{k=0}^{N} b_k x[n-k]$ and

$H(z) = \sum_{k=0}^{N} b_k z^{-k}$ respectively. Placed figures 6.14(a) and 6.14(b) picture the generalized N^{th} order FIR filter structures for the direct and the transposed direct forms respectively. To go through by an example, let us consider the FIR filter with the system function $H(z) = -3 + 4z^{-1} - 0.9z^{-2}$. The filter is a second order one and when the filter

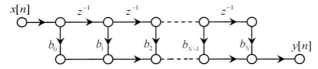

Figure 6.14(a) *Signal flow graph of the Direct form FIR digital filter of order N*

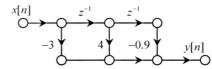

Figure 6.14(c) *Signal flow graph of the Direct form FIR representation for the given example*

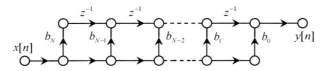

Figure 6.14(b) *Signal flow graph of the transposed Direct form FIR digital filter of order N*

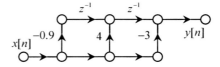

Figure 6.14(d) *Signal flow graph of the transposed Direct form FIR representation for the given example*

coefficients are placed, the filter attains the structures of the figures 6.14(c) and 6.14(d) for the direct and the transpose forms respectively. As regular, the parameter window of the Digital Filter block presented in the figure 6.12(h) holds the option for the FIR filter implementation in simulink as well. Referring to the window, you find FIR (all zero) in the popup menu under the Transfer function type. The example FIR filter is having the coefficients as [−3 4 −0.9] and we

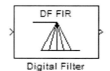

Figure 6.14(e) *Icon outlook of the Direct Form (DF) FIR filter*

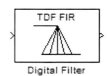

Figure 6.14(f) *Icon outlook of the Transposed Direct Form (TDF) FIR filter*

enter this coefficients in the Numerator coefficients slot in the same window. Regarding the type of filter, the reader can choose it by going through the popup menu of the Filter structure in the aforementioned window. However, as soon as you enter the FIR filter coefficients, simulink responds for the direct and the transposed direct types as shown in the figures 6.14(e) and 6.14(f) respectively. Exploring other types of filters available in the block is left as an exercise for the reader and for which the references [25]

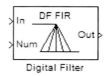

Figure 6.14(g) *Icon outlook of the Direct Form (DF) FIR filter when filter coefficients are fed externally*

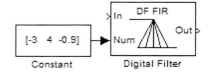

Figure 6.14(h) *Passing the Direct Form (DF) FIR filter coefficients externally*

and [26] can be checked with for the theory behind the filter structures.

So far we entered the filter coefficients in the parameter window as provided in the figure 6.12(h). Referring to the same figure, you find there a slot by name Coefficients source whose popup menu possesses $\begin{Bmatrix} \text{Specify via dialog} \\ \text{Input port(s)} \end{Bmatrix}$. We used the first type. If you select the second type for the last FIR filter, you see the icon outlook as shown in the figure 6.14(g). The filter coefficients [−3 4 −0.9] are first fed in a Constant bock and then imported to the Digital Filter block (see figure 6.14(h)). This sort of coefficient passing is also possible for the other filter structures.

Signal processing is a vast subject. It is not possible to include all relevant topics of signal processing in this short context. We guess the reader at least has had some escorting procedure. We close the section by presenting the block links.

6.16 Block links used in this chapter

Our objective in the chapter is to highlight the modeling for some signal processing problems in simulink for which we employed the blocks as presented in the table 6.B.

Table 6.B   Necessary blocks for modeling the signal processing problems as found in simulink library (not arranged in the alphabetical order)

Block name	Representative Symbol/Function	Icon Outlook	Block name	Representative Symbol/Function	Icon Outlook
Constant	Generates matrices of constant value or values	Constant [1]	Discrete Filter	Characterizes a discrete filter in terms of the Z transform transfer function as the polynomial of z^{-1}	$\frac{1}{1+0.5z^{-1}}$ Discrete Filter
Link: *Simulink → Sources → Constant*			Link: *Simulink → Discrete → Discrete Filter*		
Display	Shows the instantaneous output of the concern functional line	Display [0]	Discrete Zero-Pole	Models Z transform transfer function when it is given in the gain-zero-pole form	$\frac{(z-1)}{z(z-0.5)}$ Discrete Zero-Pole
Link: *Simulink → Sinks → Display*			Link: *Simulink → Discrete → Discrete Zero-Pole*		
Discrete Transfer Fcn	Models Z transform transfer function when it is given in the factored form	$\frac{1}{z+0.5}$ Discrete Transfer Fcn	Polynomial Stability Test	Determines whether the roots of a polynomial are inside the unit circle	|roots(u)| < 1 Polynomial Stability Test
Link: *Simulink → Discrete → Discrete Transfer Fcn*			Link: *DSP Blockset → Math Functions → Polynomial Functions → Polynomial Stability Test*		
MATLAB Fcn	Executes any MATLAB function in simulink either standard or user defined	MATLAB Function MATLAB Fcn	Convolution	Performs convolution of two discrete signals	CONV Convolution
Link: *Simulink → User-Defined Functions → MATLAB Fcn*			Link: *DSP Blockset → Signal Operations → Convolution*		
FFT	Computes the fast forward Fourier transform of a sequence only for the sequence length of the power of 2	FFT FFT	IFFT	Computes the inverse fast Fourier transform of a sequence only for the sequence length of the power of 2	IFFT IFFT
Link: *DSP Blockset → Transforms → FFT*			Link: *DSP Blockset → Transforms → IFFT*		
DCT	Computes the discrete cosine transform of a sequence only for the sequence length of the power of 2	DCT DCT	IDCT	Computes the inverse discrete cosine transform of a sequence only for the sequence length of the power of 2	IDCT IDCT
Link: *DSP Blockset → Transforms → DCT*			Link: *DSP Blockset → Transforms → IDCT*		
Quantizer	Quantizes an analog signal with specific quantizer interval	Quantizer	Sum	Σ or	$(+\ +)$
Link: *DSP Blockset → Quantizers → Quantizer*			Link: *Simulink → Math Operations → Sum*		
Sine Wave	Generates continuous sinusoidal signal of different amplitude and frequency	Sine Wave	Scope	Displays the functions or signals to which it is connected	Scope
Link: *Simulink → Sources → Sine Wave*			Link: *Simulink → Sinks → Scope*		
Digital Clock	It generates discrete independent variable at user defined sampling period	12:34 Digital Clock	To Workspace	It sends simulink functional data to MATLAB workspace	simout To Workspace
Link: *Simulink → Sources → Digital Clock*			Link: *Simulink → Sinks → To Workspace*		
Pad	It pads a sequence by the user defined constant values	Pad	Complex to Real-Imag	It separates a complex function to its real and imaginary parts	Re(u) Im(u) Complex to Real-Imag
Link: *DSP Blockset → Signal Operations → Pad*			Link: *Simulink → Math Operations → Complex to Real-Imag*		
Zero Pad	It pads a sequence by the zero values	Zero Pad	Downsample	It downsamples a sequence by an integer factor, for example, turning $x[n]$ to $x[2n]$	↓ 2 Downsample
Link: *DSP Blockset → Signal Operations → Zero Pad*			Link: *DSP Blockset → Signal Operations → Downsample*		

Continuation of previous table:

Block name	Representative Symbol/Function	Icon Outlook	Block name	Representative Symbol/Function	Icon Outlook
Upsample	It upsamples a sequence by an integers factor, for example, turning $x[n]$ to $x[\frac{n}{3}]$	Upsample	From Workspace	It imports MATLAB workspace data to simulink for analysis	simin / From Workspace
Link: *DSP Blockset → Signal Operations → Upsample*			Link: *Simulink → Sources → From Workspace*		
Window Function	It generates and applies various window functions to signals	hamming / Window Function	Complex Phase Difference	It returns the phase difference of two complex values or signals in radians	Complex Phase Difference / Complex Phase Difference
Link: *DSP Blockset → Signal Operations → Window Function*			Link: *Communications Blockset → Basic Comm Functions → Sequence Operations → Complex Phase Difference*		
Unit Delay	It delays $x[n]$ to $x[n-1]$ or $x[n+1]$ to $x[n]$	$\frac{1}{z}$ / Unit Delay	Normalization	It normalizes signal by taking sum of squares or square root of the sum of squares on all elements	$\frac{u}{\|u\|^2}$ / Normalization
Link: *Simulink → Discrete → Unit Delay*			Link: *DSP Blockset → Math Functions → Math Operations → Normalization*		
Radians to Degrees	It transforms angle entered in radians to degrees	R2D / Radians to Degrees	Clock	It simulates the continuous signal like $f(t)=t$	Clock
Link: *Simulink Extras → Transformations → Radians to Degrees*			Link: *Simulink → Sources → Clock*		
Gain	It multiplies the input function or signal by a scalar	1 / Gain	Abs	It takes the absolute value of the signal entering to its input port	|u| / Abs
Link: *Simulink → Math Operations → Gain*			Link: *Simulink → Math Operations → Abs*		
MinMax	It finds the minimum or maximum of the signal entering to its input port from the constant generation	min / MinMax	Assignment	It picks up elements from some signal line and assign those to some other signal line	U1 -> Y / U2 -> Y{E} / Y / Assignment
Link: *Simulink → Math Operations → MinMax*			Link: *Simulink → Math Operations → Assignment*		
Fcn	Any user defined function can map the input signal to other considering the independent variable as u	f(u) / Fcn	Product	It multiplies two signals entering to its input ports on a common independent variable	× / Product
Link: *Simulink → User-Defined Functions → Fcn*			Link: *Simulink → Math Operations → Product*		
dB Conversion	It converts input signal x to dB by performing operation $20\log_{10} x$	dB (1 ohm) / dB Conversion			
Link: *DSP Blockset → Math Functions → Math Operations → dB Conversion*					

Modeling Common Matrix Algebra Problems

We intend here in this chapter to introduce the simulink approach of handling the common matrix algebra problems. Matrix algebra problems usually appear in some specific caption such as matrix generation, matrix manipulation, matrix arithmetic, and functional analysis of matrices. The first category includes the generation of row-column-rectangular matrices either by the user requirement or by the matrix algebra rules. Matrix manipulation, the second category, can be explained as the resizing an existing matrix, picking up some rows, columns, or submatrix, overwriting some rows, columns, or submatrix, forming larger matrix from smaller ones ... etc. The third kind concerns the matrix arithmetic operations such as addition, subtraction, multiplication, powering ... etc according to the matrix algebra rules. The last sub caption basically analyzes the matrix-organized data, for example, eigenvector, singular value decomposition, or norm analysis. We highlight in a nutshell their simulink way of handling.

7.1 Matrix algebra and simulink

With the easy accessibility to computer, it is convenient to study in detail all kinds of functions and their applications numerically. Functions when discretized acquire the form of matrices, row or column type of which represents one-dimensional function or signal and rectangular kind of which represents two-dimensional function. Simulink is built on the wing of MATLAB in contrast MATLAB is effective and advantageous for matrix-oriented computation and manipulation. One can interpret that simulink portrayal of matrix algebra problems is just an added convenience in MATLAB replacing the MATLAB codes of the matrix algebra by specific block for specific problem. Simulink as well as MATLAB offers a great flexibility to numerical computing and handling of different data like real, complex, floating-point, or integer. The matrix data can be imported in simulink or exported to MATLAB once some modeling and analysis is performed in simulink.

7.2 Constant, diagonal, and identity matrices modeling in simulink

Constant matrix can be generated in simulink by employing the Constant block found in the link *'Simulink → Sources → Constant'*. Matrix generation command in MATLAB also applies here. For example, the matrices

$$A = \begin{bmatrix} 8 \\ -5 \\ 6 \end{bmatrix} \text{ and } B = \begin{bmatrix} 67 & 0 & 5 \\ 7 & -5 & -3 \end{bmatrix}$$ can be generated by writing the commands [8;−5;6] and [67 0 5;7 −5 −3]

7.1(a) Icon outlook of the Constant block

7.1(b) Icon outlook when the description fits out the block

7.1(c) Outlook of the block 7.1(b) when enlarged

7.1(d) Outlook of the block when matrix B fits out of the block

respectively in which the rows are separated by the operator ;, the elements in a row are separated by one space, and all elements are enclosed by the third brace [ ]. Bring the block in a new simulink model file whose outlook is the figure 7.1(a), doubleclick the block to see the block parameter window as shown in the figure 7.1(f), type the code of the matrix *A* in the slot of Constant value under Parameters, and click OK. You see the block appearance as shown in figure 7.1(b) because the matrix description fits out the block size. Enlarge the block to see the contents according to the figure

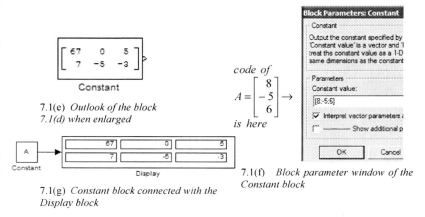

7.1(e) Outlook of the block 7.1(d) when enlarged

code of $A = \begin{bmatrix} 8 \\ -5 \\ 6 \end{bmatrix} \rightarrow$ is here

7.1(g) Constant block connected with the Display block

7.1(f) Block parameter window of the Constant block

Figure 7.1.a-g *Generation and display of constant matrices in simulink*

7.1(c). Similarly type the code of the matrix *B* in the block parameter window and click OK. Simulink displays the block of figure 7.1(d) in which you see the dimension of the matrix (because *B* contains two rows and three columns). The outlook can be like the figure 7.1(e) following the enlargement. As we generate in MATLAB, a long row matrix whose value starts at −2 and lasts to 5 with the step 0.01 can be formed by writing the code [-2:0.01:5]. Also existing matrix in MATLAB can be imported through the block. For example, go to MATLAB command window, type A=[67 0 5;7 −5 −3] in the command prompt of MATLAB, press enter, come back in the simulink model, type just A in the Constant value of the block parameter window, and click OK to see the action. But in that case you see just A inside the block. If you are interested to see the values, bring one Display block (the link is *'Simulink → Sinks → Display'*) and connect that with the Constant block. Run the simulink model and enlarge the Display block to see the contents of A as indicated in the figure 7.1(g).

Let us say we want to model a constant diagonal matrix $A = \begin{bmatrix} 9 & 0 & 0 \\ 0 & -2 & 0 \\ 0 & 0 & -1 \end{bmatrix}$ in simulink. Bring a

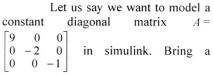

Figure 7.1(h) Icon outlook of the Constant Diagonal Matrix

Constant Diagonal Matrix block (the outlook is shown in the figure 7.1(h)) in a new simulink model file following the link *'DSP Blockset → Math Functions → Matrices and Linear Algebra → Matrix Operations → Constant Diagonal Matrix'*. Doubleclick the block and the diagonal elements [9 −2 −1] of *A* are typed in the block parameter window as shown in the figure 7.1(i). Check the Show additional parameters

Diagonal elements [9 − 2 −1]

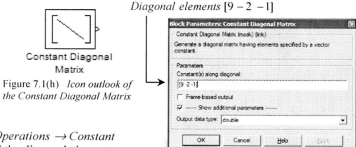

Figure 7.1(i) Block parameter window of the Constant Diagonal Matrix

button, set the output data type as double, and click OK. If you want to see the contents of the block, bring a Display block, connect the block with the Constant Diagonal matrix, and run the model as we did in figure 7.1(g). Any MATLAB workspace row or column matrix name can also be put in the block parameter window.

An identity matrix has all ones as the diagonal, for example, A

$$= \begin{bmatrix} 1 & 0 & 0 \\ 0 & 1 & 0 \\ 0 & 0 & 1 \end{bmatrix}$$ of dimension 3×3. So the A has the size 3. If A were

7×7, we would say the size of A is 7. The location of the Identity Matrix block is '*DSP Blockset → Math Functions → Matrices and Linear Algebra → Matrix Operations → Identity Matrix*'. Bring the block in a new simulink model file (the icon outlook is in the figure 7.1(j)), doubleclick the block to see the

Identity Matrix

Figure 7.1(j) *Icon outlook of Identity Matrix*

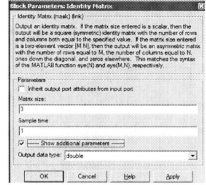

Figure 7.1(k) *Block parameter window of the Identity Matrix*

block parameter window in the figure 7.1(k), type the matrix size as 3 in the block parameter window for the A, check the button Show additional parameters, set the output data type as double, and click OK.

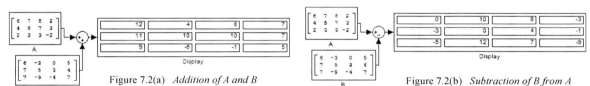

Figure 7.2(a) *Addition of A and B*　　　　Figure 7.2(b) *Subtraction of B from A*

7.3 Modeling the matrix addition, subtraction, and multiplication

Matrix addition and subtraction both can be accomplished by the block Sum found in the link '*Simulink → Math Operations → Sum*'. Let us consider the matrices $A = \begin{bmatrix} 6 & 7 & 8 & 2 \\ 4 & 5 & 7 & 3 \\ 2 & 3 & 3 & -2 \end{bmatrix}$ and $B = \begin{bmatrix} 6 & -3 & 0 & 5 \\ 7 & 5 & 3 & 4 \\ 7 & -9 & -4 & 7 \end{bmatrix}$, their

sum and subtraction are given by $A + B = \begin{bmatrix} 12 & 4 & 8 & 7 \\ 11 & 10 & 10 & 7 \\ 9 & -6 & -1 & 5 \end{bmatrix}$ and $A - B = \begin{bmatrix} 0 & 10 & 8 & -3 \\ -3 & 0 & 4 & -1 \\ -5 & 12 & 7 & -9 \end{bmatrix}$ respectively.

Bring the Constant block in a new simulink model file, doubleclick the block to enter the description of A as [6 7 8 2;4 5 7 3;2 3 3 –2], rename the block as A and enlarge the block, copy the block in the clipboard, paste it, doubleclick the block to enter the description of B as [6 –3 0 5;7 5 3 4;7 –9 –4 7], bring one Sum and one Display blocks, place the blocks relatively as in figure 7.2(a), connect them, and run the model to display the addition (you may need to enlarge the Display block). For the subtraction, what we need is doubleclick the Sum block, change its List of signs to +- ('-' corresponds to the functional line which is to be subtracted), and execute the model to see the subtraction as presented in the figure 7.2(b). The addition and subtraction necessitate that both the matrices must be of the identical order.

Matrix multiplication can happen through the block Matrix Multiply found in '*DSP Blockset → Math Functions → Matrices and Linear Algebra → Matrix Operations → Matrix Multiply*'. Let us say we have two matrices

$A = \begin{bmatrix} 5 & 6 & 9 \\ 6 & 7 & 8 \end{bmatrix}$ and $B = \begin{bmatrix} 6 & 7 \\ 7 & 9 \\ 1 & 2 \end{bmatrix}$, their matrix or vector

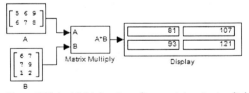

Figure 7.2(c) *Multiplication of two matrices in simulink*

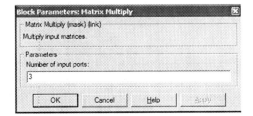

Figure 7.2(d) *Parameter window of the Matrix Multiply block*

Figure 7.2(e)
Multiplication of three matrices in simulink

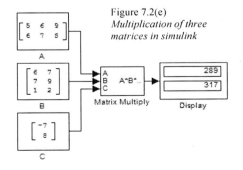

multiplication is $AB = \begin{bmatrix} 5 & 6 & 9 \\ 6 & 7 & 8 \end{bmatrix} \begin{bmatrix} 6 & 7 \\ 7 & 9 \\ 1 & 2 \end{bmatrix} = \begin{bmatrix} 81 & 107 \\ 93 & 121 \end{bmatrix}$. In order to implement the multiplication, bring one

Constant block in a new simulink model file, doubleclick the block to enter its Constant value as [5 6 9;6 7 8] for the code of the matrix A, rename the block as A, enlarge the block to see its contents, bring another Constant block in the model file, doubleclick the block to enter the code for B as [6 7;7 9;1 2], rename the block as B, enlarge the block to see its contents, bring one Matrix Multiply block in the model file following previous mentioned link, bring one Display block, place the blocks as relatively in figure 7.2(c), connect them, and run the model. Enlarge the Display block to see the complete matrix multiplication.

Thus the multiplication of three or more matrices is also possible, for example, let us multiply previous matrices A and B and $C = \begin{bmatrix} -7 \\ 8 \end{bmatrix}$. Their vector multiplication is given by $\begin{bmatrix} 5 & 6 & 9 \\ 6 & 7 & 8 \end{bmatrix} \times$

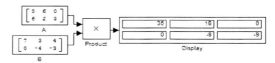

Figure 7.2(f) *Scalar multiplication of the matrices A and B*

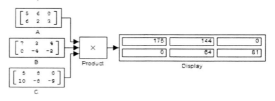

Figure 7.2(g) *Scalar multiplication of the matrices A, B, and C*

$\begin{bmatrix} 6 & 7 \\ 7 & 9 \\ 1 & 2 \end{bmatrix} \times \begin{bmatrix} -7 \\ 8 \end{bmatrix} = \begin{bmatrix} 289 \\ 317 \end{bmatrix}$. For the implementation, doubleclick the Matrix Multiply block to see its parameter window as shown in the figure 7.2(d). In that window change its Number of input ports to 3 making it enable for three matrices A, B, and C. Bring one more Constant block in the model file, doubleclick the block to enter the code for C as [-7;6], rename the block as C, uncheck the Interpret the vector parameter

as 1-D in the parameter window of Constant for displaying the matrix inside the C as a column one, and enlarge the block to display the contents inside C. Connect the output port of the block C to the third input port of the Matrix Multiply to have the model like the figure 7.2(e) and finally run the model to view the required output. It is important to mention that the vector or matrix multiplication of matrix A of order $M \times N$ (M rows and N columns) with matrix B is only possible if B has N rows.

Scalar multiplication of the matrices can also be performed in simulink and the multiplication is basically the element by element multiplication of two or more identical matrices. Let us consider the matrices $A = \begin{bmatrix} 5 & 6 & 0 \\ 6 & 2 & 3 \end{bmatrix}$ and $B = \begin{bmatrix} 7 & 3 & 4 \\ 0 & -4 & -3 \end{bmatrix}$, their scalar multiplication is given by $\begin{bmatrix} 35 & 18 & 0 \\ 0 & -8 & -9 \end{bmatrix}$. We can take the help

of the Product block for the scalar multiplication found in '*Simulink* → *Math Operations* → *Product*'. To continue with the procedure, let us bring a Constant block in a new simulink model file, doubleclick the block to enter the code of A as [5 6 0;6 2 3] in the parameter window, rename the block as A, enlarge the block to see its contents, bring another Constant block, doubleclick the block to enter the code of B as [7 3 4;0 –4 –3] in the parameter

window, rename the block as B and enlarge it, bring one Product block, doubleclick the Product block and make sure that the multiplication mode is set as Element-wise, bring one Display block, connect them as shown in the figure 7.2(f), run the model, and see the output as shown in the same figure. To perform the scalar multiplication of three matrices, let us choose another matrix $C = \begin{bmatrix} 5 & 8 & 0 \\ 10 & -8 & -9 \end{bmatrix}$. If you multiply the matrices A, B, and C element by element, you obtain the resultant matrix as $\begin{bmatrix} 175 & 144 & 0 \\ 0 & 64 & 81 \end{bmatrix}$.

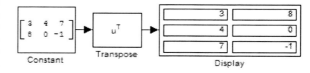

Figure 7.3(a) *Transposition of a matrix*

Figure 7.3(b) *Transposition of a complex matrix*

Now let us go to simulink and form the matrix C via another Constant block as we did for A and B, doubleclick the Product block and change its number of inputs to 3, connect the

Figure 7.3(c) *Transposition of the complex matrix with the conjugate*

block as displayed in the figure 7.2(g), run the model, and see the output in Display as presented in the same figure.

7.4 Modeling the matrix transposition

We know that the transpose of the matrix $\begin{bmatrix} 3 & 4 & 7 \\ 8 & 0 & -1 \end{bmatrix}$ is $\begin{bmatrix} 3 & 8 \\ 4 & 0 \\ 7 & -1 \end{bmatrix}$ and the transposition can be implemented through the block Transpose found in '*DSP Blockset → Math Functions → Matrices and Linear Algebra → Matrix Operations → Transpose*'. Bring a Constant block in a new simulink model file, doubleclick the block to enter the code of $\begin{bmatrix} 3 & 4 & 7 \\ 8 & 0 & -1 \end{bmatrix}$ as [3 4 7;8 0 –1], enlarge the block to display the contents, bring the Transpose block, bring one Display block, connect the three blocks as shown in the figure 7.3(a), run the model, and enlarge the Display block to see the output. The block can also handle the complex numbers, for example, we wish to transpose the complex number matrix $\begin{bmatrix} 3j & 4-j & 7 \\ 8-j & 0 & -1 \end{bmatrix}$ so that our transposed matrix should look like $\begin{bmatrix} 3j & 8-j \\ 4-j & 0 \\ 7 & -1 \end{bmatrix}$. All we need is enter the code of the complex matrix as [3i 4–i 7;8–i 0 –1] in the Constant block parameter window. Still the elements are separated by one space and the complex number $3j$ can be written as 3i. However, the modeling is shown in the figure 7.3(b), and the transposition is without the complex conjugate. Occasionally one may need to have the transpose with the complex conjugate that is we want to have the matrix $\begin{bmatrix} -3j & 8+j \\ 4+j & 0 \\ 7 & -1 \end{bmatrix}$ for the example at hand. Doubleclick the block Transpose, click the Hermitian in the block parameter window, run the model, and the action is displayed in the figure 7.3(c). The block input of the Transpose can also be any workspace complex matrix.

7.5 Some matrix arithmetic in simulink

During programming for the matrix related problems in simulink, a number of mathematical operations on the matrix elements might be necessary. Our objective here is to familiarize the icons and functions about the matrix arithmetic.

Suppose you have to take the reciprocal of all elements in a matrix, for example, $A = \begin{bmatrix} 0.1 & 0.4 \\ 1 & 0.2 \\ 0.2 & 0.25 \end{bmatrix}$. So one would get $\begin{bmatrix} 10 & 2.5 \\ 1 & 5 \\ 5 & 4 \end{bmatrix}$ following the reciprocal computation. To perform this operation in simulink, bring one Constant block in a new simulink model file, doubleclick the block to enter the code of A as [0.1 0.4;1 0.2;0.2 0.25] in the parameter window, enlarge the block to see its contents, rename the block as A, bring one Math Function block following the link '*Simulink → Math Operations → Math Function*', doubleclick the block to see its popup menu under the Function as indicated in the figure 7.4(a), pick up the function reciprocal from the menu, see that the icon outlook is changed, bring one Display block, connect the three blocks as shown in the figure 7.4(b), run the model, and enlarge the Display block to see the complete result.

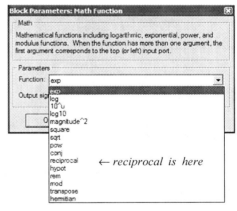

Figure 7.4(a) *Popup menu of the Math Function block*

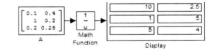

Figure 7.4(b) *Taking reciprocal of all elements of matrix A*

The appearance of icon outlook for the popup menu function including functional detail is provided in the table 7.C. Now we explain in a nutshell what the popup functions of the figure 7.4(a) would perform:

If A were $\begin{bmatrix} 0 & 1 \\ 2 & -1 \end{bmatrix}$, the exp (raising power on the e) of the popup menu would perform $\begin{bmatrix} e^0 & e^1 \\ e^2 & e^{-1} \end{bmatrix} = \begin{bmatrix} 1 & 2.718 \\ 7.389 & 0.3679 \end{bmatrix}$,

If A were $\begin{bmatrix} 4 & 1 \\ 2 & 8 \end{bmatrix}$, the log (natural logarithm of the matrix elements) of the popup menu would perform

$\begin{bmatrix} \ln 4 & \ln 1 \\ \ln 2 & \ln 8 \end{bmatrix} = \begin{bmatrix} 1.386 & 0 \\ 0.6931 & 2.069 \end{bmatrix}$ (elements of A can not be 0 or negative with the auto setting),

If A were $\begin{bmatrix} 4 & -1 \\ 2 & 0 \end{bmatrix}$, the 10^u (raising power on 10) of the popup menu would perform $\begin{bmatrix} 10^4 & 10^{-1} \\ 10^2 & 10^0 \end{bmatrix}$,

If A were $\begin{bmatrix} 7 & 1 \\ 5 & 8 \end{bmatrix}$, the log10 (common logarithm of the matrix elements) of the popup menu would

perform $\begin{bmatrix} \log_{10} 7 & \log_{10} 1 \\ \log_{10} 5 & \log_{10} 8 \end{bmatrix}$ (elements of A can not be 0 or negative with the auto setting),

If A were $\begin{bmatrix} 7+j & 1 \\ 5j & 8+2j \end{bmatrix}$, the magnitude^2 (taking the square of the magnitude of the complex matrix

elements) of the popup menu would perform $\begin{bmatrix} 7^2+1^2 & 1^2 \\ 5^2 & 8^2+2^2 \end{bmatrix}$ (elements of A are complex),

If A were $\begin{bmatrix} -7 & 0 \\ -5 & 8 \end{bmatrix}$, the square (squaring the matrix elements) of the popup menu would perform

$\begin{bmatrix} 49 & 0 \\ 25 & 64 \end{bmatrix}$,

If A were $\begin{bmatrix} 7.5 & 1.4 \\ 5.6 & 8.4 \end{bmatrix}$, the sqrt (taking the square root of the matrix elements) of the popup menu would

perform $\begin{bmatrix} \sqrt{7.5} & \sqrt{1.4} \\ \sqrt{5.6} & \sqrt{8.4} \end{bmatrix}$, and

If A were $\begin{bmatrix} 8+j & 1 \\ -35j & 8-2j \end{bmatrix}$, the conj (complex conjugate of the matrix elements) of the popup menu

would perform $\begin{bmatrix} 8-j & 1 \\ 35j & 8+2j \end{bmatrix}$ (elements of A are complex).

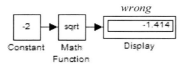

Figure 7.4(c) *Output of the simulink when the output signal type is taken as auto or real*

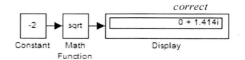

Figure 7.4(d) *Output of the simulink when the output signal type is taken as complex*

The information and range of the input and output blocks in a simulink model is very important for not having the simulation time error. For example, we know that the square root of any positive number is positive but the square root of any negative number is complex. If you doubleclick the Math Function block, you see one more option in the parameter window called the Output signal type and you have three options there – auto, real, and complex. Setting the output signal type as real or auto would provide the square root of -2 as -1.414, which is wrong. But setting the signal type as complex can provide the correct output $j1.414$.

Figure 7.4(e) *Powering of a matrix elements by the elements of another matrix*

The discrepancy that can cause simulation time error due to the improper selection of the output signal type is depicted in the figures 7.4(c) and 7.4(d) respectively.

However, some mathematical functions need two inputs, for example, the pow of the popup menu. The function raises power of a number according to another number. Let us consider the number matrix $A = \begin{bmatrix} 2 & 3 & -2 \\ 5 & 2 & 7 \end{bmatrix}$ and we want to raise the power of each element of the matrix according to $B = \begin{bmatrix} 6 & 4 & 2 \\ 2 & -1 & 3 \end{bmatrix}$

whence we should get the resultant matrix as $\begin{bmatrix} 2^6 & 3^4 & (-2)^2 \\ 5^2 & 2^{-1} & 7^3 \end{bmatrix} = \begin{bmatrix} 64 & 81 & 4 \\ 25 & 0.5 & 343 \end{bmatrix}$. Bring one constant block in a

new simulink model file, doubleclick the block to enter the code of A as [2 3 –2;5 2 7], enlarge the block, rename the block as Number, bring another Constant block, doubleclick the block to enter the code of B as [6 4 2;2 –1 3], enlarge the block, rename the block as Power, bring one Math Function block, doubleclick the block to change its Function to pow in the parameter window, bring one Display block, and connect the blocks as shown in the figure 7.4(e), and run the model to see the output according to the same figure. If the elements of the Power matrix were a single constant (let us say 3), the output would be $\begin{bmatrix} 2^3 & 3^3 & (-2)^3 \\ 5^3 & 2^3 & 7^3 \end{bmatrix}$ and the elements of the Number matrix can even be complex.

Now we concentrate on the function rem of the popup menu, which is the abbreviation of the remainder after integer division. Remembering the fact that the integer division of 2 by 3 is 0 and remainder is 2, 1 by 2 is 0 and remainder is 1...etc, the remainder of the integer division of two integer matrices is given by the function rem. For example, we have two matrices of integer numbers, $A = \begin{bmatrix} 12 & 3 & 52 \\ 5 & 42 & 17 \end{bmatrix}$ and $B = \begin{bmatrix} 2 & 3 & 2 \\ 3 & 2 & 7 \end{bmatrix}$. When the integers 12, 3, and 17 of A are divided by the integers 2, 3, and 7 of B, they leave the remainders 0, 0, and 3 respectively.

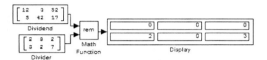

Figure 7.4(f) *Remainder after integer division of the dividend A by the divider B*

Figure 7.4(g) *Hypotenuse function on the elements of the matrices A and B*

Continuing the element by element operation for the dividend A and the divider B, one would obtain the resultant matrix as $\begin{bmatrix} 0 & 0 & 0 \\ 2 & 0 & 3 \end{bmatrix}$ – this is what we expect from simulink. Anyhow bring two Constant blocks in a new simulink model file, enter their constant values as [12 3 52;5 42 17] and [2 3 2;3 2 7] for the dividend and the divider respectively, rename the blocks as Dividend and Divider, bring one Math Function block, doubleclick the block to set its Function as rem in the popup menu of the parameter window, connect them bringing one Display block like the figure 7.4(f), run the model, and see the expected remainder as presented in the same figure. In the popup menu of the figure 7.4(a) you find another function called mod, which also performs the operation similar to the rem. The function rem has the same sign as the dividend whereas the function mod has the same sign as the divider. Notice that both the divider and the dividend matrices are of identical order or if the divider is a single scalar, the remainder operation is performed on all elements of the dividend matrix.

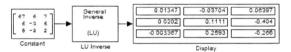

Figure 7.5(a) *Inverse of a square matrix*

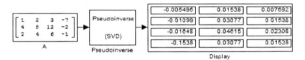

Figure 7.5(b) *Pseudoinverse of the rectangular matrix A*

Figure 7.5(c) *Icon outlook of LDL Inverse block*

Figure 7.5(d) *Icon outlook of Cholesky Inverse block*

We addressed separately the transpose and Hermitian of matrices in the article 7.4. Le us simulate the function hypot (abbreviation of hypotenuse). Suppose we have the matrices $A = \begin{bmatrix} 6 & 3 & 5 \\ 3 & 2 & 7 \end{bmatrix}$ and $B = \begin{bmatrix} -2 & 3 & 2 \\ 3 & 0 & 7 \end{bmatrix}$, their hypotenuse output should be

$$\begin{bmatrix} \sqrt{6^2 + (-2)^2} & \sqrt{3^2 + 3^2} & \sqrt{5^2 + 2^2} \\ \sqrt{3^2 + 3^2} & \sqrt{2^2 + 0^2} & \sqrt{7^2 + 7^2} \end{bmatrix} = \begin{bmatrix} 6.3246 & 4.2426 & 5.3852 \\ 4.2426 & 2 & 9.8995 \end{bmatrix}.$$ The reader can construct the model as

presented in the figure 7.4(g) and run the model to see the hypotenuse output returned in Display by simulink.

7.6 Modeling the inverse matrices

Matrix inverse A^{-1} of a square matrix A can be simulated by employing the block LU Inverse. Matrix inverse is only possible for a square and full rank matrix not for a rectangular matrix. The product of matrix A and its inverse A^{-1} is always an identity matrix. Let us consider the square matrix $A = \begin{bmatrix} 67 & 6 & 7 \\ 6 & -3 & 6 \\ 5 & -3 & 2 \end{bmatrix}$ which has the inverse $A^{-1} = \begin{bmatrix} \frac{4}{297} & -\frac{1}{27} & \frac{19}{297} \\ \frac{2}{99} & \frac{1}{9} & -\frac{40}{99} \\ -\frac{1}{297} & \frac{7}{27} & -\frac{79}{297} \end{bmatrix} = \begin{bmatrix} 0.0135 & -0.0370 & 0.0640 \\ 0.0202 & 0.1111 & -0.4040 \\ -0.0034 & 0.2593 & -0.2660 \end{bmatrix}$. Our objective is to find the inverse A^{-1} of

the square matrix A in simulink. The block resides in the link '*DSP Blockset $\rightarrow$ Math Functions $\rightarrow$ Matrices and Linear Algebra $\rightarrow$ Matrix Inverses $\rightarrow$ LU Inverse*'. Bring the LU Inverse block in a new simulink model file, bring one Constant block in the model, doubleclick the block to enter the expression of the matrix A as [67 6 7;6 –3 6;5 – 3 2] in the constant value slot of the parameter window, enlarge the block to see the complete matrix, bring one Display block, and connect them as shown in the figure 7.5(a). Run the model and enlarge the Display block to see the inverse values of the matrix.

When we have non square matrix, the matrix inverse is termed as the Pseudoinverse (also called Moore-Penrose inverse). Any non null matrix A of order $M \times N$ is said to have a Pseudoinverse G if it follows the

properties $\begin{cases} AGA = A \\ GAG = G \\ GA \text{ is symmetriv} \\ AG \text{ is symmetric} \end{cases}$, where G has the order $N \times M$ and it is unique. The block we need for the

Pseudoinverse resides in the link '*DSP Blockset $\rightarrow$ Math Functions $\rightarrow$ Matrices and Linear Algebra $\rightarrow$ Matrix*

Inverses $\rightarrow$ Pseudoinverse'. It is given that $G = \begin{bmatrix} -\frac{1}{182} & \frac{65}{65} & \frac{1}{130} \\ -\frac{1}{91} & \frac{2}{65} & \frac{1}{65} \\ -\frac{3}{182} & \frac{3}{65} & \frac{3}{130} \\ -\frac{2}{13} & \frac{2}{65} & \frac{1}{65} \end{bmatrix} = \begin{bmatrix} -0.0055 & 0.0154 & 0.0077 \\ -0.0110 & 0.0308 & 0.0154 \\ -0.0165 & 0.0462 & 0.0231 \\ -0.1538 & 0.0308 & 0.0154 \end{bmatrix}$ is the

Pseudoinverse of $A = \begin{bmatrix} 1 & 2 & 3 & -7 \\ 4 & 8 & 12 & -2 \\ 2 & 4 & 6 & -1 \end{bmatrix}$. Let us bring the Pseudoinverse block in a new simulink model file, bring a

Constant block, doubleclick the block to enter the matrix description of A as [1 2 3 –7;4 8 12 –2;2 4 6 –1], rename the block as A, enlarge the block to display its contents, bring a Display block, and connect the three blocks as shown in the figure 7.5(b). Run the model and enlarge the Display block to see the Pseudoinverse elements of G.

However, two more inverse blocks called the LDL Inverse (finds the matrix inverse using LDL factorization) and the Cholesky Inverse (finds the matrix inverse using Cholesky factorization) are also there in the same link whose icon outlooks are presented in the figures 7.5(c) and 7.5(d) respectively. Suppose you have the square or rectangular matrix by the name A residing in the workspace of MATLAB, the modeling is still operation if the Constant block contains just the name of the matrix as A.

7.7 Modeling the matrix square of a matrix

Any matrix A square or rectangular of order $M \times N$ can be converted to a square matrix of order $N \times N$

following the operation $A^T \times A$, where A^T is the transpose of matrix A. Let us say $A = \begin{bmatrix} 3 & 7 & -1 & 0 \\ 8 & 5 & 2 & 3 \end{bmatrix}$ so

$A^T \times A = \begin{bmatrix} 3 & 8 \\ 7 & 5 \\ -1 & 2 \\ 0 & 3 \end{bmatrix} \times \begin{bmatrix} 3 & 7 & -1 & 0 \\ 8 & 5 & 2 & 3 \end{bmatrix} = \begin{bmatrix} 73 & 61 & 13 & 24 \\ 61 & 74 & 3 & 15 \\ 13 & 3 & 5 & 6 \\ 24 & 15 & 6 & 9 \end{bmatrix}$ (the order of A says that M =2 and N =4). The

operation can be simulated through the block Matrix Square obtained from '*DSP Blockset $\rightarrow$ Math Functions $\rightarrow$ Matrices and Linear Algebra $\rightarrow$ Matrix Operations $\rightarrow$ Matrix Square*'. Bring a Constant block in a new simulink model file, doubleclick the block to enter the code of A as [3 7 –1 0;8 5 2 3], enlarge the block to see its contents, rename the block as A, bring one Matrix Square block, bring one Display block, connect them as shown in the figure 7.6(a), run the model, and enlarge the Display block to see the matrix product. If you have a complex matrix, the Hermitian of the matrix is considered for the multiplication. For example, the matrix $\begin{bmatrix} 5 - j \\ 7 + j6 \end{bmatrix}$ should

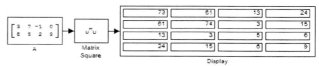

Figure 7.6(a) *Matrix square of the matrix A*

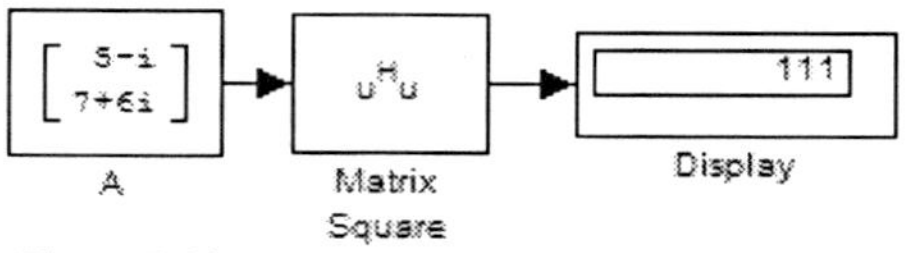

Figure 7.6(b) *Matrix square of the complex matrix*

provide the block output as $[5 + j \quad 7 - j6] \times \begin{bmatrix} 5 - j \\ 7 + j6 \end{bmatrix} = 111$. So doubleclick the block A, enter the complex matrix

as [5-i;7+6i] in the parameter window, uncheck the button Interpret vector parameter as 1-D, enlarge the block to see the contents, run the model to see the output as presented in the figure 7.6(b). You can also operate the block on the workspace variables.

7.8 Modeling the matrix concatenation

Sometimes you may need to form a larger matrix from the smaller ones. Suppose we have two matrices: $A = \begin{bmatrix} 5 \\ 8 \end{bmatrix}$ and $B = \begin{bmatrix} 6 & 3 \\ 0 & 11 \end{bmatrix}$. We want to place them side by side so that we have the matrix $\begin{bmatrix} 5 & 6 & 3 \\ 8 & 0 & 11 \end{bmatrix}$. This kind of matrix placement is called the horizontal concatenation for which one can use the block Matrix Concatenation found in the location '*DSP Blockset $\rightarrow$ Math Functions $\rightarrow$ Matrices and Linear Algebra $\rightarrow$ Matrix Operations $\rightarrow$ Matrix Concatenation*'. Let us bring one Constant block in a new simulink model file, doubleclick the block to enter the code of A as [5;8] in the parameter window, uncheck the button Interpret vector parameter as 1-D, rename the block as A, bring another Constant block, doubleclick the block to enter the code of B as [6 3;0 11], enlarge the block to see its contents, rename the block as B, bring one Matrix Concatenation and one Display blocks, connect them according to the figure 7.7(a), run the model, and see the output in the Display block as shown in the same figure. But the important point is the numbers of rows of both the matrices must have to be identical.

Let us say we have one more matrix $C = \begin{bmatrix} 2 & 5 \\ -2 & 8 \end{bmatrix}$ and we want to place C on the top of the matrix B so that our new matrix becomes $\begin{bmatrix} 2 & 5 \\ -2 & 8 \\ 6 & 3 \\ 0 & 11 \end{bmatrix}$. This operation is termed as the vertical concatenation. The Matrix Concatenation block needs little clicking operation for the Vertical counterpart. In the last model, we doubleclick the block A, enter the matrix code of C as [2 5;-2 8] in the parameter window, change the name of the block from A to C, doubleclick the Matrix Concatenation block, change the setting from the Horizontal to the Vertical in the Concatenation method in the parameter window (due to the change, the icon outlook becomes different as shown in the figure 7.7(b)), run the model, and see the output as presented in the figure 7.7(b). The condition to be fulfilled for the vertical concatenation is the numbers of the columns of the matrices C and B must be the same.

The block can also concatenate multiple matrices, for example, placing $A = \begin{bmatrix} 7 \\ 5 \end{bmatrix}$,

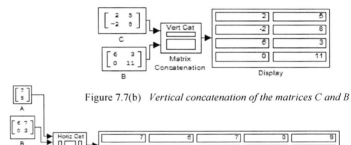

Figure 7.7(a) *Horizontal concatenation of the matrices A and B*

Figure 7.7(b) *Vertical concatenation of the matrices C and B*

Figure 7.7(c) *Horizontal concatenation of A, B, C, and D*

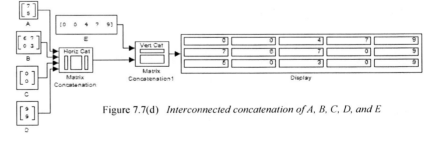

Figure 7.7(d) *Interconnected concatenation of A, B, C, D, and E*

$B = \begin{bmatrix} 6 & 7 \\ 0 & 3 \end{bmatrix}$, $C = \begin{bmatrix} 0 \\ 0 \end{bmatrix}$, and $D = \begin{bmatrix} 9 \\ 9 \end{bmatrix}$ to obtain $\begin{bmatrix} 7 & 6 & 7 & 0 & 9 \\ 5 & 0 & 3 & 0 & 9 \end{bmatrix}$ can happen by making the number of input ports of the block four (doubleclick the block to see the option for the number of inputs) as shown in the figure 7.7(c). You can also invoke any matrix residing in the workspace through the Constant block even the complex matrix of workspace. Multiple matrices' placement is also possible for the vertical counterpart. Interconnected concatenation can easily be implemented, let us say we have the row matrix $E = [0 \quad 0 \quad 4 \quad 7 \quad 9]$. We want to place this matrix on the top of just obtained four concatenated output so that we see $\begin{bmatrix} 0 & 0 & 4 & 7 & 9 \\ 7 & 6 & 7 & 0 & 9 \\ 5 & 0 & 3 & 0 & 9 \end{bmatrix}$. The model of the figure 7.7(d) can be implemented in this regard.

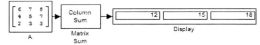

Figure 7.8(a) *Summation of only the columns of the matrix*   Figure 7.8(b) *Summation of only the rows of the matrix*

7.9 Modeling the summation and product of matrix elements

A matrix elements can be summed towards the row or column direction. One can model the summation of the matrix elements employing the block Matrix Sum placed in '*DSP Blockset → Math Functions → Matrices and Linear Algebra → Matrix Operations → Matrix Sum*'. Let us say we have the matrix $A = \begin{bmatrix} 6 & 7 & 8 \\ 4 & 5 & 7 \\ 2 & 3 & 3 \end{bmatrix}$ wherefrom the sum of only the columns and the sum of only the rows should give us $[12 \quad 15 \quad 1]$ and $\begin{bmatrix} 21 \\ 16 \\ 8 \end{bmatrix}$ respectively. Bring the Matrix Sum block in a new simulink model file (the default one is the column sum), bring one Constant block, doubleclick the block to enter the code of the matrix A as [6 7 8;4 5 7;2 3 3], rename the block as A, enlarge the block to show the contents, bring one Display block, connect them as shown in the figure 7.8(a), run the model, and enlarge the Display block to see the column sum as presented in Display. To perform the row sum, doubleclick the Matrix Sum block, change the 'sum along' from column to row in the parameter window (the icon outlook is changed due to the action), run the model, and enlarge the Display block to see the output of the simulation as shown in the figure 7.8(b). All elements in the matrix has the sum 45. If we want to sum all elements, the model of the figure 7.8(c) can be exercised wherein one Column Sum and one Row Sum blocks are placed side by side.

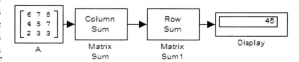

Figure 7.8(c) *Summing all elements in the matrix*

Row or column based product like the sum can be conducted with the help of the block Matrix Product found in '*DSP Blockset → Math Functions → Matrices and Linear Algebra → Matrix Operations → Matrix Product*'. Let us consider the matrix A which has the column and row directed product as $[48 \quad 105 \quad 168]$ and $\begin{bmatrix} 336 \\ 140 \\ 18 \end{bmatrix}$ respectively. Replace the Matrix Sum block of the figure 7.8(a) by the Matrix Product one (the default is Column Product) and run the model as shown in figure 7.8(d) to obtain the column directed product. Doubleclick the

Figure 7.8(d) *Product of only the columns of the matrix*   Figure 7.8(e) *Product of only the rows of the matrix*

block Matrix Product, change the 'multiply along' from the column to row in the parameter window, and run the model to see the row directed product as shown in the figure 7.8(e). All elements in the matrix A has the product 846720 and that can be simulated by placing two blocks of the Matrix Product in cascade as shown in the figure 7.8(f). As you see, the return of the simulink is in the exponential form wherein e correspond to 10 not the constant e ($846720 \equiv 8.4672 \times 10^5$).

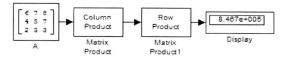

Figure 7.8(f) *Taking product of all elements in the matrix*

7.10 Modeling the matrix scaling

If the matrix elements are multiplied by specific numbers, that is termed as the matrix scaling. Scaling can happen for all elements (also called gain), only for rows, or only for columns. Let us consider the matrix $A =$

$$\begin{bmatrix} 6 & 7 & 8 & 2 \\ 4 & 5 & 7 & 3 \\ 2 & 3 & 3 & -2 \end{bmatrix}$$ which has three rows and four columns. To scale the four columns, we need four scalars and

assume that the scalars are $S = \begin{bmatrix} 6 & -3 & 0 & 5 \end{bmatrix}$. So we should obtain $\begin{bmatrix} 36 & -21 & 0 & 10 \\ 24 & -15 & 0 & 15 \\ 12 & -9 & 0 & -10 \end{bmatrix}$ following the

column scaling and the necessary model is presented in the figure 7.9(a). The block Matrix Scaling (obtained from the link '*DSP Blockset → Math Functions → Matrices and Linear Algebra → Matrix Operations → Matrix Scaling*') performs the scaling operation. As usual, bring a Constant block in a new simulink model file, doubleclick the block to enter the expression of A as [6 7 8 2;4 5 7 3;2 3 3 –2] in the parameter window, enlarge the block to see the inside matrix elements, and rename the block as A. Next bring another Constant block in the model file, doubleclick the block to enter the scalars as [6 –3 0 5] (as if a row matrix) in the parameter window, enlarge the block, and rename the block as S. Bring the Matrix Scaling block (the default is the row scaling one) in the model, doubleclick the block to change its parameters mode from Scale Rows to Scale Columns in the parameter window, bring one Display block, connect the blocks with relative positions of the figure 7.9(a), run the model, and see the outputs as presented in the Display block.

For the row scaling of A, the number of scalars we need is three and let us

assume that the scalars are $S = \begin{bmatrix} 6 \\ -2 \\ 5 \end{bmatrix}$ and the

matrix A turns to

$$\begin{bmatrix} 36 & 42 & 48 & 12 \\ -8 & -10 & -14 & -6 \\ 10 & 15 & 15 & -10 \end{bmatrix}$$ following the

scaling operation. Now doubleclick the block S of the model of figure 7.9(a), change

the scalars to $\begin{bmatrix} 6 \\ -2 \\ 5 \end{bmatrix}$ by writing [6;-2;5] and

uncheck the Interpret vector parameters as 1-D in the parameter window, enlarge the block, doubleclick the Matrix Scaling block, change its parameters mode to Scale Rows, run the model, and see the result in the Display block as shown in the figure 7.9(b).

Apart from the row or column scaling, all elements in a matrix can be

Figure 7.9(a) *Column scaling of the matrix A*

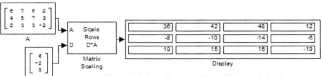

Figure 7.9(b) *Row scaling of the matrix A*

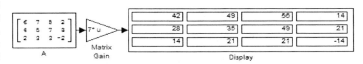

Figure 7.9(c) *Gain of the matrix A by 7*

scaled up or down by some scalar constant, for instance, let us scale up the matrix A by a factor 7 hence A should

turn to $\begin{bmatrix} 42 & 49 & 56 & 14 \\ 28 & 35 & 49 & 21 \\ 14 & 21 & 21 & -14 \end{bmatrix}$. The block Matrix Gain located in the link '*Simulink → Math Operations → Matrix*

Gain' can perform the scaling so bring the block in the model, connect it as shown in the figure7.9(c), doubleclick the block to set its Gain as 7 in the parameter window, enlarge the block, and finally run the model to see the Display output like the figure 7.9(c).

Figure 7.10(a) *Icon outlook of the Reshape block depending on the popup menu selection*

7.11 Matrix manipulation in simulink

A lot of matrix manipulations such as reshaping to different dimensions, forming sub matrices, overwriting values to some existing ones ... etc can be performed in simulink. Broadly speaking, the manipulation can be divided into four classes: changing the matrix dimension from one to another, forming smaller matrix or submatrix, changing the orientation of the matrix elements, and replacing the values of existing matrices by another values or matrix. We address them in the following.

151

7.11.1 Reshaping the matrix elements

Let us say we have the matrix $A = \begin{bmatrix} 5 & 6 & 2 \\ 8 & 9 & 3 \end{bmatrix}$, we want to turn the matrix as column vector $C = \begin{bmatrix} 5 \\ 8 \\ 6 \\ 9 \\ 2 \\ 3 \end{bmatrix}$ by

placing column after column. The block Reshape can be helpful in this regard. Let us bring one Constant block in a

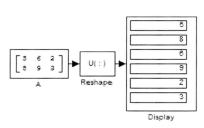

Figure 7.10(b)   *Reshaping a rectangular matrix to form a column matrix*

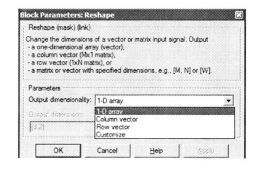

Figure 7.10(c)   *Block parameter window of the Reshape block*

new simulink model file, doubleclick the block to enter the matrix code of A as [5 6 2;8 9 2], rename the block as A, enlarge the block to see its contents, bring one Reshape block looking into the link *Simulink* → *Math Operations* → *Reshape'*, bring one Display block, and connect the blocks like the figure 7.10(b), and run the model to see the output. You need to enlarge the Display block to see all in it. Now doubleclick the Reshape block (figure 7.10(c) is parameter window) and click the Output dimensionality popup menu,

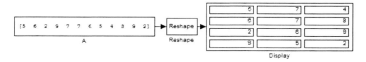

Figure 7.10(d)   *Turning the 12 element row matrix A to the 4×3 rectangular matrix placing the elements column after column*

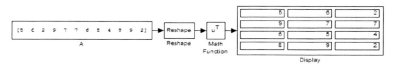

Figure 7.10(e)   *Turning the 12 element row matrix A to the 4×3 rectangular matrix placing the elements row after row*

you see some options there in the popup menu. The option Column vector and 1-D array would give you the same output as we found in figure 7.10(b). The option Row vector would show you the elements of C as a row matrix instead of the column one.

There is one more option in the popup menu called Customize. There are six elements in the matrix A and columnwise they take the shape of the matrix C. What if we form a matrix from the elements of C of order 3×2, then we should have the matrix $B = \begin{bmatrix} 5 & 9 \\ 8 & 2 \\ 6 & 3 \end{bmatrix}$. So in the Output dimensionality popup menu, select Customize and the

window prompts you by giving the option for the output dimension. You enter there [3,2] for the required order 3×2, and run the model to see the output like the matrix B.

As another example, let us say now we have a twelve element row matrix $A =$[5 6 2 9 7 7 6 5 4

8 9 2] from which we form another matrix $\begin{bmatrix} 5 & 7 & 4 \\ 6 & 7 & 8 \\ 2 & 6 & 9 \\ 9 & 5 & 2 \end{bmatrix}$ of order 4×3 placing the elements sequentially in

columns. The necessary change you need is select the Customize in the Output dimensionality popup menu and enter [4,3] in the Output dimensions of the parameter window and the simulation is presented in the figure 7.10(d). Rowwise orientation of the elements of A can happen by passing the Reshape block output through a Transpose block as presented in the model 7.10(e) but we have to enter [3,4] (interchanging the row and column number) in lieu of [4,3] in the Output dimensions slit. When you reshape some matrix, make sure that the number of the elements in the matrix you start with is equal to the product of the row and column numbers in the required matrix. Depending on the selection of the popup menu of figure 7.10(c), the icon outlook of the Reshape block becomes different as presented in the figure 7.10(a).

7.11.2 Flipping the matrix elements

Flipping the elements can happen for the row, column, or rectangular matrices. Suppose we have the row matrix $A =[7 \quad 8 \quad 9 \quad 0 \quad -3 \quad -1]$, flipping the elements of matrix A from the left to right should impart us the row matrix $[-1 \quad -3 \quad 0 \quad 9 \quad 8 \quad 7]$. The flipping just we did in A is called the flipping along rows. To simulate the problem in simulink, let us bring one Constant block in a new simulink model file, doubleclick the block to enter the code of A as [7 8 9 0 –3 –1] in the parameter window, uncheck Interpret vector parameter as 1-D in the same window, rename the block as A, enlarge the block horizontally, bring one Flip block from the link *'DSP Blockset →*

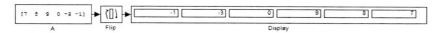

Figure 7.11(a) *Flipping the row matrix A*

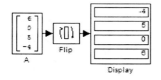

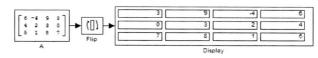

Figure 7.11(b) *Flipping the column matrix A*   Figure 7.11(c) *Flipping the rectangular matrix in row direction*

Figure 7.11(d) *Flipping the rectangular matrix in column direction*

Signal Management → Indexing → Flip', doubleclick the Flip block, change the flipping direction to row in the parameter window of Flip, bring one Display block, and connect the blocks like the figure 7.11(a), run the model, and see the output in the Display.

For the column matrix $A = \begin{bmatrix} 6 \\ 0 \\ 5 \\ -4 \end{bmatrix}$, the up to down flipping

should result the matrix $\begin{bmatrix} -4 \\ 5 \\ 0 \\ 6 \end{bmatrix}$ (called flipping along the column

direction). To have the simulation done, enter the code of A in the Constant block of previous model by writing [6;0;5;–4], resize the block A, doubleclick the Flip block, change its Flip direction from the row to column in the parameter window, run the model, resize the Display block and see the output as indicated in the figure 7.11(b).

As another example, the rectangular matrix $A = \begin{bmatrix} 6 & -4 & 9 & 3 \\ 4 & 2 & 3 & 0 \\ 5 & 1 & 8 & 7 \end{bmatrix}$ when flipped along the row and column directions,

we obtain the matrix $\begin{bmatrix} 3 & 9 & -4 & 6 \\ 0 & 3 & 2 & 4 \\ 7 & 8 & 1 & 5 \end{bmatrix}$ and $\begin{bmatrix} 5 & 1 & 8 & 7 \\ 4 & 2 & 3 & 0 \\ 6 & -4 & 9 & 2 \end{bmatrix}$

respectively. The models and the simulation results are presented in the figures 7.11(c) and 7.11(d) for the row and column directions in which you need the code for A as [6 –4 9 3;4 2 3 0;5 1 8 7] and the

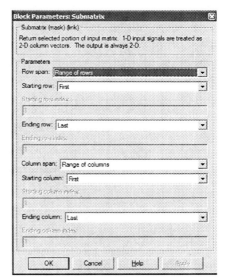

Figure 7.11(e) *Parameter window of the block Submatrix*

parameter window of the Flip block should have the flipping setting in the act of row and column respectively.

7.11.3 Forming sub matrices

Single row or column and particular rows or columns can be picked up to form sub matrix from some existing matrix in simulink. One can obtain a modular matrix employing the block Submatrix found in the link *DSP Blockset → Math Functions → Matrices and Linear Algebra → Matrix Operations → Submatrix'*. Let us choose the

test matrix as $A = \begin{bmatrix} 6 & -4 & 9 & 3 \\ 4 & 2 & 3 & 0 \\ 5 & 1 & 8 & 7 \\ 0 & 0 & 0 & 0 \\ -9 & -8 & -5 & -4 \end{bmatrix}$, enter

the matrix code for A as [6 –4 9 3;4 2 3 0;5 1 8 7;0 0 0 0;–9 –8 –5 –4] in the Constant block parameter

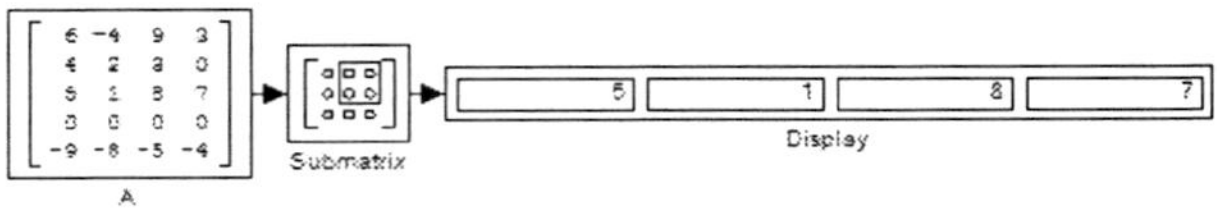

Figure 7.11(f) *Middle row of A is selected by the block Submatrix*

window bringing the block in a new simulink model file, rename the block as A, enlarge the block, bring the Submatrix and one Display blocks, run the model to see the output as the whole matrix A (because of the default setting). Now doubleclick the block Submatrix to see the parameter window like the figure 7.11(e). In that window you see the popup dialog for row and column spans. These span dialogs ask you the submatrix row and column information residing in A, which is to be picked up.

Let us say we want to pick up only the middle row of A, which is [5 1 8 7]. It goes without saying that that we need to choose all columns because we are taking all elements in the row. When we click the Row span drop down menu in the parameter window, we see three options – All rows, One row, and Range of rows. For sure we must choose One row for the row we are interested in. As soon as we select One row in the parameter window,

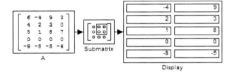

Figure 7.11(g) *Picking up the second and the third columns of matrix A*

Figure 7.11(h) *Picking up the third and fourth rows of the matrix A*

right down the Row span we see the parameter Row asking for specific row number to be picked up. Again if you click the Row popup menu, you see the drop down option as First, Index, Offset from last, Last, Offset from middle, and Middle. Since the row we started with is the middle one in the matrix A, we should choose the Middle. Now let us change our focus to Column span. Column span of the parameter window provides us similar options as the Row Span does. For the example at hand, we should choose all columns. However, successful construction and simulation of the model should show you the output like the figure 7.11(f). A number of examples are presented in the following to feel you master in the modeling of the sub matrices.

The second and third columns of the matrix A take the submatrix form $\begin{bmatrix} -4 & 9 \\ 2 & 3 \\ 1 & 8 \\ 0 & 0 \\ -8 & -5 \end{bmatrix}$. We are taking all

rows so the settings in the parameter window should be $\begin{cases} \text{Row span : All Rows} \\ \text{Column span : Range of columns} \\ \text{Starting column : Index(choose)} \\ \text{Starting column index : 2} \\ \text{Ending column : Index(choose)} \\ \text{Ending column index : 3} \end{cases}$. The model and the output

are shown in the figure 7.11(g).

Let us form another submatrix taking the third and the fourth rows of the matrix A so we should get the

submatrix as $\begin{bmatrix} 5 & 1 & 8 & 7 \\ 0 & 0 & 0 & 0 \end{bmatrix}$. One should set the parameters as $\begin{cases} \text{Row span : Range of rows} \\ \text{Starting row : Index(choose)} \\ \text{Starting row index : 3} \\ \text{Ending row : Index(choose)} \\ \text{Ending row index : 4} \\ \text{Column span : All columns} \end{cases}$ in the

dialog window of the block Submatrix and the figure 7.11(h) shows the simulation action.

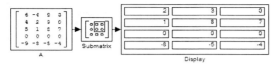

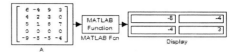

Figure 7.11(i) *Submatrix from the intersection of rows and columns*

Figure 7.11(j) *Submatrix from the intersection of the discrete rows and columns of A*

How if we select the elements in A taken from the intersection of the second through fifth rows and the second through fourth columns that is we pick up the submatrix $\begin{bmatrix} 2 & 3 & 0 \\ 1 & 8 & 7 \\ 0 & 0 & 0 \\ -8 & -5 & -4 \end{bmatrix}$ from A. The necessary setting

in the parameter window is $\begin{cases} \text{Row span : Range of rows} \\ \text{Starting row : Index (choose)} \\ \text{Starting row index : 2} \\ \text{Ending row : Index (choose)} \\ \text{Ending row index : 5} \\ \text{Column span : Range of columns} \\ \text{Starting column : Index (choose)} \\ \text{Starting column index : 2} \\ \text{Ending column : Index (choose)} \\ \text{Ending column index : 4} \end{cases}$ and

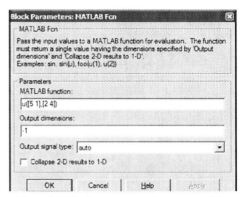

Figure 7.11(k) *Parameter window setting of the block MATLAB Fcn*

the simulation is depicted in the figure 7.11(i).

Problems of varying complicity might be encountered in real situations. The problem with the Submatrix block is the range of the rows or columns must have to be consecutive. What if we need to form the submatrix from the discrete rows and columns, for example, we want to form a submatrix in which the elements are taken from the intersection of the fifth and first rows (note that the

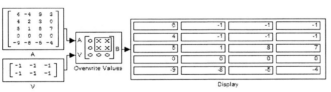

Figure 7.12(a) *Overwriting some elements of A by the elements of the matrix V*

$$A = \begin{bmatrix} 6 & -4 & 9 & 3 \\ 4 & 2 & 3 & 0 \\ 5 & 1 & 8 & 7 \\ 0 & 0 & 0 & 0 \\ -9 & -8 & -5 & -4 \end{bmatrix}$$

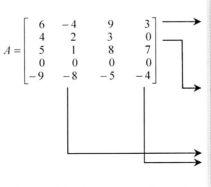

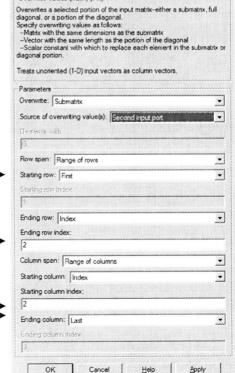

Figure 7.12(b) *Parameter window of the block Overwrite Values*

fifth row comes first, then does the first) and the second and fourth columns. Let us put the fifth row on the top of the first row of A to have $\begin{bmatrix} -9 & -8 & -5 & -4 \\ 6 & -4 & 9 & 3 \end{bmatrix}$ from which we expect to form the submatrix $\begin{bmatrix} -8 & -4 \\ -4 & 3 \end{bmatrix}$. For this problem we replace the Submatrix block by the MATLAB Fcn one located in *'Simulink → User-Defined Functions → MATLAB Fcn'*. In a word we can say that in the row direction the needed indexes of A are 5 and 1 and in the column direction they are 2 and 4. One can best put them as A([5 1],[2 4]) in the MATLAB code but the block definition is for u that is why we bring about the command as u([5 1],[2 4]) in the parameter window of the MATLAB Fcn block as shown in the figure 7.11(k). To prevent from keeping the matrix as a long vector, we also uncheck

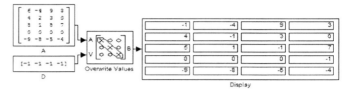

Figure 7.12(c) *Overwriting the diagonal of A by the elements of D*

the Collapse 2-D results to 1-D in the parameter window of MATLAB Fcn. However, the simulink model and the output held in the Display block are shown in the figure 7.11(j).

7.11.4 Overwriting new values to some existing ones

This kind of manipulation is imperative when we want to modify our existing matrix in simulink. Let us consider the same test matrix A as we chose for the last section. Suppose we have another matrix called $V = \begin{bmatrix} -1 & -1 & -1 \\ -1 & -1 & -1 \end{bmatrix}$ and we are going to replace the elements intersecting the first and second rows and the second through fourth columns of the matrix A so that our desired matrix appears as $\begin{bmatrix} 6 & -1 & -1 & -1 \\ 4 & -1 & -1 & -1 \\ 5 & 1 & 8 & 7 \\ 0 & 0 & 0 & 0 \\ -9 & -8 & -5 & -4 \end{bmatrix}$.

The block Overwrite Values found in the link '*DSP Blockset → Math Functions → Matrices and Linear Algebra → Matrix Operations → Overwrite Values*' can perform the operation. The model is given in the figure 7.12(a) in which you see that some elements of block A are to be replaced by the elements of V. Doubleclick the block Overwrite Values and we see the parameter window of the block in the figure 7.12(b). From the last section the reader is familiar with different drop down concepts. Similar popup or dropdown options are also applicable here in the parameter window. Indicatory arrows of the figure 7.12(b) show the specific row and column indexes where exactly the submatrix V is to be overwritten.

As another overwriting example, suppose we want to replace the diagonal [6 2 8 0] of the matrix A by the elements of $D =$[-1 -1 -1 -1] whence the expected matrix appears as $\begin{bmatrix} -1 & -4 & 9 & 3 \\ 4 & -1 & 3 & 0 \\ 5 & 1 & -1 & 7 \\ 0 & 0 & 0 & -1 \\ -9 & -8 & -5 & -4 \end{bmatrix}$. The model

of the figure 7.12(c) includes the necessary modifications for the diagonal replacement. The modifications we engaged are changing the Overwrite options from the Submatrix to Diagonal in the parameter window of the figure 7.12(b) and forming the matrix D in a Constant block. Run the model and Display the output. We selected the number of elements in D as that of the diagonal elements in A but different options are also there in the parameter window. Also notice that the icon outlook of the block is changed as soon as the overwriting option is changed.

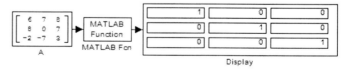

Figure 7.13(a) *Finding the reduced row echelon form of the matrix A*

7.12 Modeling the reduced row echelon form, rank, and determinant of a matrix

The reduced row echelon form of a matrix is necessary to find the rank of a matrix or the linear dependence of the vectors. Elementary row transformations are required to transform a matrix to the reduced row echelon form. It is given that the matrix $A = \begin{bmatrix} 6 & 7 & 8 \\ 8 & 0 & 7 \\ -2 & -7 & 3 \end{bmatrix}$ has the reduced row echelon form $\begin{bmatrix} 1 & 0 & 0 \\ 0 & 1 & 0 \\ 0 & 0 & 1 \end{bmatrix}$ and our objective is to implement that in simulink. Let us bring one Constant block in a new simulink model file, doubleclick the block to enter the matrix description of *A* as [6 7 8;8 0 7;–2 –7 3], rename the block as A, resize the block to display the contents, bring one MATLAB Fcn block, doubleclick the block to enter the function as rref (abbreviation for the reduced row echelon form) in the parameter window, uncheck the Collapse 2-D results to 1-D in the same parameter

window otherwise you see all elements in a column matrix and no separation between the rows would exist, bring one Display block, connect them according to the figure 7.13(a), run the model, and resize the Display block to see the reduced row echelon form.

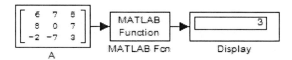

Figure 7.13(b) *Finding the rank of the matrix A in simulink*

The rank of a matrix is defined as the number of the linearly independent rows or columns of the matrix. If the matrix *A* is of order $M \times N$, the rank of $A \le$ minimum between *M* and *N*. It is given that the rank of just mentioned matrix *A* is 3. We wish to implement that in simulink. In the model of the figure 7.13(a), doubleclick the MATLAB Fcn block, change the function to rank (employed for finding the ranks of matrices) in the parameter window, run the model, and resize the Display block to see the rank as shown in the figure 7.13(b).

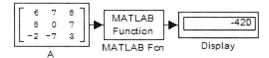

Figure 7.13(c) *Finding the determinant of the matrix A*

In a similar fashion you can also find the determinant of a square matrix. The determinant of the matrix *A* is –420. Doubleclick the MATLAB

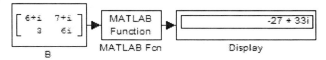

Figure 7.13(d) *Finding the determinant of the complex matrix B*

Fcn block, change the function to det (abbreviation for finding the determinant of a square matrix), run the model, and see the output as shown in the figure 7.13(c). A square complex matrix is no different from the modeling, let us

consider the complex matrix $B = \begin{bmatrix} 6+j & 7+j \\ 3 & 6j \end{bmatrix}$ whose determinant is $-27 + j33$. The model and its outcome are

presented in the figure 7.13(d) for which you need the matrix code of B as [6+i 7+i;3 6i].

7.13 Modeling the characteristic polynomial, eigenvalues, eigenvectors, and singular values

Any square matrix *A* of order $N \times N$ has the characteristic equation $| \lambda I - A | = 0$ where λ is a scalar, *I* is an identity matrix of the same order as that of *A*, and the sign | .. | indicates the determinant. Let us consider the square matrix $A = \begin{bmatrix} 6 & 5 & 9 \\ 0 & 2 & 8 \\ 8 & 0 & 3 \end{bmatrix}$ which has the characteristic

polynomial $\lambda^3 - 11\lambda^2 - 36\lambda - 212$. The polynomial coefficients become [1   –11   –36   –212] as the descending power of λ. Our aim of the simulation is to obtain that. As usual, bring a Constant block in a new simulink model file, doubleclick the block to enter the matrix code of *A* as [6 5 9;0 2 8;8 0 3] in the parameter window, rename the block as A, resize the block, bring one MATLAB Fcn block, doubleclick the block to enter the function as poly (which computes the characteristic polynomial), uncheck the Collapse 2-D results to 1-D, bring one Display block, and connect them in accordance with the figure 7.14(a), run the model, and resize the Display block to see the characteristic

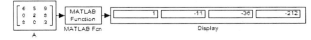

Figure 7.14(a) *Finding the characteristic polynomial of the square matrix A*

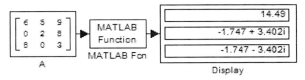

Figure 7.14(b) *Finding the eigenvalues of the matrix A*

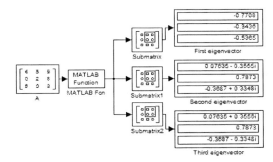

Figure 7.14(c) *Finding the eigenvectors of the matrix A in simulink*

polynomial coefficients.

The roots of the characteristic equation are the eigenvalues and we have the eigenvalues 14.4932, $-1.7466 + j\,3.4025$, and $-1.7466 - j\,3.4025$ for the example of A. Let us doubleclick the block MATLAB Fcn, change its function to eig (abbreviation for eigenvalues) in the parameter window, change its output signal type to complex because of the complex eigenvalues, run the model, and see the output like the figure 7.14(b).

The next rational question is how the eigenvectors can be found in simulink. Little bit functional programming can comfort us bring the eigenvectors in simulink. Just to mention the theoretical issue of the eigenvectors, there is a matrix V of order $N \times 1$ (basically a column matrix) that satisfies the matrix equation $AV = \lambda V$ for every eigenvalue λ. The vector V is called the eigenvector of the square matrix A. The number of the eigenvectors is equal to the number of eigenvalues. It is given that the matrix A has the eigenvectors $\begin{bmatrix} -0.7708 \\ -0.3436 \\ -0.5365 \end{bmatrix}$,

$\begin{bmatrix} 0.0763 - j0.3555 \\ 0.7873 \\ -0.3687 + j0.3348 \end{bmatrix}$, and $\begin{bmatrix} 0.0763 + j0.3555 \\ 0.7873 \\ -0.3687 - j0.3348 \end{bmatrix}$, and we aim to simulate them in simulink. As a first step, we need to

write an M file function called evalue (it can be any name you select, we selected evalue). Write the following in an M file:

> function V=evalue(A)
> [E1,E2]=eig(A);
> V=E1;

and save the file by the name evalue. As you see, the number of the output arguments of the MATLAB function eig is two and they are named as E1 and E2. The eigenvectors of the matrix A are returned to the first output argument E1. The matrix E1 is a square matrix and each column of the E1 represents the eigenvector of the matrix A. The eigenvectors so returned correspond to the eigenvalues as found in the model 7.14(b). Anyhow we assign the eigenvectors to V for the functional return in the next line. Since we do not require E2 for the problem, we are ignoring its description. Now let us concentrate in simulink and the model is presented in the figure 7.14(c). Doubleclick the MATLAB Fcn block of the figure 7.14(c) to change its function to evalue in the parameter window, uncheck the Collapse 2-D results to 1-D in the parameter window of Fcn, bring one Submatrix block to pick up the

first column of the return, doubleclick the block to set its setting as $\begin{cases} \text{Row span : All rows} \\ \text{Column span : One column} \\ \text{Column : Index (choose)} \\ \text{Column index : 1} \end{cases}$, copy the Submatrix

block in the clipboard, and paste it in the model to see the block Submatrix1 for the picking up of the second column,

doubleclick the Submatrix1 to set its setting as $\begin{cases} \text{Row span : All rows} \\ \text{Column span : One column} \\ \text{Column : Index (choose)} \\ \text{Column index : 2} \end{cases}$, obtain the block Submatrix2 in a similar

way but with the setting as $\begin{cases} \text{Row span : All rows} \\ \text{Column span : One column} \\ \text{Column : Index (choose)} \\ \text{Column index : 3} \end{cases}$, rename the Display block as the First eigenvector, copy it in

the clipboard, paste it in the model, rename the block as Second eigenvector, and obtain the other Display block similarly. Finally, connect various blocks and run the model to display the eigenvectors as shown in the same figure.

The terminologies eigenvalue and eigenvector administer the spectral decomposition only for the square matrix. The spectral decomposition theorem says that any symmetric matrix can be decomposed into the product of three matrices. Let A be a rectangular matrix of order $M \times N$, A can be expressed as $A = U \times D \times V^T$, where the multiplication sign indicates the matrix multiplication and V^T is the transpose of V and the U and V are the unitary matrices. The orders of U, D, and V are $M \times M$, $M \times M$, and $N \times M$ respectively. This sort of decomposition is termed as the singular value decomposition of A. The decomposition exists for general matrices – square or rectangular. The D is a diagonal matrix and the diagonal elements of D are called the singular values of A. The singular values are obtained by taking the positive square roots of the eigenvalues of $A \times A^T$ (A^T is the transpose of A). The matrix $A^T \times A$ or $A \times A^T$ is symmetric and their eigenvalues are nonnegative. The columns of U and V are the eigenvectors of $A \times A^T$ and $A^T \times A$ respectively.

Let us clarify the decomposition through the numerical example of $A = \begin{bmatrix} 4 & 1 & 0 & 9 \\ 5 & 7 & -1 & 0 \\ 6 & 9 & 4 & 2 \end{bmatrix}$ where $M = 3$

and $N = 4$. The matrix $A \times A^T = \begin{bmatrix} 98 & 27 & 51 \\ 27 & 75 & 89 \\ 51 & 89 & 137 \end{bmatrix}$ has the eigenvalues 73.1662, 10.9423, and 225.8915. The singular

values are the positive square roots of these eigenvalues, which are 8.5537, 3.3079, and 15.0297 respectively. The diagonal matrix $D = \begin{bmatrix} 15.0297 & 0 & 0 \\ 0 & 8.5537 & 0 \\ 0 & 0 & 3.3079 \end{bmatrix}$ is formed from

these singular values placing them in descending order. The columns in matrix $U = \begin{bmatrix} 0.4090 & -0.9062 & 0.1073 \\ 0.5168 & 0.3269 & 0.7912 \\ 0.7521 & 0.2682 & -0.6020 \end{bmatrix}$ is formed from

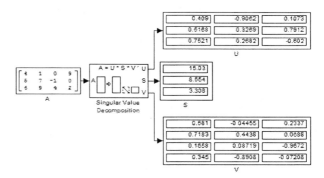

Figure 7.14(d) *Singular value decomposition of the rectangular matrix A*

the normalized eigenvectors (the magnitude of the vectors is 1) of $A \times A^T$ and the magnitude of the determinant of U is 1. On the contrary the

columns in the matrix $V = \begin{bmatrix} 0.5810 & -0.0445 & 0.2337 \\ 0.7183 & 0.4438 & 0.0688 \\ 0.1658 & 0.0872 & -0.9672 \\ 0.3450 & -0.8908 & -0.0721 \end{bmatrix}$ is formed from the normalized eigenvectors of

$A^T \times A$. Let us simulate the singular value decomposition in simulink with the given matrices quadruple $\{ A, U, D, V \}$. Let us bring one Constant block in a new simulink model file, doubleclick the block to enter the code of A as [4 1 0 9;5 7 –1 0;6 9 4 2] in the parameter window of the Constant, rename the block as A, resize the block to show the whole matrix, bring one Singular Value Decomposition block following the link '*DSP Blockset → Math Functions → Matrices and Linear Algebra → Matrix Factorizations → Singular Value Decomposition*', bring three Display blocks and rename them as U, S, and V respectively, connect various blocks as shown in the figure 7.14(d), run the model, and resize the Display blocks to show the complete matrices. The block has one input port that intakes the matrix A and three output ports that return the matrix U, the diagonal elements of the matrix D or the singular values, and the matrix V from the top respectively.

7.14 Orthonormalization, null space basis, and Jordan form of matrices

A matrix Q is said to be an orthonormal basis for the range of the matrix A if it satisfies $Q^T Q = I$, where I is the identity matrix of order $r \times r$ and r is the rank of A. If the matrix A has the order $M \times N$, the matrix Q will be of order $M \times r$.

Let us consider the matrix $A = \begin{bmatrix} 2 & 2 & -3 \\ 2 & 2 & 2 \\ -2 & -2 & 0 \\ 0 & 0 & 2 \end{bmatrix}$ for the

Figure 7.15(a) *Orthonormalization of the matrix A*

orthonotrmalization whose rank is 2 whence our matrix Q should be of order 4×2 because of $M = 4$ and $r = 2$. It is given that the matrix Q is $\begin{bmatrix} -0.7333 & 0.4714 \\ -0.4 & -0.7071 \\ 0.5333 & 0.2357 \\ 0.1333 & -0.4714 \end{bmatrix}$ and let us carry out that

in simulink. Bring one Constant block in a new simulink model file, doubleclick the block to enter the code of A as [2 2 –3;2 2 2;–2 –2 0;0 0 2], rename the block as A, resize the block to show its contents, bring one MATLAB Fcn block, doubleclick the block to enter orth (function employed for the orthonormalization) in the slit of MATLAB function, uncheck the button Collapse 2-D results to 1-D, and enter the output Dimension as [4,2] in the parameter window of MATLAB Fcn, bring one Display block, connect the blocks, and run the model to see the output as shown in the figure 7.15(a) with resizing the Display block. The simulation indicates that the rank of A must be determined first in order to enter the output dimension in the parameter window of MATLAB Fcn.

Our next agenda is to implement the basis for the null space of a matrix. The basis vectors Y for the null space of the matrix A are obtained from the equation $A Y = 0$. That is from the matrix A we find some Y which satisfies the equation $A Y = 0$ and the matrix Y is not unique. The nullity of a matrix A is defined as the difference between the number of columns of A and the rank of A. If the order of A is $M \times N$ and the nullity of A is v, the basis matrix for the null space of A should have the order $N \times v$. It is given that the

matrix $A = \begin{bmatrix} 2 & 3 & 1 & 2 & 7 \\ 8 & 8 & 8 & -3 & 5 \end{bmatrix}$ has the rank 2

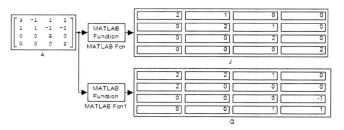

Figure 7.15(b) *Basis for the null space of the matrix A*

hence the nullity is 5−2=3 (because $N = 5$). On account of that the null space matrix has the order 5×3 and is given

by $\begin{bmatrix} -0.1738 & -0.1917 & -0.7981 \\ -0.5877 & 0.4909 & 0.3587 \\ 0.7518 & 0.2170 & 0.1775 \\ 0.2017 & 0.7217 & -0.4178 \\ 0.1365 & -0.3928 & 0.1683 \end{bmatrix}$. As usual, let us bring one Constant block in a new simulink model file,

doubleclick the block to enter the code of A as [2 3 1 2 7;8 8 8 −3 5], resize the block to show the complete A, rename the block as A, bring one MATLAB Fcn block, doubleclick the block to enter MATLAB Function as null (utilized for the null space matrix finding), enter the Output Dimension as [5,3] for the order of the null space matrix, and uncheck the button Collapse 2-D results to 1-D in the parameter window of the Fcn, bring one Display block, connect the blocks like the figure 7.15(b), run the model, and Display the expected output with resizing as shown in the same figure. The implementation also emphasizes that we know the nullity beforehand.

If the eigenvalues of a square matrix A are all distinct, the matrix A has the diagonal representation with the eigenvalues on the diagonal by choosing the set of eigenvectors as a basis. For the repeated eigenvalues, it is not always possible to find a diagonal matrix representation. However, it is possible to find some basis vectors so that the representation $Q^{-1}AQ = J$ is almost a diagonal form, J is called the Jordan form. The form has the eigenvalues of A in the diagonal and either 0 or 1 in the superdiagonal or subdiagonal. For example, if A has an eigenvalue λ with multiplicity 3, then J (assuming superdiagonal) takes one of the following forms − $\begin{bmatrix} \lambda & 0 & 0 \\ 0 & \lambda & 0 \\ 0 & 0 & \lambda \end{bmatrix}$,

$\begin{bmatrix} \lambda & 1 & 0 \\ 0 & \lambda & 0 \\ 0 & 0 & \lambda \end{bmatrix}$, and $\begin{bmatrix} \lambda & 0 & 0 \\ 0 & \lambda & 1 \\ 0 & 0 & \lambda \end{bmatrix}$. Which

Jordan form the matrix A assumes depends on the nature of A. The form is applicable only for the square matrix and not unique. The Jordan form just mentioned is the upper one, there can be the lower Jordan form as well.

It is given that the square matrix $A =$

$\begin{bmatrix} 3 & -1 & 1 & 1 \\ 1 & 1 & -1 & -1 \\ 0 & 0 & 2 & 0 \\ 0 & 0 & 0 & 2 \end{bmatrix}$ has the eigenvalue 2 of

Figure 7.15(c) *Jordan form decomposition of the matrix A*

multiplicity 4 and the expressible Jordan form is conferred as $Q^{-1} = \begin{bmatrix} 0 & \frac{1}{2} & 0 & 0 \\ \frac{1}{2} & -\frac{1}{2} & -\frac{1}{2} & -\frac{1}{2} \\ 0 & 0 & 1 & 1 \\ 0 & 0 & -1 & 0 \end{bmatrix}$, $Q =$

$\begin{bmatrix} 2 & 2 & 1 & 0 \\ 2 & 0 & 0 & 0 \\ 0 & 0 & 0 & -1 \\ 0 & 0 & 1 & 1 \end{bmatrix}$, and $J = \begin{bmatrix} 2 & 1 & 0 & 0 \\ 0 & 2 & 1 & 0 \\ 0 & 0 & 2 & 0 \\ 0 & 0 & 0 & 2 \end{bmatrix}$ so that $Q^{-1}AQ = J$ is maintained. To mention about the simulink

approach, the model of the figure 7.15(c) presents the implementation. The MATLAB Fcn block conceives the

160

setting $\begin{cases} \text{MATLAB function : jordan} \\ \text{Output dimensions} : -1 \\ \text{Output signal type : Auto} \\ \text{uncheck Collapse 2 - D results to 1 - D} \end{cases}$ in the parameter window. The function jordan is utilized for the

form computation but the problem is the return of the function is only the J not along with the Q. We need to write a MATLAB function file (M file) with the following codes:

```
function y=jform(A)
[Q,J]=jordan(A);
y=Q;
```

Let us save above MATLAB function file by the name jform. The function jordan actually has two output arguments: Q and J, simulink only handles the J from the output because that is the default return if no output argument is mentioned. The user defined function jform provides the control on the return for the matrix Q. However, the settings in the MATLAB Fcn1 block of the figure 7.15(c) are the following $\begin{cases} \text{MATLAB function : jform} \\ \text{Output dimensions} : -1 \\ \text{Output signal type : Auto} \\ \text{uncheck Collapse 2 - D results to 1 - D} \end{cases}$. We also renamed the Display blocks associated in the model as Q and J.

7.15 Matrix norms and special matrix generation

The determinant is defined only for the square matrices but the concept is not defined for the rectangular matrices. The matrix norm in some sense measures the magnitude of a rectangular matrix by a single number. There are several types of matrix norms such as Frobenius norm, 1-norm (L_1 norm), 2-norm (L_2 norm), ∞-norm (L_∞ norm), ... etc. But in simulink we find only the block for 1-norm (location is *'DSP Blockset $\rightarrow$ Math Functions $\rightarrow$ Matrices and Linear Algebra $\rightarrow$ Matrix Operations $\rightarrow$ Matrix 1–Norm')*. The 1-norm of a matrix is defined as the

Figure 7.16(a) *Finding the 1-norm of the matrix A in simulink*

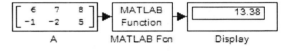

Figure 7.16(b) *Finding the Frobenius norm of the matrix A in simulink*

largest column sum of the absolute values of the elements in the matrix. For instance, the matrix $A = \begin{bmatrix} 7 & 8 & 8 \\ 0 & -3 & 1 \end{bmatrix}$

has the largest column sum or 1-norm as 11. We can employ the model of the figure 7.16(a) for the simulation.

Table 7.A   Matrix norm functional description for implemenation in simulink

Norm name	Functional setting of the MATLAB Fcn block
2-norm or L_2 norm	norm(u,2)
∞-norm (L_∞ norm)	norm(u,inf)
$-\infty$-norm ($L_{-\infty}$ norm)	norm(u,-inf)
p-norm or L_p norm for example, L_4 norm	norm(u, p) for example, norm(u,4)

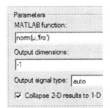

Figure 7.16(c) *MATLAB Fcn parameter window setting of the figure 7.16(b)*

For the other kinds of norms, we can take the help of the MATLAB Fcn block because of their unavailability in the library. The Frobenius norm of any matrix A of order $M \times N$ is defined by $\|A\|_F = \sqrt{\sum_{j=1}^{M} \sum_{i=1}^{N} A_{ij}^2}$ or the norm of the matrix is the positive square root of the sum of its squared elements. For example, the matrix $\begin{bmatrix} 6 & 7 & 8 \\ -1 & -2 & 5 \end{bmatrix}$ has the Frobenius norm 13.3790 whose model and MATLAB Fcn block window settings are shown in the figures 7.16(b) and 7.16(c) respectively. The function norm(u,'fro') of the figure 7.16(c) has two input arguments, the first of which is the control to the simulink block input (must be u because of the simulink block design), and the second one indicates the norm to be considered (fro for Frobenius). In table 7.A we mention the settings of the functional code in the MATLAB Fcn block for different kinds of norms.

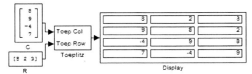

Figure 7.16(d) *Toeplitz matrix formation from the row matrix R*

Figure 7.16(e) *Non symmetric toeplitz from the matrices R and C*

Table 7.B Some matrix generation in simulink referring to the figure 7.16(f)

Matrix Name	MATLAB Fcn settings	Contents of the Constant block
Hilbert matrix of order $N \times N$ for example, 3×3	MATLAB function : hilb Output dimensions : [3,3] Output signal type: Auto uncheck Collapse 2 – D results to 1 – D	3
Companion matrix from a row vector for example, $R =[7 \quad 8 \quad 9]$	MATLAB function : compan Output dimensions : – 1 Output signal type: Auto uncheck Collapse 2 – D results to 1 – D	[7 8 9]
Vandermonde matrix from a column vector for example, $C = \begin{bmatrix} 8 \\ 5 \\ 2 \end{bmatrix}$	MATLAB function : vander Output dimensions : – 1 Output signal type: Auto uncheck Collapse 2 – D results to 1 – D	[8 5 2]
Discrete cosine transform matrix of order $N \times N$ for example, 4×4	MATLAB function : dctmtx Output dimensions : [4,4] Output signal type: Auto uncheck Collapse 2 – D results to 1 – D	[4]
Discrete Fourier transform matrix of order $N \times N$ For example, 7×7	MATLAB function : dftmtx Output dimensions : [7,7] Output signal type: Complex uncheck Collapse 2 – D results to 1 – D	[7]

Now we concentrate on the formation of some special matrices in simulink platform. Let us say we want to generate the toeplitz matrix. By definition, a toeplitz matrix is the one whose entries are constant along each diagonal and the matrix is symmetric as well. The matrix can be generated either from a row or from a column matrix. For example, the row matrix $R =[8 \quad 9 \quad -4 \quad 7]$ can generate the toeplitz $\begin{bmatrix} 8 & 9 & -4 & 7 \\ 9 & 8 & 9 & -4 \\ -4 & 9 & 8 & 9 \\ 7 & -4 & 9 & 8 \end{bmatrix}$. Let us

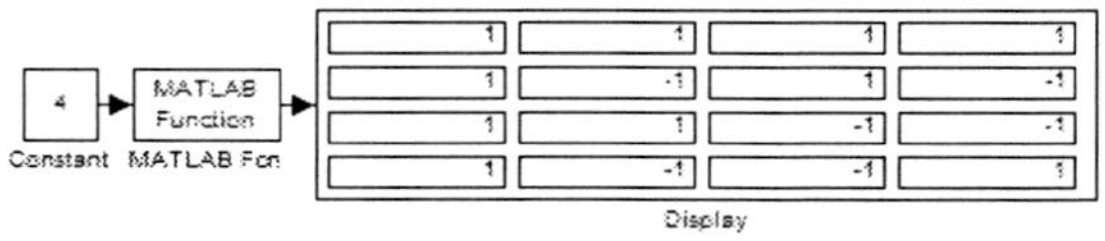

Figure 7.16(f) *Hadamard matrix formation by MATLAB Fcn block*

bring one Constant block in a new simulink model file, doubleclick the block to enter the code of R as [8 9 –4 7], rename the block as R, resize the block, bring one Toeplitz block following the link *'DSP Blockset → Math Functions → Matrices and Linear Algebra → Matrix Operations → Toeplitz'*, bring one Display block, and connect them as shown in the figure 7.16(d), run the model, and see the output as presented in the same figure. The matrix so formed is a symmetric toeplitz. You can also have the non-symmetric toeplitz from one column and one row matrices. For example, $C = \begin{bmatrix} 8 \\ 9 \\ -4 \\ 7 \end{bmatrix}$ and $R = [8 \quad 2 \quad 3]$ should provide us the non symmetric toeplitz $\begin{bmatrix} 8 & 2 & 3 \\ 9 & 8 & 2 \\ -4 & 9 & 8 \\ 7 & -4 & 9 \end{bmatrix}$

for which the model of the figure 7.16(e) applies. Doubleclick the block Toeplitz and uncheck the button symmetric. The moment you alter that, you see the block appearance of the Toeplitz as shown in the figure 7.16(e).

Not all the matrices have the dedicated blocks in simulink library. Unavailable matrices can be generated employing the MATLAB Fcn block. Let us consider the Hadamard matrix. The number of rows or columns of a Hadamard matrix (N) must be an integer which is 2^m, where $m = 1, 2, 3...$etc. The lowest order Hadamard matrix is

of order 2×2 ($N = 2$ and $m = 1$), and given by $H_{2 \times 2} = \begin{bmatrix} 1 & 1 \\ 1 & -1 \end{bmatrix}$. Letting H_N represent the Hadamard matrix of order

$N \times N$, the matrix can be obtained recursively from $H_N = \begin{bmatrix} H_{N/2} & H_{N/2} \\ H_{N/2} & -H_{N/2} \end{bmatrix}$ that is how $H_4 = \begin{bmatrix} 1 & 1 & 1 & 1 \\ 1 & -1 & 1 & -1 \\ 1 & 1 & -1 & -1 \\ 1 & -1 & -1 & 1 \end{bmatrix}$.

The model of the figure 7.16(f) can simulate the matrix generation in which the Constant block contains the order of

the Hadamard matrix and the MATLAB Fcn block has the settings $\left\{ \begin{array}{l} \text{MATLAB function : hadamard} \\ \text{Output dimensions : [4,4]} \\ \text{Output signal type : Auto} \\ \text{uncheck Collapse } 2-D \text{ results to } 1-D \end{array} \right\}$. Of

coarse, the function hadamard is a built-in one and its dimension must be known beforehand. However, we include few examples of matrices that can be generated like the hadamard as depicted in table 7.B.

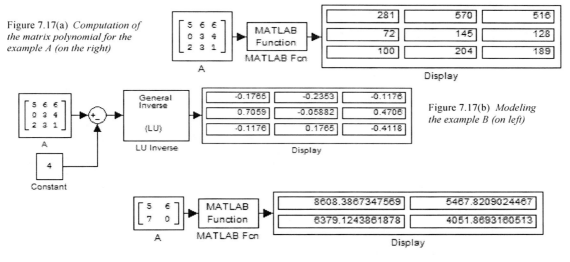

Figure 7.17(a) *Computation of the matrix polynomial for the example A (on the right)*

Figure 7.17(b) *Modeling the example B (on left)*

Figure 7.17(c) *Model for the matrix exponentiation for the square matrix A*

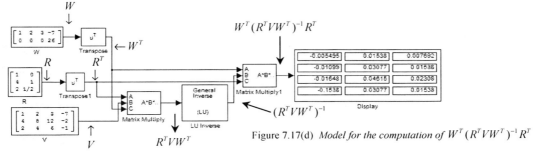

$W^T (R^T V W^T)^{-1} R^T$

$\leftarrow W^T$

R^T

$R^T V W^T$

$(R^T V W^T)^{-1}$

Figure 7.17(d) *Model for the computation of $W^T (R^T V W^T)^{-1} R^T$*

7.16 Some matrix algebraic computation

Here and now we present some examples of the matrix algebraic numeration carried out in simulink employing the blocks we mentioned so far. Let us compute the following in simulink:

⌗⌗ Example A

Compute the matrix polynomial $A^3 - 5A + I$ where $A = \begin{bmatrix} 5 & 6 & 6 \\ 0 & 3 & 4 \\ 2 & 3 & 1 \end{bmatrix}$.

Solution:

The I in the matrix polynomial is an identity matrix or order 3×3. With that one can eventually obtain the computation as $\begin{bmatrix} 281 & 570 & 516 \\ 72 & 145 & 128 \\ 100 & 204 & 189 \end{bmatrix}$ for which the model of the figure 7.17(a) is presented, and the MATLAB Fcn block in the figure has the setting $\begin{cases} \text{MATLAB function}: u^\wedge 3 - 5 * u + eye(3) \\ \text{Output dimensions}: -1 \\ \text{Output signal type}: \text{Auto} \\ \text{uncheck Collapse 2} - \text{D results to 1} - \text{D} \end{cases}$. The function eye(3) generates the identity matrix of order 3×3.

⌗⌗ Example B

Compute the matrix $(A-4)^{-1}$ where A is taken from the example A.

Solution:

Subtracting 4 from each element of A, followed by the matrix inverse, provides us $\begin{bmatrix} -0.1765 & -0.2353 & -0.1176 \\ 0.7059 & -0.0588 & 0.4706 \\ -0.1176 & 0.1765 & -0.4118 \end{bmatrix}$, and the model of the figure 7.17(b) can be applicable for this.

⌗⌗ Example C

Compute the matrix exponential for the square matrix $A = \begin{bmatrix} 5 & 6 \\ 7 & 0 \end{bmatrix}$.

Solution:

The matrix exponentiation does not mean that we raise the power of every element in A according to the exponent e. For the explanation, the reader is referred to [2]. However, it is given that the matrix exponential of the matrix A is $\begin{bmatrix} 8608.3867 & 5467.8209 \\ 6379.1243 & 4051.8693 \end{bmatrix}$ and the figure 7.17(c) implements the simulation with MATLAB Fcn setting $\begin{cases} \text{MATLAB function}: expm \\ \text{Output dimensions}: -1 \\ \text{Output signal type}: \text{Auto} \\ \text{uncheck Collapse 2} - \text{D results to 1} - \text{D} \end{cases}$. Also we changed the numeric format of the Display block (doubleclick the block) from the short to long one.

⌗⌗ Example D

Compute the matrix expression $W^T (R^T V W^T)^{-1} R^T$ for $W = \begin{bmatrix} 1 & 2 & 3 & -7 \\ 0 & 0 & 0 & 26 \end{bmatrix}$, $V = \begin{bmatrix} 1 & 0 \\ 4 & 1 \\ 2 & \frac{1}{2} \end{bmatrix}$, and $R = \begin{bmatrix} 1 & 2 & 3 & -7 \\ 4 & 8 & 12 & -2 \\ 2 & 4 & 6 & -1 \end{bmatrix}$.

Solution:

The computation involves several matrix functions: W^T is the transpose of the matrix W, $R^T V W^T$ is the matrix product of R^T, V, and W^T, $(R^T V W^T)^{-1}$ is the matrix inverse of $R^T V W^T$. However, the entire matrix computation is given by $\begin{bmatrix} -0.005494 & 0.015384 & 0.007692 \\ -0.010989 & 0.030769 & 0.015384 \\ -0.016483 & 0.046153 & 0.023076 \\ -0.153846 & 0.030769 & 0.015383 \end{bmatrix}$. The model of the figure 7.17(d) presents the simulation and its output. The matrix flow in various functional lines is also attached for the reader's convenience.

7.17 Block links used in this chapter

We discussed in this chapter common matrix related problems to the context of simulink for which the blocks of the table 7.C are utilized.

Table 7.C   Necessary blocks for modeling the matrix algebra problems as found in simulink library (not arranged in the alphabetical order)

Block name	Representative Symbol/Function	Icon Outlook	Block name	Representative Symbol/Function	Icon Outlook
Constant	Generates matrices of constant value	Constant	LU Inverse	Performs matrix inverse using LU factorization for a square matrix	General Inverse (LU) LU Inverse
Link: *Simulink → Sources → Constant*			Link: *DSP Blockset → Math Functions → Matrices and Linear Algebra → Matrix Inverses → LU Inverse*		
Display	Shows the output of the concern functional line	Display	Matrix Multiply	Multiplies two or more matrices according to matrix algebra rules	A A*B B Matrix Multiply
Link: *Simulink → Sinks → Display*			Link: *DSP Blockset → Math Functions → Matrices and Linear Algebra → Matrix Operations → Matrix Multiply*		
Constant Diagonal Matrix	Generates matrix from the diagonal description	Constant Diagonal Matrix	Identity Matrix	Generates identity matrices of different size	eye(5) Identity Matrix
Link: *DSP Blockset → Math Functions → Matrices and Linear Algebra → Matrix Operations → Constant Diagonal Matrix*			Link: *DSP Blockset → Math Functions → Matrices and Linear Algebra → Matrix Operations → Identity Matrix*		
Pseudoinverse	Finds Pseudoinverse for a rectangular matrix	Pseudoinverse (SVD) Pseudoinverse	Cholesky Inverse	Finds matrix inverse using Cholesky factorization	Sym. Pos. Def. Inverse (Chol) Cholesky Inverse
Link: *DSP Blockset → Math Functions → Matrices and Linear Algebra → Matrix Inverses → Pseudoinverse*			Link: *DSP Blockset → Math Functions → Matrices and Linear Algebra → Matrix Inverses → Cholesky Inverse*		
LDL Inverse	Finds matrix inverse using LDL factorization	Sym. Pos. Def. Inverse (LDL) LDL Inverse	Transpose	Finds the transpose of a real or complex matrix	u^T Transpose
Link: *DSP Blockset→ Math Functions → Matrices and Linear Algebra → Matrix Inverses → LDL Inverse*			Link: *DSP Blockset → Math Functions → Matrices and Linear Algebra → Matrix Operations → Transpose*		
Matrix Square	Finds the matrix square of a real or complex matrix	u^{H}u Matrix Square	Matrix Concatenation (Horizontal)	Places matrices of identical row numbers side by side	Horiz Cat Matrix Concatenation
Link: *DSP Blockset →Math Functions → Matrices and Linear Algebra → Matrix Operations → Matrix Square*			Link: *DSP Blockset → Math Functions → Matrices and Linear Algebra → Matrix Operations → Matrix Concatenation*		
Matrix Concatenation (Vertical)	Places matrices of identical columns on top of the others	Vert Cat Matrix Concatenation	Matrix Sum (Column)	Sums only the elements in the column direction of a matrix	Column Sum Matrix Sum
Link: *DSP Blockset → Math Functions → Matrices and Linear Algebra → Matrix Operations → Matrix Concatenation*			Link: *DSP Blockset → Math Functions → Matrices and Linear Algebra → Matrix Operations → Matrix Sum*		
Matrix Sum (Row)	Sums only the elements in the row direction of a matrix	Row Sum Matrix Sum	Matrix Product (Column)	Products only the elements in the column direction of a matrix	Column Product Matrix Product
Link: *DSP Blockse → Math Functions → Matrices and Linear Algebra → Matrix Operations → Matrix Sum*			Link: *DSP Blockset → Math Functions → Matrices and Linear Algebra → Matrix Operations → Matrix Product*		
Matrix Product (Row)	Products only the elements in the row direction of a matrix	Row Product Matrix Product	Sum	Sums or subtracts two or more matrices of identical order	
Link: *DSP Blockset →Math Functions → Matrices and Linear Algebra → Matrix Operations → Matrix Product*			Link: *Simulink → Math Operations → Sum*		
Matrix Gain	Scales up or down all elements in a matrix	K*u Matrix Gain	Product	Can perform scalar multiplication for two or more matrices	× Product
Link: *Simulink → Math Operations → Matrix Gain*			Link: *Simulink → Math Operations → Product*		

Continuation of previous table:

Block name	Representative Symbol/Function	Icon Outlook	Block name	Representative Symbol/Function	Icon Outlook
Matrix Scaling	Scales up or down each row separately according to some constant		Matrix Scaling	Scales up or down each column separately according to some constant	
Link: *DSP Blockset → Math Functions → Matrices and Linear Algebra → Matrix Operations → Matrix Scaling*			Link: *DSP Blockset → Math Functions → Matrices and Linear Algebra → Matrix Operations → Matrix Scaling*		
Math Function (log)	Takes natural logarithm on all elements in a matrix		Math Function (square)	Squares all elements in a matrix	
Math Function (exp)	Raises power on *e* according to the elements in a matrix		Math Function (rem)	Can perform the remainder after integer division, the upper input dividend and the lower input divider	
Math Function (10^u)	Raises power on 10 according to the elements in a matrix		Math Function (mod)	Can perform the remainder after integer division, the upper input dividend and the lower input divider	
Math Function (reciprocal)	Takes reciprocal on all elements in a matrix		Math Function (conj)	Takes complex conjugates on all elements in a complex matrix	
Math Function (sqrt)	Takes square root on all elements in a matrix		Math Function (magnitude^2)	Takes square of the magnitudes on all elements in a complex matrix	
Math Function (pow)	Raises power on elements of a matrix according to another of identical dimension (upper input base and lower input power)		Math Function (hypot)	Finds the hypotenuse values of two identical dimension matrices	
All Math Functions presented above are found in the link: *Simulink → Math Operations*					
Reshape	Turns a row or column matrix to rectangular matrix or vice versa		Flip	Flips in general a rectangular matrix from left to right or up to down	
Link: *Simulink → Math Operations → Reshape*			Link: *DSP Blockset → Signal Management → Indexing → Flip*		
Submatrix	Picks up smaller matrix blocks from a rectangular matrix		Overwrite Values	Replace some row, column, or submatrix block in a rectangular matrix	
Link: *DSP Blockset → Math Functions → Matrices and Linear Algebra → Matrix Operations → Submatrix*			Link: *DSP Blockset → Math Functions → Matrices and Linear Algebra → Matrix Operations → Overwrite Values*		
Toeplitz	Find symmetric or nonsymmetric toeplitz matrix from row or column vector		MATLAB Fcn	Perform execution of MATLAB functions in simulink	
Link: *DSP Blockset → Math Functions → Matrices and Linear Algebra → Matrix Operations → Toeplitz*			Link: *Simulink → User-Defined Functions → MATLAB Fcn*		
Matrix 1-Norm	Finds 1-norm of a rectangular matrix in general		Singular Value Decomposition	Finds the singular values and unitary matrices from a rectangular matrix	
Link: *DSP Blockset → Math Functions → Matrices and Linear Algebra → Matrix Operations → Matrix 1-Norm*			Link: *DSP Blockset → Math Functions → Matrices and Linear Algebra → Matrix Factorizations → Singular Value Decomposition*		

Modeling Common Statistics and Conversion Problems

As the caption implies, there are two main headings in this chapter – the first one is the simulink approach of simulating and computing the basic statistical problems and the second section presents some conversion regarding units and other quantities. The statistical and probabilistic analyses are getting more and more importance in the fields of communications engineering, network theory, meteorological prediction, demographic prediction, artificial intelligence, ... etc. The scope and implementation of the statistics are vast but here we outline only the first level topics or the commonly seen problems often studied and required in most science and engineering disciplines. The conversion of units for the basic physical quantities such as density, mass, or temperature are often tabulated but the simulink block is very convenient for the conversion, we wish to focus that in the second section of the chapter.

8.1 Statistical functions

In this section our principal focus is to bring about the simulation of the common statistical functions mainly to the context of the descriptive statistics. Since our objective is to give an introduction to simulink usage, the advanced statistical problems are not covered in the section.

8.1.1 Mean, variance, and standard deviation of some sample

One can arrange the sample values of some observations in terms of row, column, or rectangular matrices. The statistical computation can then be carried on the matrix oriented data. Let us jump into the simulation considering that the random variable data is arranged as the row matrix $R = [7 \quad 9 \quad 3 \quad 7 \quad 3 \quad 0]$ whose mean is

$\dfrac{\sum\limits_{i=1}^{N} R_i}{N} = \dfrac{29}{6} = 4.8333$. The N is the number of the observations and here it is the number of the elements in the matrix R. Let us bring one Constant block in a new simulink model file, doubleclick the block to enter the code of R as [7 9 3 7 3 0], resize the block to see its contents, rename the block as R, bring one Mean block following the link *'DSP Blockset → Statistics → Mean'* in the model, bring one Display block from the link *'Simulink → Sinks → Display'*, connect the blocks as shown in the figure 8.1(a), run the model, and see the output in the Display block. The data of

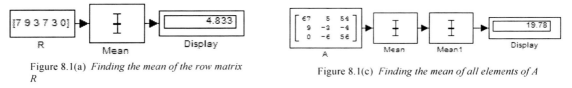

Figure 8.1(a) *Finding the mean of the row matrix R*

Figure 8.1(c) *Finding the mean of all elements of A*

the rectangular matrix $A = \begin{bmatrix} 67 & 5 & 54 \\ 9 & -3 & -4 \\ 0 & -6 & 56 \end{bmatrix}$ has the means 25.3333, −1.3333, and 35.3333 for the first, second, and third columns of A

Figure 8.1(b) *Mean on columns of A*

respectively. In the model of the figure 8.1(a) we doubleclick the block R, change the description of the constant value to [67 5 54;9 −3 −4;0 −6 56] for the code of the matrix A, rename the block as A, run the model, and see the output in the resized block of the model 8.1(b). All elements in the matrix A has the mean 19.7778 and the simulation is performed with the help of two Mean blocks like the model in the figure 8.1(c).

The variance of a random variable X with the mean $m = \dfrac{\sum\limits_{i=1}^{N} X_i}{N}$ is defined as $V(X) = \dfrac{\sum\limits_{i=1}^{N}(X_i - m)^2}{N-1}$ hence

working on the elements of R should provide us the variance as $V(X) = \dfrac{\sum\limits_{i=1}^{N}(X_i - m)^2}{N-1} =$

$\dfrac{(7-4.8333)^2 + (9-4.8333)^2 + (3-4.8333)^2 + (7-4.8333)^2 + (3-4.8333)^2 + (0-4.8333)^2}{6-1} = 11.3667$. The block

Figure 8.1(d) *Finding the variance for the elements of R*

Figure 8.1(e) *Finding the variance for the columns of A*

Variance placed in the link *'DSP Blockset → Statistics → Variance'* can compute the variance and we replace the block Mean of the model 8.1(a) by the Variance, run the model, and see the variance returned by simulink as shown in the figure 8.1(d). The previous mentioned matrix A have the variances 1322.3333, 32.3333, and 1161.3333 for the first, second, and third columns respectively and whose model is presented in the figure 8.1(e). Since the variance is a nonlinear computation, placing two blocks of Variance does not provide the variance of all elements in a rectangular matrix. It is given that the variance of all elements in the rectangular matrix A is 898.4444. To simulate that, we

Figure 8.1(f) *Finding the variance of all elements of A*

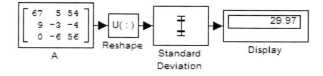

Figure 8.1(g) *Finding the standard deviation of all elements of A*

turn the rectangular matrix A first to a column or row one with the aid of the block Reshape (found in '*Simulink* → *Math Operations* → *Reshape*') and then apply the block Variance on the output of the Reshape. The model and its simulated output are shown in the figure 8.1(f).

Since the standard deviation is equal to the square root of the variance, replacing the block Variance by the Standard Deviation (also located in the link '*DSP Blockset* → *Statistics* → *Standard Deviation*') can easily compute the standard deviation of previous mentioned matrices. For example, all elements in the matrix A has the standard deviation $\sqrt{898.4444}$ =29.9741 for which we presented the model of the figure 8.1(g).

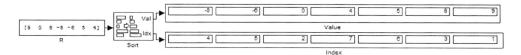

Figure 8.2(a) *Sorting the elements of the matrix R in ascending order*

Figure 8.2(b) *Sorting the elements of the matrix R in the descending order*

8.1.2 Sorting, minimum, and maximum of a sample

We can sort the elements of a sample arranged in matrices by dint of the block Sort discovered in the link '*DSP Blockset* → *Statistics* → *Sort*'. Let us say we have the row matrix R =[9 0 8 −8 −6 5 4]. If the elements of R are sorted in ascending order, we end up with the row matrix [−8 −6 0 4 5 8 9] and the elements in the last matrix occurs at the index [4 5 2 7 6 3 1] in R. Let us bring one Constant block in a new simulink model file, doubleclick the block to enter the code of R as [9 0 8 −8 −6 5 4], uncheck the button Interpret vector parameter as 1-D in the parameter window for displaying the output as the row matrix, rename the block as R, resize the block, bring one Sort block (the block has one input and two output ports and the default setting is for the ascending order), one Display block, rename the block as Value, bring another Display block, rename the block as Index, connect the blocks as shown in the figure 8.2(a), run the model, and see the sorted values and indexes in the resized blocks in the same figure.

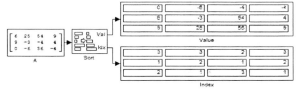

Figure 8.2(c) *Block Sort operated on the rectangular matrix A*

Doubleclick the block Sort and you see there output port options for only the Value, for only the index, and for both the Value and Index under mode in the parameter window. Also you find there the Sort order options from there one can choose the descending one. The model of the figure 8.2(b) is run with the descending order option. Note that the icon outlook is changed because of the

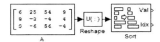

Figure 8.2(d) *Model for sorting all elements of A*

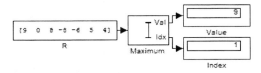

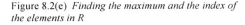

Figure 8.2(e) *Finding the maximum and the index of the elements in R*

Figure 8.2(f) *Finding the maxima and their indexes along the columns of A*

change of the ordering option. For the rectangular matrix $A = \begin{bmatrix} 6 & 25 & 54 & 9 \\ 9 & -3 & -4 & 4 \\ 0 & -6 & 56 & -4 \end{bmatrix}$ (whose code is [6 25 54 9;9 −3 −4 4;0 −6 56 −4]) the block operates on the columns of A and for which we presented the model of the figure 8.2(c) considering the ascending order. If you need to sort all elements in a rectangular matrix, bring one Reshape block and connect the block in series with the matrix A as shown in the figure 8.2(d).

169

The aforementioned row matrix R has the maximum value 9 and it happens at index 1. The block Maximum (the link is '*DSP Blockset → Statistics → Maximum*') finds the maximum value of any applied input in the form of a matrix to its input port and the model for it is shown in the figure 8.2(e). The columns of the last rectangular matrix A has the maximum values and indexes 9, 25, 56, and 9 and 2, 1, 3, and 1 for the first, second, third, and fourth columns respectively which can be implemented by the model of the figure 8.2(f). The block Maximum has the output options for only the Value, for only the Index, and for both the Value and Index. You can verify that by doubleclicking the block

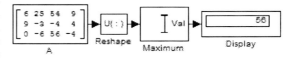

Figure 8.2(g) *Finding the maximum of all elements of A*

and checking the Parameters mode in the dialog window. The maximum in the whole matrix A is 56 for which the model of the figure 8.2(g) can be implemented. We employed the Reshape block to convert the rectangular matrix to a column one and selected the Value mode in the parameter widow of the Maximum block.

The computation pertaining to the minimum can be performed in a similar fashion with the help of the block Minimum whose link and icon outlook are presented in the table 8.C.

8.1.3 Median and RMS values of a sample

Upon sorted the middle value in a sample is the median, for example, the row matrix $R =[23 \quad -1 \quad 98 \quad 0 \quad 2 \quad 10 \quad 101 \quad 28 \quad 34]$ has the ascending order sorted form $[-1 \quad 0 \quad 2 \quad 10 \quad 23 \quad 28 \quad 34 \quad 98 \quad 101]$. So the median 23 can easily be found and we provided the model 8.3(a) for the implementation. One can obtain the block Median from the link is '*DSP Blockset → Statistics → Median*'. As a procedural step, bring a Constant block in a new simulink model file, doubleclick the block to enter the code of R as [23 –1 98 0 2 10 101 28 34], rename the block as R, resize the block, bring one Median and one Display blocks, rename the Display block as Median Value, connect the blocks as presented in the figure 8.3(a) and run the model. For the even number of elements, the block returns the average of the two middle elements in the sample.

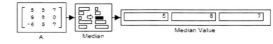

Figure 8.3(a) *Finding the median from the row matrix R*

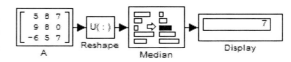

Figure 8.3(b) *Block Median functions on the columns of A*

Figure 8.3(c) *Median in the entire matrix A*

For the rectangular matrix $A = \begin{bmatrix} 5 & 8 & 7 \\ 9 & 8 & 0 \\ -6 & 5 & 7 \end{bmatrix}$, the block functions on the columns as depicted in the figure 8.3(b). If we are interested to find the

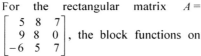

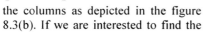

Figure 8.3(d) *Finding the RMS value for the row matrix R*

median in the entire matrix A (which is 7), the model of the figure 8.3(c) can be mentioned.

As a formulation, one can write the root mean square as RMS= $\sqrt{\dfrac{\sum\limits_{i=1}^{i=N} R_i^2}{N}}$. Let us consider the previous

mentioned R for which RMS value is $\dfrac{(-1)^2+0+2^2+10^2+23^2+28^2+34^2+98^2+101^2}{9}$ =49.8654. The block RMS

which has the link '*DSP Blockset → Statistics → RMS*' can help us perform the computation. The model and its output are presented in the figure 8.3(d). If the input of the block is a rectangular matrix, the output of the block is the RMS for each column. Also you can find the RMS value for all elements in a rectangular matrix by employing the Reshape block as we did for the Median in the figure 8.3(c).

8.1.4 Least square polynomial fit of some x-y data

In a least square fit problem, we have some x and y data and we find the equation for the best fit straight line or the curve of higher degree passing through the data with the minimal error sense. For the mathematical derivation, reference [2] can be mentioned.

Let us say that our data is $x = \begin{bmatrix} 7 \\ 9 \\ 12 \\ 13 \\ 15 \\ 17 \end{bmatrix}$ and $y = \begin{bmatrix} -5 \\ 6 \\ 0 \\ 3 \\ 7 \\ 8 \end{bmatrix}$ and we want to fit the data in a straight line such that the

minimal error is occurred due to the straight-line assumption (because practically the data may not follow the straight-line variation) and which should have the equation $y = ax + b$. So our objective is to find the unknown constants starting with the data for x and y.

Figure 8.4(a) *Finding the least squares fit straight line parameters*

The straight-line coefficients are given by the expressions

$$a = \frac{\sum x \sum y - N \sum xy}{(\sum x)^2 - N \sum x^2} \quad \text{and} \quad b = \frac{\sum x \sum xy - \sum y \sum x^2}{(\sum x)^2 - N \sum x^2}.$$ Easy computation

on the given x and y data shows that $\sum x = 73$, $\sum y = 19$, $\sum y^2 = 183$,

$\sum x^2 = 957$, and $\sum xy = 299$ from which $a = \dfrac{73 \times 19 - 6 \times 299}{73^2 - 6 \times 957} = 0.9855$

and $b = \dfrac{73 \times 299 - 19 \times 957}{73^2 - 6 \times 957} = -8.8232$. Now let us concentrate on

simulink model of figure 8.4(a), bring one Constant block in a new simulink model file, doubleclick the block to enter the y data as a row matrix [–5 6 0 3 7 8] in the parameter window, keep the Interpret vector parameters as 1-D checked in the parameter window, resize the block, rename the block as y values, bring one Least Squares Polynomial Fit block in the model file from the link '*DSP Blockset* → *Math Functions* → *Polynomial Functions* → *Least Squares Polynomial Fit*', doubleclick the block to see the parameter window like the figure 8.4(b), enter the x data as a row matrix by writing [7 9 12 13 15 17] under the slot of the Control points in the last parameter window, enter the Polynomial order as 1 in that window because straight line has the order 1, bring one Display block, connect the blocks as shown in the figure 8.4(a), run the model, and verify the coefficients as indicated in the figure.

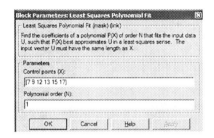

Figure 8.4(b) *Parameter window for the block Least Squares Polynomial Fit*

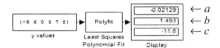

Figure 8.4(c) *Finding the least squares fit parabola parameters*

What if the assumed polynomial is of degree 2 or parabola that is the equation of the assumed curve is now $y = ax^2 + bx + c$. The coefficients related to the equation are more suitably expressed in the matrix form as

$$\begin{bmatrix} a \\ b \\ c \end{bmatrix} = \begin{bmatrix} \sum x^2 & \sum x & N \\ \sum x^3 & \sum x^2 & \sum x \\ \sum x^4 & \sum x^3 & \sum x^2 \end{bmatrix}^{-1} \begin{bmatrix} \sum y \\ \sum xy \\ \sum x^2 y \end{bmatrix}.$$ Considering the aforementioned data for x and y, we have $\sum x = 73$,

$\sum x^2 = 957$, $\sum x^3 = 13285$, $\sum x^4 = 192405$, $\sum y = 19$, $\sum xy = 299$, and $\sum x^2 y = 4635$ whence $\begin{bmatrix} a \\ b \\ c \end{bmatrix} =$

$\begin{bmatrix} 957 & 73 & 6 \\ 13285 & 957 & 73 \\ 192405 & 13285 & 957 \end{bmatrix}^{-1} \begin{bmatrix} 19 \\ 299 \\ 4635 \end{bmatrix} = \begin{bmatrix} -0.0213 \\ 1.4931 \\ -11.6041 \end{bmatrix}$. So we expect to obtain these polynomial coefficients from

simulink. Let us change the polynomial order in the parameter window from 1 to 2 and run the model. The simulink response is shown in the figure 8.4(c) in which the return is the polynomial coefficients in descending power of x.

Thus one can obtain any order polynomial coefficients from the given x and y data in the minimal error sense. The number of coefficients is one more than the order of the assumed polynomial. If the data is in MATLAB workspace assigned to some variable x as a row or column matrix, just type x in the parameter window instead of typing in the Constant block or control points slot of the parameter window. You will have the same effect on running the model. In real data analysis one may have hundreds or thousands of data in MATLAB workspace whose least square fit is required.

8.1.5 Covariance and correlation of the random variables

As a matter of representation, the random variables can be considered as the columns in a rectangular matrix. The number of columns then indicates the number of the random variables. The mean, median, or variance of a random variable describes the information about the variable itself but the covariance provides a relationship between the random variables or about their tendency to vary together rather than independently. The covariance of a matrix is another matrix but a symmetric one. The elements of the covariance matrix $V = \begin{bmatrix} V_{11} & V_{12} & \cdots & V_{1N} \\ V_{21} & V_{22} & \cdots & V_{2N} \\ \vdots & \vdots & \ddots & \vdots \\ V_{N1} & V_{N2} & \cdots & V_{NN} \end{bmatrix}$

of a matrix $A = \begin{bmatrix} A_{11} & A_{12} & \cdots & A_{1N} \\ A_{21} & A_{22} & \cdots & A_{2N} \\ \vdots & \vdots & \ddots & \vdots \\ A_{M1} & A_{M2} & \cdots & A_{MN} \end{bmatrix}$ is defined as $V_{ij} = \dfrac{\sum_{k=1}^{M}(A_{ki} - \overline{A_i})(A_{kj} - \overline{A_j})}{M-1}$, where $A_1 = \begin{bmatrix} A_{11} \\ A_{21} \\ \vdots \\ A_{M1} \end{bmatrix}$,

$A_2 = \begin{bmatrix} A_{12} \\ A_{22} \\ \vdots \\ A_{M2} \end{bmatrix}$ ......, and $A_N = \begin{bmatrix} A_{1N} \\ A_{2N} \\ \vdots \\ A_{MN} \end{bmatrix}$ and $\overline{A_j}$ is the mean of the jth column of A. The diagonal elements of V are the

variances and the off-diagonal elements are the covariances. The orders of A and V are $M \times N$ and $N \times N$ respectively.

Figure 8.5(a) Finding covariance of random variables placed in a matrix

Figure 8.5(b) Finding correlation coefficient matrix of random variables placed in a matrix

Considering the example of $A = \begin{bmatrix} -2 & 6 & 6 \\ 4 & 30 & 1 \\ 1 & -4 & -4 \\ 5 & 0 & 5 \end{bmatrix}$ and employing the expression for V_{ij}, we have $\overline{A_1} = 2$,

$\overline{A_2} = 8$, $\overline{A_3} = 2$, $V_{11} = \dfrac{\sum_{k=1}^{4}(A_{k1} - \overline{A_1})^2}{4-1} = \dfrac{(-2-2)^2 + (4-2)^2 + (1-2)^2 + (5-2)^2}{4-1} = 10$, $V_{21} = V_{12} = \dfrac{\sum_{k=1}^{4}(A_{k1} - \overline{A_1})(A_{k2} - \overline{A_2})}{4-1} =$

$\dfrac{(-2-2)(6-8) + (4-2)(30-8) + (1-2)(-4-8) + (5-2)(0-8)}{4-1} = 13.3333$, ... and so on and eventually one obtains $V =$

$\begin{bmatrix} 10 & 13.3333 & -1 \\ 13.3333 & 232 & 6 \\ -1 & 6 & 20.6667 \end{bmatrix}$. Let us simulate the covariance computation. Bring one Constant block in a new

simulink model file, doubleclick the block to enter the code of A as [−2 6 6;4 30 1;1 −4 −4;5 0 5], rename the block as A, resize the block, bring one MATLAB Fcn block, doubleclick the MATLAB Fcn block to enter

$\begin{cases} \text{MATLAB function : cov} \\ \text{Output dimensions : } -1 \\ \text{Output signal type : Auto} \\ \text{uncheck Collapse 2 - D results to 1 - D} \end{cases}$ (the function cov finds the covariance among the random variables), and

connect the blocks with one Display block as shown in the figure 8.5(a) and run the model to see the expected output.

The correlation coefficient gives a measure of the linear association between two random variables. The correlation of the matrix patterned by placing the random variables in columns is a matrix, which is called the correlation matrix. As a general rule, it is applicable for more than two random variables and can be expressed in terms of the variance-covariance matrix V, which is defined earlier. The correlation matrix R of the rectangular matrix A is defined as $R = D \times V \times D$, where $\times$ indicates the multiplication of matrices. From A, the diagonal matrix D is obtained by taking the reciprocals of the standard deviations of the columns. With regard to the variance-covariance matrix we discussed, the diagonal matrix is given by $D = \begin{bmatrix} \frac{1}{\sqrt{V_{11}}} & 0 & 0 & 0 \\ 0 & \frac{1}{\sqrt{V_{22}}} & 0 & 0 \\ \vdots & \vdots & \vdots & \ddots & \vdots \\ 0 & 0 & 0 & \frac{1}{\sqrt{V_{NN}}} \end{bmatrix}$. The

diagonal elements of R are unity because a variable is perfectly correlated with itself. Each element of R satisfies – $1 \leq R_{ij} \leq 1$, where R_{ij} is any element of R. Both the V and R are of order $N \times N$, and R is a symmetric matrix too.

Taking into account aforementioned A and V, one can have $D = \begin{bmatrix} \frac{1}{\sqrt{10}} & 0 & 0 \\ 0 & \frac{1}{\sqrt{232}} & 0 \\ 0 & 0 & \frac{1}{\sqrt{62/3}} \end{bmatrix}$ and

$R = D \times V \times D = \begin{bmatrix} 1 & 0.2768 & -0.0696 \\ 0.2768 & 1 & 0.0867 \\ -0.0696 & 0.0867 & 1 \end{bmatrix}$. Now doubleclick the model of the figure 8.5(a), change the

MATLAB function from cov to corrcoef (used for determining the correlation coefficient among random variables), run the model, and the figure 8.5(b) presents the output.

The functions cov and corrcoef are written in a general type and whose outputs are returned as matrices. In pure statistics, we define the covariance and correlation of two random variables as a single number not as a matrix. So to be consistent with the pure statistics, we can form a rectangular matrix in which the first and second columns occupy the observations of the first and second random variables respectively. Then we utilize the functions cov or corrcoef as mentioned and pick up only the element (1, 2) or (2, 1) from the output matrix.

8.1.6 Regression analysis

Suppose y is a dependent variable and it is a function of n independent variables $x_1, x_2, x_3, \ldots\ldots\ldots, x_n$. These independent variables are called the predictors in regression analysis. Widely used linear relationship between the dependent variable y and the predictors is as follows:

$$y = \beta_1 x_1 + \beta_2 x_2 + \beta_3 x_3 + \beta_4 x_4 + \ldots\ldots\ldots\ldots\ldots + \beta_n x_n + \varepsilon \qquad \text{Eqn. (8.1)}$$

The coefficients $\beta_1, \beta_2, \beta_3, \beta_4, \ldots\ldots\ldots\ldots\ldots, \beta_n$ are termed as the regression coefficients and ε is called an error due to the regression analysis. The equation 8.1 is referred to as the linear regression model, and the model represents one dependent variable. If we have many dependent variables, say, m, then the equation for the ith dependent variable can be written as follows:

$$y_i = \beta_1 x_{i1} + \beta_2 x_{i2} + \beta_3 x_{i3} + \beta_4 x_{i4} + \ldots\ldots\ldots\ldots\ldots + \beta_n x_{in} + \varepsilon_i \qquad \text{Eqn. (8.2)}$$

where $i = 1, 2, 3, \ldots, m$

In matrix form, we can organize the variables as $\begin{bmatrix} y_1 \\ y_2 \\ y_3 \\ \vdots \\ y_m \end{bmatrix} = \begin{bmatrix} x_{11} & x_{12} & x_{13} & & x_{1n} \\ x_{21} & x_{22} & x_{23} & & x_{2n} \\ x_{31} & x_{32} & x_{33} & \cdots & x_{3n} \\ & & \vdots & & \\ x_{m1} & x_{m2} & x_{m3} & & x_{mn} \end{bmatrix} \begin{bmatrix} \beta_1 \\ \beta_2 \\ \beta_3 \\ \vdots \\ \beta_n \end{bmatrix} + \begin{bmatrix} \varepsilon_1 \\ \varepsilon_2 \\ \varepsilon_3 \\ \vdots \\ \varepsilon_m \end{bmatrix}$ or $Y = X\beta + \varepsilon$,

where $Y = \begin{bmatrix} y_1 \\ y_2 \\ y_3 \\ \vdots \\ y_m \end{bmatrix}$, $X = \begin{bmatrix} x_{11} & x_{12} & x_{13} & & x_{1n} \\ x_{21} & x_{22} & x_{23} & & x_{2n} \\ x_{31} & x_{32} & x_{33} & \cdots & x_{3n} \\ & & \vdots & & \\ x_{m1} & x_{m2} & x_{m3} & & x_{mn} \end{bmatrix}$, $\beta = \begin{bmatrix} \beta_1 \\ \beta_2 \\ \beta_3 \\ \vdots \\ \beta_n \end{bmatrix}$, and $\varepsilon = \begin{bmatrix} \varepsilon_1 \\ \varepsilon_2 \\ \varepsilon_3 \\ \vdots \\ \varepsilon_m \end{bmatrix}$. The objective of the regression analysis

is to estimate the regression coefficients $\beta = \begin{bmatrix} \beta_1 \\ \beta_2 \\ \beta_3 \\ \vdots \\ \beta_n \end{bmatrix}$ from the given $Y = \begin{bmatrix} y_1 \\ y_2 \\ y_3 \\ \vdots \\ y_m \end{bmatrix}$ and $X = \begin{bmatrix} x_{11} & x_{12} & x_{13} & & x_{1n} \\ x_{21} & x_{22} & x_{23} & & x_{2n} \\ x_{31} & x_{32} & x_{33} & \cdots & x_{3n} \\ & & \vdots & & \\ x_{m1} & x_{m2} & x_{m3} & & x_{mn} \end{bmatrix}$.

The order for the Y, X, β, and ε are $m \times 1$, $m \times n$, $n \times 1$, and $m \times 1$ respectively. The error due to the regression is $\varepsilon = Y - X\beta$ and the solution for β is $(X^T X)^{-1} X^T Y$ (for the derivation the reader is referred to [2]). The regression analysis is applicable only for the inconsistent equations. For inconsistent equations, the rank of $[A \quad B]$ is greater than that of A. We may seek for an approximate solution of $A\,X = B$ that minimizes the distance between $A\,X$ and B in spite of their inconsistency.

To exemplify this, let us take $X = \begin{bmatrix} 5 & 1 & 2 \\ 4 & -2 & 1 \\ -1 & 3 & -1 \\ 0 & 2 & -3 \end{bmatrix}$ and $Y = \begin{bmatrix} 7 \\ -2 \\ 8 \\ 0 \end{bmatrix}$. Some related computations are

$$X^T = \begin{bmatrix} 5 & 4 & -1 & 0 \\ 1 & -2 & 3 & 2 \\ 2 & 1 & -1 & -3 \end{bmatrix}, \quad X^T X = \begin{bmatrix} 42 & -6 & 15 \\ -6 & 18 & -9 \\ 15 & -9 & 15 \end{bmatrix}, \quad \beta = \begin{bmatrix} \beta_1 \\ \beta_2 \\ \beta_3 \end{bmatrix} = (X^T X)^{-1} X^T Y = \begin{bmatrix} \dfrac{7}{184} & -\dfrac{5}{552} & -\dfrac{1}{23} \\ -\dfrac{5}{552} & \dfrac{15}{184} & \dfrac{4}{69} \\ -\dfrac{1}{23} & \dfrac{4}{69} & \dfrac{10}{69} \end{bmatrix} \times$$

$$\begin{bmatrix} 5 & 4 & -1 & 0 \\ 1 & -2 & 3 & 2 \\ 2 & 1 & -1 & -3 \end{bmatrix} \begin{bmatrix} 7 \\ -2 \\ 8 \\ 0 \end{bmatrix} = \begin{bmatrix} 0.2319 \\ 2.9130 \\ 1.7826 \end{bmatrix}, \text{ and } \varepsilon = Y - X\beta = \begin{bmatrix} -0.6377 \\ 1.1159 \\ 1.2754 \\ -0.4783 \end{bmatrix}.$$ Our expectation from simulink is to obtain the

β and ε values. Since there is no specific block for the regression analysis, we can implement the matrix expression for the computation of the $\beta = (X^T X)^{-1} X^T Y$ through the model of the figure 8.6(a). The matrix flow in the figure will make the reader perceive the computational algorithm. However, bring one Constant block in a new simulink model file, doubleclick the block to enter the code of the X as [5 1 2;4 −2 1;−1 3 −1;0 2 −3], rename the block as X, resize the block to see its contents, bring another Constant block, doubleclick the block to enter the code of Y as [7 −2 8 0]', uncheck the vector parameter as 1-D in the parameter window, rename (as Y) and resize the block, bring one Matrix Square (to compute $X^T X$), one Transpose (to compute X^T), two Matrix Multiply (one for the multiplication of X and β

and the other for the multiplication of $(X^T X)^{-1}$, X^T, and Y), one LU Inverse (to compute $(X^T X)^{-1}$), one Sum, and two Display blocks, rename the Display blocks as Beta and Error, doubleclick the Sum block, enter +- in the list of signs in the parameter window of the Sum (for the subtraction of $X\beta$ from Y), doubleclick the Matrix Multiply block and

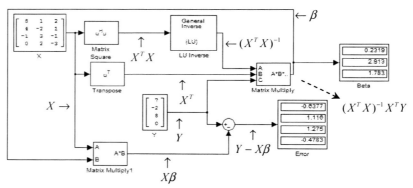

Figure 8.6(a) *Simulink model for the regression analysis*

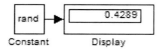

Figure 8.6(b) *Generation of a uniformly distributed continuous random variable from 0 to 1*

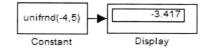

Figure 8.6(c) *Generation of a uniformly distributed continuous random variable from −4 to 5*

change its number of inputs from the default 2 to 3), place and connect various blocks relatively according to the figure 8.6(a), run the model, and finally enlarge the Beta and Error blocks to see the expected results. The reader is referred to the tables 7.C and 8.C for the links of the employed blocks.

8.1.7 Random variable and signal generation

Random number and signal generation is the primary step in many statistical analyses. Now we mention how one can generate the random numbers and signals in simulink.

⊟ **Example 1**

Let us say we want to generate the uniformly distributed continuous random variable between 0 and 1. Bring one Constant block in a new simulink model file, doubleclick the block to enter the function rand (used for the generation of the uniformly distributed random number between 0 and 1) in the constant value slot of the parameter window, bring one Display block, connect the blocks as shown in figure 8.6(b), and run the model to see the random

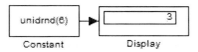

Figure 8.6(d) *Generation of a uniformly distributed discrete random variable from 1 to 6*

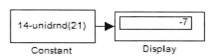

Figure 8.6(e) *Generation of a uniformly distributed discrete random variable from –7 to 13*

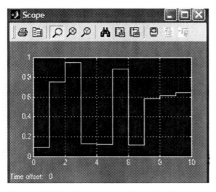

Figure 8.6(h) *Scope output from the Random Source block of the figure 8.6(f)*

Figure 8.6(f) *Generation of a random signal*

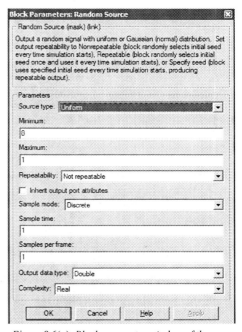

Figure 8.6(g) *Block parameter window of the Random Source*

variable. The Display block of the model 8.6(b) is showing the value 0.4289 but when you run the model, you see another value. It is obvious that the value will be any fractional value from 0 to 1. The meaning of the continuous is that we have the fractional return from the generator.

⊟ **Example 2**

When a uniformly distributed continuous random variable on the domain $-4 \leq X \leq 5$ is to be generated, the function unifrnd can be utilized inside the Constant block. Figure 8.6(c) shows the modeling in which the function unifrnd possesses two input arguments – the first and second of which are the lower and upper bounds of the interval $-4 \leq X \leq 5$ respectively. When you run the model, you do not see the Display block value -3.417 but the generation is from -4 to 5 that is for sure.

Table 8.A   Random number generators that can be used in collaboration with the Constant block in simulink
(not in alphabetical order)

Name of the generator	Description	Example	Contents of the Constant block
exprnd	Exponentially distributed continuous random variable with the mean θ where $\theta > 0$	Generate an exponentially distributed continuous random number of parameter 4	exprnd(4)
betarnd	Beta random numbers with parameters α and β	Generate a beta random number with the parameters 3 and 5	betarnd(3,5)
binornd	Binomial random numbers with parameters N and p	Generate a binomial random number with the parameters 7 and 0.55	binornd(7,0.55)
normrnd	Normal or Gaussian random numbers with parameters μ and σ	Generate a normal random number with the parameters 0 and 1	normrnd(0,1)
poissrnd	Poisson random numbers with the parameter λ	Generate a Poisson random number with the parameters 3.5	poissrnd(3.5)

⏻ Example 3

The last two generations are on the continuous or fractional number ones. The discrete number generation also occurs in a similar fashion. As a discrete example, the uniformly distributed integers from 1 to 6 (symbolically $1 \leq X \leq 6$, where X is a discrete random variable) can be generated by the function unidrnd(6). Figure 8.6(d) shows the model and its output.

The defined input argument (let us N) in the function unidrnd says that the generation is from 1 to N. If we need to generate any other discrete integers within $A \leq X \leq B$ (where $B > A$), the formula $B + 1 - unidrnd(B - A + 1)$ can be applied. Let us say we want to generate uniform discrete random variable from -7 to 13, then the function inside the Constant block should be 14–unidrnd(21). The reader is referred to the figure 8.6(e) for the implementation.

⏻ Example 4

A single binary number (0 or 1) can be generated by setting the function as 2-unidrnd(2).

However, aforementioned generations are not the ones we find in simulink. There are so many found in the statistical toolbox. Few more random number generators are presented in the table 8.A. The reader is referred to [2] and the statistical toolbox reference guide for more available random number generators in simulink.

So far we concentrated only on the generation of a single random number. Circumstances might be seen when the generation of the random numbers requires over time. On that occasion, the random variable viewed or observed in different time is called the random signal or random process.

⏻ Example 5

Let us say that we want to generate a random signal whose instantaneous random value is uniformly distributed and we want to see that at one sample per second for 10 seconds.

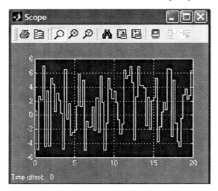

Figure 8.6(i) *Scope output for the uniformly distributed random signal from –5 to 7 of sample time 0.25 sec and of existent for 20 sec*

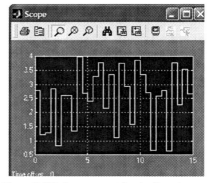

Figure 8.6(j) *Scope output for the Gaussian random signal of mean 2.5 and variance 1.5 and of sample time 0.5 sec and of existent for 15 sec*

Let us bring one Random Source block following the link '*DSP Blockset → DSP Source → Random Source*' in a new simulink model file and one Scope block, connect the blocks as shown in the figure 8.6(f), run the model, and doubleclick the Scope block to see the output. Our trial produced the

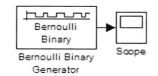

Figure 8.6(k) *Generation of a random binary signal*

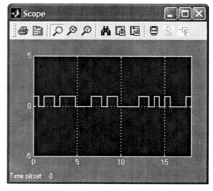

Figure 8.6(l) *Scope output from the model of the figure 8.6(k)*

output like the figure 8.6(h) with the autoscale setting. Referring to the Scope output, the horizontal axis corresponds to the 10 seconds and the random signal values are separated by 1 second. Doubleclick the Random Source block to see the parameter window of the figure 8.6(g). In that window you find the Sample time as 1 that is why the samples of the signal are separated by 1 second. The window also shows that the maximum and minimum values as 0 and 1 respectively. For this reason the signal generated is supposed to be from 0 and 1 in the figure 6.8(h).

⎘ Example 6

Let us say we want to generate a uniformly distributed random signal of value from –5 to 7 and of sample time 0.25 seconds and for the time 20 seconds. The settings for the parameter window of the figure 8.6(g) should be

$$\begin{cases} \text{Minimum}:-5 \\ \text{Maximum}:7 \\ \text{Sample time}:0.25 \end{cases}$$ keeping the others as default. But the time interval 20 seconds can not be entered in the

parameter window. From the model menu bar, click *Simulation → Simulation parameters → Solver* and in the slot of Stop time you enter 20. After running the model, our trial from the Scope returns the figure 8.6(i) with the autoscale setting.

⎘ Example 7

Anyhow, there is one more type of the random signal that can be generated from the block Random Source. Referring to the figure 8.6(g), click the popup menu of the Source type, you see another option called Gaussian. Let us say we want to generate a Gaussian random signal whose instantaneous random value has a mean 2.5 and a variance 1.5. The random signal should exist for 15 sec and the samples should be taken at every 0.5 sec. In the

parameter window of the figure 8.6(g), the change you need is $\begin{cases} \text{Source type}:\text{Gaussian} \\ \text{Mean}:2.5 \\ \text{Variance}:1.5 \\ \text{Sample time}:0.5 \end{cases}$ keeping the other settings

unchanged. The moment you change the source type, the parameters window prompts by giving the slot for the mean and variance. For the interval setting, change the solver stop time to 15 as we did for the last example. Our session with simulink provided the Scope output as shown in the figure 8.6(j) with the autoscale setting for the Gaussian signal generation.

⎘ Example 8

As another random generation, let us generate a random binary signal (all values either 0 or 1 but equally likely) which will exist for 18 seconds and the signal sample values should be 0.6 seconds apart. We can utilize the block Bernoulli Binary Generator found in the link '*Communications Blockset → Comm Sources → Data Sources → Bernoulli Binary Generator*'. Let us implement the model of the figure 8.6(k). Doubleclick the block Bernoulli Binary Generator and enter the sample time as 0.6 in the parameter window of the block. Also change the Solver stop time as 18 from the model menu bar. The output of our trial is displayed in the figure 8.6(l).

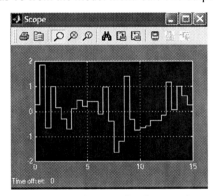

Figure 8.6(m) *Gaussian Random signal of* P_{avg} *=0.3 watts,* T_s *=0.5 seconds, and* T *=15 seconds*

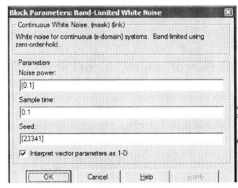

Figure 8.6(n) *Parameter window for the block Band-Limited White Noise*

Sometimes the random signals are generated from the power specifications. By power we mean the electrical power which is essentially the product of voltage and current. In such cases the generation is mostly for the additive white Gaussian because it contains almost all spectrums. Let us review the definition of the average power in a continuous random signal X which exists for T seconds. The average power is given by $\frac{1}{T}\int_0^T X^2 dt$. In

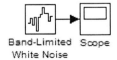

Band-Limited White Noise   Scope

Figure 8.6(o) *Gaussian Random signal of fixed power, fixed sampling period, and fixed duration generation*

calculation of the average power in a random signal it is assumed that the power is taken by a one-ohm resistor.

Hence it does not matter that X represents a voltage or current wave. The instantaneous continuous power is always given by X^2. Bu we take the samples of such signal from the continuous one. Let us say Y represents the discrete signal in which the samples are taken from the X with a sampling frequency f_s (hence period $T_s = \dfrac{1}{f_s}$ sec). So we have the relationship from the continuous to discrete as $Y_n = X(nT_s)$ where n is any integer or you can say the sample number. Now assuming constant value between the consecutive samples (also called the zero order hold), the average power in a time interval T seconds from the discrete counterpart is given by $P_{avg} = \dfrac{\sum_{n=0}^{N} Y_n^2 T_s^2}{T}$ where $N+1$ is the number of samples in the discrete signal. We generate our random signal from the given P_{avg}. Let us proceed with the following example.

⊟ Example 9

Generate a Gaussian signal whose average power is given by 0.3 watts and samples are 0.5 seconds apart, and existence should be for 15 seconds.

The given random signal specification says that P_{avg} =0.3 watts, T_s =0.5 seconds, and T =15 seconds. The block Band-Limited White Noise located in the link '*Simulink → Sources → Band-Limited White Noise*' can easily generate this kind of random signal for which the model of the figure 8.6(o) is implemented. Doubleclick the block to see the parameter window for the Band-Limited White Noise as shown in the figure 8.6(n). In that window let us enter $\begin{cases} \text{Noise power:}[0.3] \\ \text{Sample time:}[0.5] \end{cases}$ for the P_{avg} and T_s. For the duration T, change the solver stop time to 15 from the model menu bar as we did before. Our trial is presented in the figure 8.6(m) with the autoscale setting.

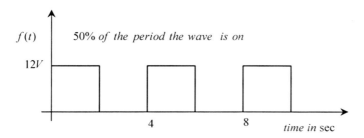

Figure 8.7(a) *A rectangular pulse of amplitude 12V, 50% duty cycle, and period 4 seconds*

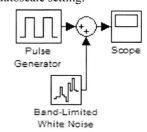

Figure 8.7(b) *The rectangular pulse of the figure 8.7(a) is corrupted with a additive white Gaussian noise*

8.1.8 Corrupted signal generation

Sometimes we corrupt artificially an ideal wave by adding some specific noise signals. We do it intentionally to study the behavior of the noise and its effect on the signal. Any unexpected disturbance signal can be termed as noise. Noise signals are nothing but the random variables of specific type and specific characteristics. Very briefly we present here few examples of corrupted signal generation. For various wave generation the reader is referred to chapter 2.

⊟ Example 1

Let us say we have a rectangular pulse wave whose amplitude $12V$, time period 4 seconds, and duty cycle 50% and the plot of the pulse is given in the figure 8.7(a). We aim here to corrupt the wave of the figure 8.7(a) by additive white Gaussian Noise (AWGN) with $\begin{cases} \text{Noise power : 0.1 watts} \\ \text{Noise samples taken} \\ \text{at every 0.1 sec} \end{cases}$. Let us also consider the simulation on the default time interval 10 sec. The

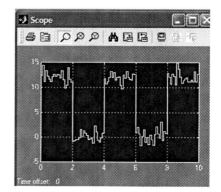

Figure 8.7(c) *The rectangular pulse corrupted with additive white Gaussian noise*

model of the figure 8.7(b) applies to the problem for which you need to bring one Pulse Generator (link: '*Simulink → Sources → Pulse Generator*'), one Scope, one Band-Limited White Noise, and one Sum blocks in a new simulink model file. Doubleclick the block Pulse Generator to enter the wave parameters as $\left\{\begin{array}{l} \text{Ampliude}:12 \\ \text{Period}:4 \\ \text{Pulse width}:50 \end{array}\right\}$, doubleclick

the block Band-Limited White Noise and make sure the parameter settings as $\left\{\begin{array}{l} \text{Noise power}:[0.1] \\ \text{Sample time}:[0.1] \end{array}\right\}$ (default one),

place and connect the blocks relatively as shown in the figure 8.7(b), and run the model. The figure 8.7(c) is the outcome of our trial from the Scope with the autoscale setting.

⊟ **Example 2**

Customarily noise is described by the signal to noise ratio (SNR). Suppose we have a sine wave of amplitude $3\,V$ and frequency $1\,KHz$. The wave is to be corrupted by an AWGN whose SNR is 10 dB and noise samples should be 0.05 $m\sec$ apart. Let us see the simulation over a time interval of $3m\sec$.

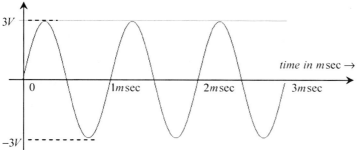

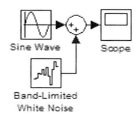

Figure 8.7(e) *The model for addition Gaussian noise with the sine wave*

Figure 8.7(d) *The sine wave of the example 2 which has amplitude 3V and frequency 1 KHz*

First one needs to compute the time average power held by the sine wave which is $\dfrac{(\text{amplitude})^2}{2}$. The given wave tells us that the average power with the sine wave is $\dfrac{3^2}{2}$ =4.5 watts. The SNR in dB is defined as $10\log_{10}\dfrac{Signal\ power}{Noise\ power}$ hence the noise power is obtained from $10\log_{10}\dfrac{4.5}{Noise\ power}$ =10 and which is 0.45 watts. The time period of the sine wave is $\dfrac{1}{1KHz}=10^{-3}\sec=1m\sec$. The sine wave should have 3 cycles in the required time interval $3m\sec$ as it is plotted in the figure 8.7(d). The necessary model for the problem is presented in the figure 8.7(e). You need to bring one Sine Wave (link: '*Simulink $\to$ Sources $\to$ Sine Wave*'), one Band-Limited White Noise, one Sum, and one Scope blocks in a new simulink model file and connect them as shown in the figure 8.7(e). Doubleclick the block Sine

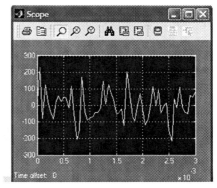

Figure 8.7(f) *The sine wave corrupted by the AWGN*

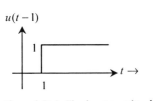

Figure 8.7(g) *The function* $u(t-1)$

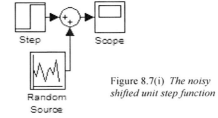

Figure 8.7(h) *The model for the uniform noise addition*

Figure 8.7(i) *The noisy shifted unit step function*

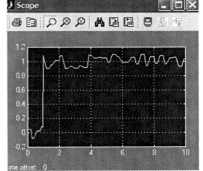

Wave, enter the settings as $\begin{Bmatrix} \text{Amplitude}:3 \\ \text{Frequency (rad/sec)}:2*pi*1e3 \end{Bmatrix}$ in the parameter window keeping the others as default,

doubleclick the block Band-Limited White Noise, and enter the settings as $\begin{Bmatrix} \text{Noise Power}:0.45 \\ \text{Sample time}:0.05e-3 \end{Bmatrix}$ in the parameter window keeping the others as default. Since the frequency intaking of the Sine Wave block happens in terms of the angular one, we have to enter the frequency as $2\pi f$. The frequency $1\ KHz$ can be coded as $10^3 = 1e3$. The sample time separation $0.05\ m\sec$ can be coded as 0.05×10^{-3} sec or 0.05e-3 sec. Still we are left with the solver stop time setting, which should be 3e-3 to take the wave interval $3m\sec$. Now you are in a position to run the model. However, figure 8.7(f) presents the random signal we had following the simulation.

☐ Example 3

The last two examples only consider taking AWGN. We can add any other noise in a similar fashion. Let us assume that we have the shifted unit step function $u(t-1)$, we want to add a uniformly distributed noise ranging from the –0.1 to 0.1. The noise samples should be 0.3 seconds apart and the noisy signal should continue for 10 sec.

Figures 8.7(g), 8.7(h), and 8.7(i) show the function $u(t-1)$, the model for the simulation, and one Scope output from the model respectively. The block Step (link: '*Simulink → Sources → Step*') has the default setting as $u(t-1)$. The parameter setting in the Random Source block you enter is $\begin{Bmatrix} \text{Minimum}:-0.1 \\ \text{Maximum}:0.1 \\ \text{Sample time}:0.3 \end{Bmatrix}$ leaving the others as default. Since the default solver stop time is 10, there is no need to change the stop time of the solver.

Anyhow we presented some examples of noisy signal implementation. Once again when you run the model, you may get different shape of the wave but the wave follows undoubtedly the specified characteristics. With this we close the sub section.

8.1.9 Discrete cross and auto correlation of random signals

The correlation of two random variables does not bear the time information. A random signal or process is not only a function of the random variable but also a function of time. A more informative approach is the cross correlation of two random signals indicating the similarities of the two random signals regarding both the variable value and the time. The crosscorrelation can be biased and unbiased. We compute the biased crosscorrelation of the discrete random signals $x[n]$ and $y[n]$ as presented in the following table:

$x[n]$	−1	6	−4	−5
$y[n]$	9	3	−10	8

The biased crosscorrelation $C_{xy}[m]$ is the convolution of $x[n]$ and $y[-n]$ computed from the polynomial multiplication of the given $x[n]$ and $y[n]$, and $C_{xy}[m]$ is given as [−8   58   −95   9   92   −51   −45] which we want to implement in simulink.

Let us bring one Constant block in a new simulink model file, doubleclick the block to enter the code of $x[n]$ as [−1 6 −4 −5]' (we wish to enter the vector as column matrix) and uncheck the button Interpret vector parameter as 1-D in the parameter window, rename the block as x[n], resize the block to show its contents, and model the sequence $y[n]$ in a similar way. We bring one Correlation (the block computes

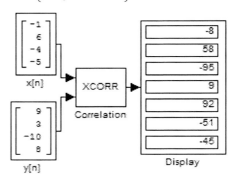

Figure 8.8(a) *The biased crosscorrelation of two discrete random signals*

Figure 8.8(b) *The unbiased crosscorrelation of the sequences x[n] and y[n]*

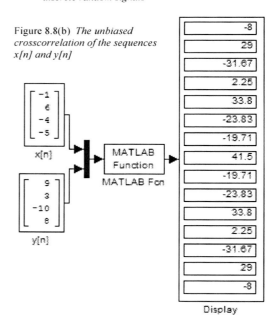

180

the biased cross correlation) and one Display blocks in the model file, where the link for the Correlation block is '*DSP Blockset → Statistics → Correlation*'. Place and connect various blocks relatively as shown in the figure 8.8(a), change the solver option type from the Variable step to the Fixed step (from the model menu bar: *Simulation → Simulation parameter→ Solver*), and run the model to see the expected output in the resized Display block as shown in the same figure.

However, the unbiased discrete crosscorrelation $C_{xy}[n]$ of two discrete random signals $x[n]$ and $y[n]$ is defined as $C_{xy}[m] = E\{x[n]y[n-m]\}$, where E is the expectation operator and the (n, m) are the integer sample indexes. It is given that the unbiased crosscorrelation of the above mentioned tabular sequences $x[n]$ and $y[n]$ is $C_{xy}[n] = [-8 \quad 29 \quad -31.6667 \quad 2.25 \quad 33.8 \quad -23.8333 \quad -19.7143 \quad 41.5]$ which is to be implemented in

Figure 8.8(c) *The autocorrelation of the sequences x[n]*

simulink. But the problem is the unbiased crosscorrelation is not defined in the block Correlation. The model of the figure 8.8(b) can be applied in this regard. The Mux block (link: '*Simulink → Signal Routing → Mux*') mulplexes the sequences $x[n]$ and $y[n]$ and then passes the output to the MATLAB Fcn block (link: '*Simulink → User-Defined Functions → MATLAB Fcn*') for crosscorrelation computation. The MATLAB Fcn has the setting $\begin{cases} \text{MATLAB Function} : \text{xcorr(u,'unbiased')} \\ \text{Output Dimensions} : -1 \end{cases}$ in which the function xcorr is applied for the computation. The first argument of xcorr hands over the control of the block input and the second one is for the type of the crosscorrelation indication. The return of the model 8.8(b) says that the function is a symmetric one (identical value at the same samples apart before and after the value 41.5).

The crosscorrelation becomes autocorrelation when the sequences $x[n]$ and $y[n]$ are identical. It is given that the autocorrelation of the sequence $x[n] = [7 \quad 8 \quad 9 \quad -1]$ is $[195 \quad 119 \quad 55 \quad -7]$ for which the model of the figure 8.8(c) can be employed. The block Autocorrelation resides in the same link as Correlation does. Doubleclick the block to see various autocorrelation options available in the block.

As an introductory text of simulink, we do not layout to cover all statistical problems found in practice. With the implementation of the crosscorrelation we bring an end to the section.

8.2 Conversion of various quantities

This section is devoted to an introduction to the conversion of various quantities that are easily carried out in simulink. Let us consider the conversion from the radian to degree, for instance, π radians are 180^0. Bring one Constant block in a new simulink model file, doubleclick the block to enter the code of π as pi, bring one Radians to Degrees block following the link '*Simulink Extras → Transformations → Radians to Degrees*', bring one Display block, connect the blocks as shown in the figure 8.9(a), run the model, and see the expected output inside the Display block.

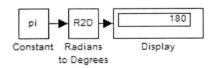

Figure 8.9(a) *Conversion from radians to degrees*

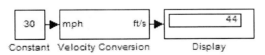

Figure 8.9(c) *Conversion of velocity from miles/hour to ft/second*

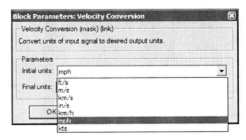

Figure 8.9(b) *Parameter window of the Velocity Conversion block*

As another example, 30 miles per hour becomes $\dfrac{30 \times 1760 \times 3}{60 \times 60} = 44$ ft/second. The necessary model for the conversion is presented in the figure 8.9(c) in which the block Velocity Conversion has the link '*Aerospace Blockset → Transformations → Units → Velocity Conversion*'. Doubleclick the block Velocity Conversion, you see a number of velocity options as a popup menu are under the slot Initial units in the parameter window as presented in the figure 8.9(b). Not only that, similar popup also resides in the slot of the Final units that means conversion from any

of the unit in $\begin{Bmatrix} ft/\sec \\ m/\sec \\ km/\sec \\ in/\sec \\ km/h \\ mph \\ kts \end{Bmatrix}$ to another is possible. The velocity conversion is not the only one that simulink offers, a

comprehensive list of conversion blocks is furnished in the tables 8.B and 8.D.

Table 8.B   Blocks for the conversion of various quantities in simulink. All of them are found in the Link: *Simulink Extras* → *Transformations*

Block name	Representative Symbol/Function		Icon Outlook
Cartesian to Polar	It has two input ports for taking (x , y) and two output ports for returning (r , θ) and converts the Cartesian coordinates to polar ones as attached.	$x \to$ $y \to$ Cartesian to Polar $\leftarrow r = \sqrt{x^2 + y^2}$ $\leftarrow \theta = \tan^{-1} \dfrac{y}{x}$ θ in radians	Cartesian to Polar
Cartesian to Spherical	It has three input ports for taking (x , y , z) and three output ports for returning (r , θ , φ) and converts the Cartesian coordinates to the spherical ones as formulated by $r = \sqrt{x^2 + y^2 + z^2}$, $\theta = \tan^{-1} \dfrac{y}{x}$, and $\varphi = \tan^{-1} \dfrac{\sqrt{x^2 + y^2}}{z}$.	$x \to$ $y \to$ $z \to$ Cartesian to Spherical $\leftarrow r$ $\leftarrow \theta$ $\leftarrow \varphi$ θ and φ in radians	Cartesian to Spherical
Polar to Cartesian	It has two input ports for taking (r , θ) and two output ports for returning (x , y) and converts the polar coordinates to the Cartesian ones as attached.	$r \to$ $\theta \to$ Polar to Cartesian $\leftarrow x$ $\leftarrow y$ θ in radians $x = r\cos\theta$ $y = r\sin\theta$	Polar to Cartesian
Spherical to Cartesian	It has three input ports for taking (r , θ , φ) and three output ports for returning (x , y , z) and converts the spherical coordinates to the Cartesian ones as attached.	$r \to$ $\theta \to$ $\varphi \to$ Spherical to Cartesian $\leftarrow x = r\sin\varphi\cos\theta$ $\leftarrow y = r\sin\varphi\sin\theta$ $\leftarrow z = r\cos\varphi$ θ and φ in radians	Spherical to Cartesian

8.3 Block links used in this chapter

The subject matter in this chapter is mainly to simulate the statistics related problems for which the blocks listed in the table 8.C become useful.

Table 8.C   Necessary blocks for modeling the statistical problems as found in simulink library (not arranged in the alphabetical order)

Block name	Representative Symbol/Function	Icon Outlook	Block name	Representative Symbol/Function	Icon Outlook
Sort	Sorts the elements in a sample and returns the sorted value with the index	Val Idx Sort	Bernoulli Binary Generator	Generates binary signal	Bernoulli Binary Bernoulli Binary Generator
Link: *DSP Blockset → Statistics → Sort*			Link: *Communications Blockset → Comm Sources → Data Sources → Bernoulli Binary Generator*		
Minimum	Finds the minimum in a matrix	Val Idx Minimum	Variance	Computes the variance from a sample	VAR Variance
Link: *DSP Blockset → Statistics → Minimum*			Link: *DSP Blockset → Statistics → Variance*		

Continuation of previous table:

Block name	Representative Symbol/Function	Icon Outlook	Block name	Representative Symbol/Function	Icon Outlook
Constant	Generates matrices of constant value	1 Constant	Display	Shows instantaneous output of the concern functional line	0 Display
Link: *Simulink → Sources → Constant*			Link: *Simulink → Sinks → Display*		
Mean	Computes the mean from a sample	Mean	Standard Deviation	Computes the standard deviation from a sample	Standard Deviation
Link: *DSP Blockset → Statistics → Mean*			Link: *DSP Blockset → Statistics → Standard Deviation*		
Maximum	Finds the maximum in a matrix	Val Idx Maximum	RMS	Computes the root mean square value from some matrix	RMS RMS
Link: *DSP Blockset → Statistics → Maximum*			Link: *DSP Blockset → Statistics → RMS*		
Autocorrelation	Computes the autocorrelation of a signal	ACF Autocorrelation	Correlation	Computes the cross correlation of two signals	XCORR Correlation
Link: *DSP Blockset → Statistics → Autocorrelation*			Link: *DSP Blockset → Statistics → Correlation*		
Random Source	Generates uniform or Gaussian random signal	Random Source	Least Squares Polynomial Fit	Finds the polynomial coefficients in a minimal error sense from the given x and y data	Polyfit Least Squares Polynomial Fit
Link: *DSP Blockset → DSP Sources → Random Source*			Link: *DSP Blockset → Math Functions → Polynomial Functions → Least Squares Polynomial Fit*		
MATLAB Fcn	Executes MATLAB functions from simulink	MATLAB Function MATLAB Fcn	Reshape	In general, it changes the dimension of a matrix	U(:) Reshape
Link: *Simulink → User-Defined Functions → MATLAB Fcn*			Link: *Simulink → Math Operations → Reshape*		
Median	Finds the median from a matrix or from a vector	Median	Scope	Function viewer	Scope
Link: *DSP Blockset → Statistics → Median*			Link: *Simulink → Sinks → Scope*		
Band-Limited White Noise	Generates additive white Gaussian noise	Band-Limited White Noise	Pulse Generator	Generates periodic pulse train	Pulse Generator
Link: *Simulink → Sources → Band-Limited White Noise*			Link: *Simulink → Sources → Pulse Generator*		
Sine Wave	Generates sinusoidal waves	Sine Wave	Step	Generates unit step function	Step
Link: *Simulink → Sources → Sine Wave*			Link: *Simulink → Sources → Step*		
Sum	Sums two or more input functions, constants, or signals	+ +	Mux	Multiplexes two or more signals or functions	
Link: *Simulink → Math Operations → Sum*			Link: *Simulink → Signal Routing → Mux*		
Radians to Degrees	Converts radians to degrees	R2D Radians to Degrees	Histogram	Performs histogram analysis of a signal	Histogram
Link: *Simulink Extras → Transformations → Radians to Degrees*			Link: *DSP Blockset → Statistics → Histogram*		

Table 8.D   Blocks for the conversion of various quantities in simulink. All of them are found in the Link:
Aerospace Blockset → Transformations → Units

Block name	Representative Symbol/Function	Icon Outlook
Acceleration Conversion	Converts from any of $\begin{Bmatrix} \text{ft/s}^2 \\ \text{m/s}^2 \\ \text{km/s}^2 \\ \text{in/s}^2 \\ \text{km/h}-\text{s} \\ \text{mph/s} \end{Bmatrix}$ to another of the same	ft/s^2 m/s^2 — Acceleration Conversion
Angle Conversion	Converts from any of $\begin{Bmatrix} \text{deg} \\ \text{rad} \\ \text{rev} \end{Bmatrix}$ to another of the same	deg rad — Angle Conversion
Angular Acceleration Conversion	Converts from any of $\begin{Bmatrix} \text{deg/s}^2 \\ \text{rad/s}^2 \\ \text{rpm/s} \end{Bmatrix}$ to another of the same	deg/s^2 rad/s^2 — Angular Acceleration Conversion
Angular Velocity Conversion	Converts from any of $\begin{Bmatrix} \text{deg/s} \\ \text{rad/s} \\ \text{rpm} \end{Bmatrix}$ to another of the same	deg/s rad/s — Angular Velocity Conversion
Density Conversion	Converts from any of $\begin{Bmatrix} \text{lbm/ft}^3 \\ \text{kg/m}^3 \\ \text{slug/ft}^3 \\ \text{lbm/in}^3 \end{Bmatrix}$ to another of the same	lbm/ft^3 kg/m^3 — Density Conversion
Force Conversion	Converts from any of $\begin{Bmatrix} \text{lbf} \\ \text{N} \end{Bmatrix}$ to another of the same	lbf N — Force Conversion
Length Conversion	Converts from any of $\begin{Bmatrix} \text{ft} \\ \text{meter} \\ \text{kilometer} \\ \text{inch} \\ \text{mile} \\ \text{Naut mile} \end{Bmatrix}$ to another of the same	ft m — Length Conversion
Mass Conversion	Converts from any of $\begin{Bmatrix} \text{lbm} \\ \text{kg} \\ \text{slug} \end{Bmatrix}$ to another of the same	lbm kg — Mass Conversion
Pressure Conversion	Converts from any of $\begin{Bmatrix} \text{psi} \\ \text{Pa} \\ \text{psf} \\ \text{atm} \end{Bmatrix}$ to another of the same	psi Pa — Pressure Conversion
Temperature Conversion	Converts from any of $\begin{Bmatrix} \text{K} \\ \text{F} \\ \text{C} \\ \text{R} \end{Bmatrix}$ to another of the same	R K — Temperature Conversion
Velocity Conversion	Converts from any of $\begin{Bmatrix} \text{ft/sec} \\ \text{meter/sec} \\ \text{kilometer/sec} \\ \text{inch/sec} \\ \text{kilometer/hour} \\ \text{mile/hour} \\ \text{Naut mile/sec} \end{Bmatrix}$ to another of the same	ft/s m/s — Velocity Conversion

Fourier Analysis Problems

This chapter elucidates the implementation of Fourier analysis problems in simulink. We present the model and its Fourier simulated results for a number of functions. The topics in the chapter are the Fourier spectrums of continuous and non periodic functions, Fourier series of continuos periodic functions, Fourier transform of discrete functions, discrete Fourier transform of discrete functions, and discrete Fourier series of the periodic discrete functions followed by the block links. Functional analysis through Fourier transform domain is very common in communications, biomedical engineering, all kinds of signal processing, and in many other disciplines. The terminology always follows one common notion that is forward and inverse or in other words analysis-synthesis approach.

9.1 What is Fourier analysis?

Fourier analysis is basically the decomposition of any function in terms of the sinusoidal functions. The reason for the analysis is the sine wave has well defined functional characteristics. The selection of the Fourier method depends on the nature and properties of the function. Again the sine wave has the different form of representation, for example, the cosine wave is basically the shifted sine wave. The complex exponential is also a sine wave with appropriate translation, rotation, or scaling. In a broad context, the functions that enfold the Fourier

analysis are $\begin{Bmatrix} perodic \\ nonperiodic \end{Bmatrix}$ and $\begin{Bmatrix} continuous \\ discrete \end{Bmatrix}$. In the subsequent sections very briefly we highlight how one can

utilize simulink to implement different class of Fourier terminology to different functions. All Fourier analysis has the transform domain and its inverse. The transform domain in most cases is complex and has the magnitude-phase or real-imaginary analysis. The inversion of the transform is just the turning to the domain of the function we start with.

185

9.2 Spectrums of the continuous and nonperiodic functions

If a function $f(t)$ is continuous, finite, and non periodic, we employ the Fourier transform to the function whose forward counterpart is formulated as $F(\omega)=\int\limits_{t=-\infty}^{t=\infty}e^{-j\omega t}f(t)dt$. The finiteness of the function $f(t)$ is assured by

the finiteness on the $\int\limits_{t=-\infty}^{t=\infty}|f(t)|\,dt$. The inverse Fourier transform is the recovery of $f(t)$ from $F(\omega)$, which is given

by $f(t)=\frac{1}{2\pi}\int\limits_{\omega=-\infty}^{\omega=\infty}e^{j\omega t}F(\omega)d\omega$. Here both the t and ω are continuous and their domain compasses from $-\infty$ to $+\infty$.

A continuous and non periodic function $f(t)$ has the forward Fourier transform $F(\omega)$ which in general complex in nature. The $F(\omega)$ can have either the real and imaginary or the magnitude and phase form. Given that we have a forward transform frequency function $F(\omega)$, we want to see the magnitude and phase responses of the transform function. It is well known that these two responses are termed as the magnitude and phase spectrums of $f(t)$ respectively. The transform function $F(\omega)$ is also continuous. In simulink we vary the ω from 0 to some finite value and plot the $F(\omega)$ on sampling some ω. Let us explore few examples on finding the spectrums.

❖❖ Example 1

Let us find the magnitude and phase spectrums of the forward Fourier transform function $F(\omega)=\frac{10}{6+5i\omega-\omega^2}$ for $0\le\omega\le50$.

⌨ Simulink Solution

The problem is basically a complex functional plotting in simulink and figure 9.1(a) shows the model to simulate the problem. Bring one Clock (for the generation of ω), one MATLAB Fcn (for the coding of $F(\omega)$), one Complex to Magnitude-Angle (for the separation of magnitude and phase angle parts of $F(\omega)$), and

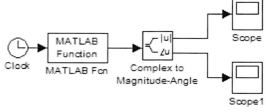

Figure 9.1(a) *Model for finding the magnitude and phase spectrums of the function $F(\omega)$*

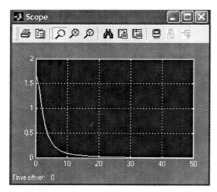

Figure 9.1(b) *Scope output for the magnitude spectrum of $F(\omega)$*

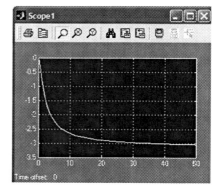

Figure 9.1(c) *Scope output for the phase spectrum of $F(\omega)$*

two Scope blocks (for displaying the magnitude and phase angle parts of $F(\omega)$) in a new simulink model file (please see the table at the end of the chapter for the links). Place the blocks relatively and connect them as shown in the figure 9.1(a). The coding of the frequency function $F(\omega)$ can be written as 10/(6+5*i*u-u^2) considering u as the independent variable. Doubleclick the block MATLAB Fcn, enter the code in the slot of MATLAB function in the parameter window of the block, also change the output signal type to complex in the same window, click from the model menu bar 'Simulation → Simulation parameters → Solver', change the solver stop time to 50 for the insertion of the interval $0\le\omega\le50$, and run the model. The Scope and Scope1 display the magnitude and phase responses as presented in the figures 9.1(b) and 9.1(c) respectively each with the autoscale setting. The horizontal axes of each Scope represent the angular frequency ω in radians/sec even though the Scope window outlook is designed originally for the time oriented analysis.

Not infrequently is wanted the magnitude $|F(\omega)|$ to be in dB ($20\log_{10}|F(\omega)|$) and the phase angle $\angle F(\omega)$ to be in degrees. The modified model of the figure 9.1(d) can be applied in this regard. We inserted two more blocks by the name dB Conversion (link: '*DSP Blockset → Math Functions → Math Operations → dB Conversion*') and Radians to Degrees (link: '*Simulink Extras → Transformations → Radians to Degrees*') in the model of the figure 9.1(a). With that inclusion, the vertical axes of the Scope and Scope1 now represent dB and degrees while the horizontal one retaining in ω.

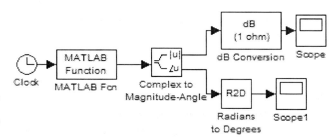

Figure 9.1(d) *Model for finding the magnitude and phase spectrums of the function $F(\omega)$ in dB and degrees respectively*

❖ ❖ Example 2

Let us find the real and imaginary components of the $F(\omega)$ in example 1 for the interval of $0 \le \omega \le 15$. The change we need in the model of the figure 9.1(a) is replace the Complex to Magnitude-Angle block by the Complex to Real-Imag (link: '*Simulink → Math Operations → Complex to Real-Imag*') one as presented in the figure 9.1(e) and change the solver stop time to 15. So now the Scope and Scope1 display the Real{$F(\omega)$} and Imaginary {$F(\omega)$} respectively which are shown in the figures 9.1(f) and 9.1(g) respectively. Reducing the ω interval from 50 to 15 can manifest the $F(\omega)$ curve variation more pronouncedly.

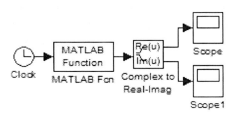

Figure 9.1(e) *Model for finding the real and imaginary parts of the function $F(\omega)$*

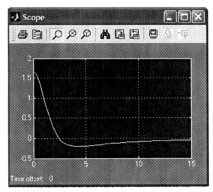

Figure 9.1(f) *Scope output for the real part of $F(\omega)$ versus ω*

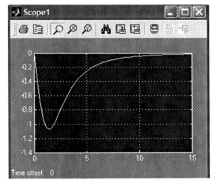

Figure 9.1(g) *Scope1 output for the imaginary part of $F(\omega)$ versus ω*

❖ ❖ Example 3

To find the spectrums of the function $F(\omega) = \dfrac{56 - j\omega - \omega^2}{5 - 6j\omega - j\omega^3}$, all we need is enter the simulink code of the frequency function as (56-i*u-u^2)/(5-6*i*u-i*u^3) in the slit of MATLAB function in the parameter window of the block MATLAB Fcn attached with previous models.

❖ ❖ Example 4

The frequency function $F(\omega) = \dfrac{5e^{j(2\omega-4)}}{7 - j(3 - \omega)}$ needs the code 5*exp(i*(2*u-4))/(7-i*(3-u)) in the MATLAB Fcn block to be simulated for different components of $F(\omega)$ as we did in the other three examples.

❖ ❖ **Example 5**

In previous examples we chose some complex frequency functions $F(\omega)$ and found their various spectrums. Sometimes starting with the function $f(t)$, viewing the spectrums might be necessary. In Fourier theory it is well known that the finite rectangular function has the sinc function frequency response. Let us speculate that on the $f(t)$ of the figure 9.1(h). The first thing is we need the mapping of $f(t)$ to $F(\omega)$ through the formula $F(\omega) = \int_{t=-\infty}^{t=\infty} e^{-j\omega t} f(t) dt$. As though the lower and upper limits are infinite but for the example of the figure 9.1(h), we have

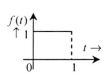

$f(t)$
↑ 1

$t \rightarrow$

0 1

Figure 9.1(h) *A finite duration function* $f(t)$

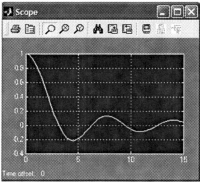

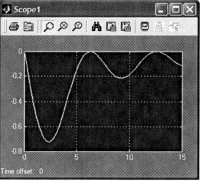

Figure 9.1(i) *Real spectrum of the finite rectangular function* $f(t)$

Figure 9.1(j) *Imaginary spectrum of the finite rectangular function* $f(t)$

$F(\omega) = \int_{t=0}^{t=1} e^{-j\omega t} dt = \dfrac{j(e^{-j\omega}-1)}{\omega}$. Since simulink intakes u dependent expression, we can write the forward transform as

$F(u) = \int_{t=0}^{t=1} e^{-jut} dt = \dfrac{j(e^{-ju}-1)}{u}$. We do not even have to integrate the $f(t)$ if we seek help from the symbolic toolbox of MATLAB. Let us conduct this integration in MATLAB Command window as follows:

MATLAB Command

```
>>syms u t ↵
>>int(exp(-j*u*t),t,0,1) ↵

ans =

i*(exp(-i*u)-1)/u
```

The first line of the command says that our independent variables for the input $f(t)$ and output $F(u)$ are symbolic. The function int in the second line integrates the symbolic expression $\int_{t=0}^{t=1} e^{-jut} dt$. The function int has four input arguments, the first, second, third, and fourth of which are the code for e^{-jut}, the declaration of the independent variable of the integration, lower limit of the integration, and upper limit of the integration respectively. As MATLAB returns, we have the $F(u)$ expression as shown above. You do not even need to memorize, write, or type the expression of $F(u)$. Select above line with your mouse and copy that to the clipboard of the computer. As far as our objective is concern, we concentrate on the model of the figure 9.1(e), doubleclick the block MATLAB Fcn, bring mouse pointer in the slit of MATLAB function, delete previous expression from the MATLAB function slit, rightclick the mouse, click paste, and you see the code for $F(u)$ in the parameter window. If you run the model, you encounter some error. Looking into the expression $\dfrac{j(e^{-ju}-1)}{u}$, one can inspect that there is one indeterminate situation (which is $\dfrac{0}{0}$) at $u = 0$ and computer can not handle this sort of situation. We apply some numerical artifice here by making the epsilon of the computer (ε) as zero. The epsilon is machine or computer system dependent, the computer we used has the $\varepsilon = 2.2204 \times 10^{-16}$. Anyhow Simulink has the code eps for the ε. Hence in the solver setting, which is accessed from the model menu bar as we did before, we set the start time of the simulation as eps. This action corresponds to $\omega = 0$ in the original expression $F(\omega) = \dfrac{j(e^{-j\omega}-1)}{\omega}$. Let us run the model with the modification we just did. The Scope and Scope1 return the real and imaginary spectrums of $f(t)$ as shown in the figures 9.1(i) and 9.1(j) respectively with the autoscale setting for each. Let us not forget that the horizontal axis corresponds to the angular frequency ω and we plotted the spectrums for $0 \leq \omega \leq 15$. Figure 9.1(i) helps us infer the transforming a rectangular function to the sinc function through the Fourier transform.

♦ ♦ Example 6

Find the Fourier real and imaginary spectrums for the function $f(t)$ of the figure 9.2(a). The procedure we executed in the example 5 is also applicable for the problem. There is no single function that can describe the function $f(t)$.

Figure 9.2(a) A short existent function $f(t)$

Piece by piece one can define the function as $f(t) = \begin{cases} t & \text{for } 0 \leq t \leq 1 \\ 1 & \text{for } 1 \leq t \leq 4 \\ 5-t & \text{for } 4 \leq t \leq 5 \end{cases}$. The

Fourier transform also applies piece by piece therefore $F(\omega) = \int_{t=0}^{t=1} te^{-j\omega t}\, dt + \int_{t=1}^{t=4} e^{-j\omega t}\, dt + \int_{t=4}^{t=5} (5-t)e^{-j\omega t}\, dt$ or $F(u) =$

$\int_{t=0}^{t=1} te^{-jut}\, dt + \int_{t=1}^{t=4} e^{-jut}\, dt + \int_{t=4}^{t=5} (5-t)e^{-jut}\, dt$ to be fed in simulink. We also wish to carry out the integration in MATLAB as follows:

MATLAB Command
```
>>syms u t ↵
>>F=int(t*exp(-j*u*t),t,0,1)+int(exp(-j*u*t),t,1,4)+int((5-t)*exp(-j*u*t),t,4,5); ↵
>>simplify(F) ↵

ans =

(exp(-i*u)-1-exp(-5*i*u)+exp(-4*i*u))/u^2
```

There are three integrations associated with $F(u)$ for that reason we employed three int functions respectively with proper coding and limits of integration. This time we assigned the MATLAB return to some variable F and latter we simplified the contents in F by the command simplify. This is just to avoid the long expressions while entering to simulink. However, copy the last expression by your mouse and paste the expression in the slit of the MATLAB function in the parameter window of the block MATLAB Fcn of the model in the figure 9.1(e). Let us see how the numerator and denominator of $F(u)$ look like by executing the following in the command prompt:

```
>>pretty(simplify(F)) ↵

    exp(-i u) - 1 - exp(-5 i u) + exp(-4 i u)
    ---------------------------------------------
                         2
                        u
```

This time the expression for $F(u)$ also takes $\dfrac{0}{0}$ form. Hence we change the solver start time to eps and stop time to 10 (let us see the spectrum for $0 \leq \omega \leq 10$). Figures 9.2(b) and 9.2(c) show the results of the simulation for the real and imaginary spectrums respectively.

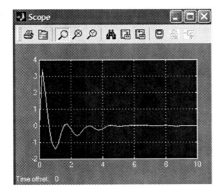

Figure 9.2(b) Real spectrum of the function in the figure 9.2(a)

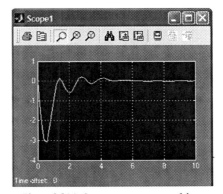

Figure 9.2(c) Imaginary spectrum of the function in the figure 9.2(a)

189

Since the inverse Fourier transform of any frequency function happens through the formula $f(t) = \frac{1}{2\pi} \int_{\omega=-\infty}^{\omega=\infty} e^{j\omega t} F(\omega) dt$, the implementation is very similar to the forward counterpart. Now we have a multiplier of $\frac{1}{2\pi}$ with the integral and the complex exponential is positive. To be implementable in simulink, the function $F(\omega)$ must be finite for some ω.

9.3 Fourier series for the periodic continuous functions

The concept of Fourier series applies to the functions which are continuous and periodic. Periodicity of the function necessitates that we must have a frequency for the function. If a function $f(t)$ is continuous and has a period T, the function can be expressed as

$$f(t) = \frac{a_0}{2} + \sum_{n=1}^{n=\infty} c_n \sin\left(n\frac{2\pi}{T}t + \varphi_n\right), \quad \text{where} \quad a_n = \frac{2}{T}\int_{t=0}^{t=T} f(t)\cos\left(n\frac{2\pi}{T}\right)dt, \quad b_n = \frac{2}{T}\int_{t=0}^{t=T} f(t)\sin\left(n\frac{2\pi}{T}\right)dt, \quad c_n =$$

$\sqrt{a_n^2 + b_n^2}$, and $\varphi_n = \tan^{-1}\left(\frac{a_n}{b_n}\right)$ with $n = 0, 1, 2, 3$, etc. The n is integer and called harmonics of the fundamental

frequency $f = \frac{1}{T}$ for which $n = 1$. The term for $n = 0$ is called the average value of the function. The expansion is termed as the Fourier series expansion of the periodic function $f(t)$. Primarily the independent variable t is assumed to be time but it can represent other quantities as well. For instance, if it represents displacement, then we have the spatial frequency. Simulink's default setting handles the functions defined over $t \geq 0$, the function to be analyzed for the series should be provided accordingly.

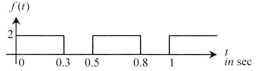

Figure 9.3(a) *A rectangular pulse of period 0.5 sec and amplitude 2*

All the while we sought the exemplary approach in the text. Let us choose the periodic wave of the figure 9.3(a). The wave has the time period $T = 0.5$ sec hence the frequency $f = 2$ Hz and the generalized coefficients are $a_n =$

$4\int_{t=0}^{t=0.3} 2\cos(4n\pi t)\, dt = \frac{2}{n\pi}\sin\left(\frac{6n\pi}{5}\right)$ and $b_n = 4\int_{t=0}^{t=0.3} 2\sin(4n\pi t)\, dt$

$= \frac{2}{n\pi}\left[1 - \cos\left(\frac{6n\pi}{5}\right)\right]$. We calculate several coefficients of the

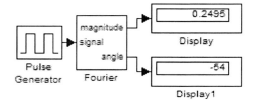

Figure 9.3(b) *Model for calculating the Fourier series coefficients for a periodic wave*

series in the following.

when $n = 0$:

$a_0 = 2.4$, $b_0 = 0$, $c_0 = \frac{a_0}{2} = 1.2$, and $\varphi_0 = 90^0$ (only for $n = 0$, $c_0 = \frac{a_0}{2}$ by definition else $c_n = \sqrt{a_n^2 + b_n^2}$),

when $n = 1$ (called the fundamental frequency):

$a_1 = -0.3742$, $b_1 = 1.1517$, $c_1 = 1.2109$, and $\varphi_1 = -18^0$,

when $n = 2$ (called the second harmonic):

$a_2 = 0.3027$, $b_2 = 0.2199$, $c_2 = 0.3742$, and $\varphi_2 = 54^0$,

when $n = 3$ (called the third harmonic):

$a_3 = -0.2018$, $b_3 = 0.1466$, $c_3 = 0.2495$, and $\varphi_3 = -54^0$, and so on for other harmonics.

The concise statement of the problem can be rephrased as

from the given periodic wave of the figure 9.3(a), we want to obtain the c_n coefficients and φ_n angles from simulink.

Figure 9.3(b) illustrates the modeling for the computation of the Fourier series coefficients. Bring one Pulse Generator (link: '*Simulink → Sources → Pulse Generator*'), one Fourier (link: '*SimPowerSystems → Extra Library → Measurements → Fourier*'), and two Display blocks in a new simulink model file. Place and connect the blocks in accordance with the figure 9.3(b). The duty cycle for the wave in the figure 9.3(a) is $\frac{0.3}{0.5} \times 100\% = 60\%$. To model the wave in the figure 9.3(a), doubleclick the block Pulse Generator and enter its setting as

$$\left.\begin{array}{l} \text{Amplitude}: 2 \\ \text{Period(secs)}: 0.5 \\ \text{Pulse Width (\% of Period)}: 60 \end{array}\right\}$$ in the parameter window of the

block keeping the others as default. The block Fourier has one input port to which we feed the signal whose Fourier series analysis is necessary. Out of the two output ports of the block, the upper and lower ones return the coefficient c_n and the phase angle φ_n in degrees respectively. The block output is one at a time that is one set of $\{c_n, \varphi_n\}$ for one simulation. The two Display blocks in the model are to show the $\{c_n, \varphi_n\}$ values for one particular harmonic. We also need to feed the wave information to the block Fourier whose block parameter window is presented in the figure 9.3(c). In the parameter window you find the slot of the fundamental frequency which is 2Hz for the wave of the figure 9.3(a). Now the slot Harmonic in the parameter window depends on which harmonic's amplitude and phase you examine for. For instance, we ran the model for $n=3$ whose Display block outputs are shown in the figure 9.3(b) confirming our computation $c_3 = 0.2495$ and $\varphi_3 = -54^0$. If you set the Harmonic as 2 in the parameter window in the figure 9.3(c), you should get $c_2 = 0.3742$ and $\varphi_2 = 54^0$ in the Display blocks.

In chapter two, we provided a number of periodic waves from where we picked up some waves for the Fourier series computation in the following.

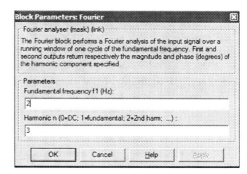

Figure 9.3(c) *Parameter window of the block Fourier*

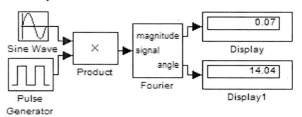

Figure 9.3(d) *Model for the Fourier series analysis of the triggered sine wave in the figure 2.8(a)*

✦ ✦ Example 1

Let us find the Fourier series magnitude and phase coefficients in simulink for the triggered sine wave of the figure 2.8(a). We connect the wave generator portion of the model in figure 2.8(b) with the Fourier block as

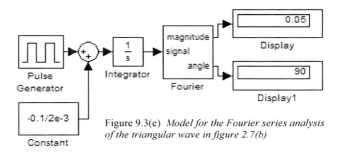

Figure 9.3(e) *Model for the Fourier series analysis of the triangular wave in figure 2.7(b)*

presented in the figure 9.3(d). To pass the frequency information to the Fourier block, doubleclick the block and enter 50 in the Fundamental frequency slot in the parameter window of the figure 9.3(c). Let us find the coefficients for the fourth harmonic. So in the parameter window we enter the Harmonic as 4, run the model, and the Display blocks should return the output as shown in the figure 9.3(d) indicating $c_4 = 0.07$ and $\varphi_4 = 14.04^0$ corresponding to the symbols we have presented earlier. The reader can verify the magnitude-angle Fourier coefficients

$$\left.\begin{array}{l} c_0 = 0.1273, \ \varphi_0 = -90^0 \\ c_1 = 0.6134, \ \varphi_1 = -11.98^0 \\ c_2 = 0.1898, \ \varphi_2 = 153.4^0 \\ c_3 = 0.1273, \ \varphi_3 = 90^0 \end{array}\right\}$$ on running the model in the figure 9.3(d).

✦ ✦ Example 2

Referring to the figure 2.7(b) of chapter 2, the wave is a triangular-derived one whose frequency is $f = \dfrac{1}{4m\sec} = 250$ Hz. Like the example 1, we connect the wave generator part of the model in the figure 2.7(f) with the block Fourier as shown in the figure 9.3(e) to perform the analysis. Change the Fundamental frequency in the parameter window of the block Fourier to 250. We ran the model for $n=0$ for which $c_0 = 0.05$ and $\varphi_0 = 90^0$ are exhibited by the Display blocks in the same model.

❖ ❖ Example 3

Once we are apt to find the Fourier series magnitude and phase coefficients, the next legitimate query is why we do not plot the spectrums for different harmonics. Let us say we wish to see the amplitude and phase spectrums of the wave in the example 2 for $0 \le n \le 7$.

To model the problem, one important point needs to be addressed. A simulink model is solver start and stop time or state dependent. Since the wave generation is also time dependent, we can not pass the harmonic information until the model has been run completely. We encountered this sort of situation in the example 2 of the section 5.6. The reader is suggested to go through the section for handling simulink model parameters from MATLAB Command window. We mention here only the procedural step for viewing the harmonics variation.

We replace the two Display blocks of the model 9.3(e) by two To Workspace (link: *'Simulink → Sinks → To Workspace'*) blocks to send the magnitude and phase data for each n to MATLAB Workspace. Figure 9.3(f) shows the model with the modification. Change the settings as

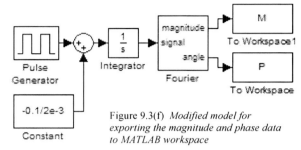

$$\begin{Bmatrix} \text{Variable name}: \text{M} \\ \text{Sample time}: 0.01 \\ \text{Save Format}: \text{Array} \end{Bmatrix} \text{ and } \begin{Bmatrix} \text{Variable name}: \text{P} \\ \text{Sample time}: 0.01 \\ \text{Save Format}: \text{Array} \end{Bmatrix} \text{ in}$$

Figure 9.3(f) *Modified model for exporting the magnitude and phase data to MATLAB workspace*

the parameter window of the To Workspace1 and To Workspace blocks respectively on doubleclicking each. Since we set the solver start and stop time as 0 and 0.01 sec respectively (from chapter 2) and also we set the sample time in To Workspace blocks as 0.01, the values of M and P exported to the MATLAB workspace happens in terms of a

two-element matrix $\begin{Bmatrix} \text{one for 0} \\ \text{another for 0.01} \end{Bmatrix}$. Let us doubleclick the block Fourier, enter the harmonic as 0, run the model,

go to MATLAB Command prompt, and execute the following:

MATLAB Command

        >>M ↵   >>P ↵

    M =   P =

             0   0   ← for t =0 sec

       0.0500   90   ← for t =0.01 sec

As we explained, we have the output now available in MATLAB Workspace for n =0 but the matter is we need only the second element M(2) or P(2) for the magnitude or phase respectively. Now doubleclick the block Fourier and just type n in the slot of Harmonic in the parameter window for enabling the block to accept variable n. We saved our simulink model of the figure 9.3(f) by the name test and run the model from the command prompt as follows:

        >>A=[]; Ph=[]; ↵
        >>for n=0:7 sim('test'); A=[A M(2)]; Ph=[Ph P(2)]; end ↵
        >>n=0:7; ↵
        >>bar(n,A) ↵
        >>figure, bar(n,Ph) ↵

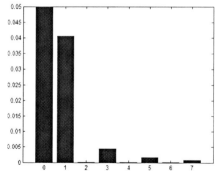

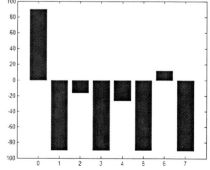

Figure 9.3(g) *Amplitude spectrum of the wave in the figure 2.7(b) of chapter 2*

Figure 9.3(h) *Phase spectrum of the wave in the figure 2.7(b) of chapter 2*

The first line is the initialization of the row matrix variable A and Ph respectively (A for amplitude and Ph for phase). A for loop must be ended by an end in MATLAB. The command A=[A M(2)]; places side by side the simulink exported M(2) for different n, so does Ph=[Ph P(2)];. The command for n=0:7 changes the integer n from 0 to 7 and complies with the n of the Fourier block. The command sim('test') executes simulink model test from the command window. Once the for loop is executed, the n values are not retained hence we generate n's again after the for loop. The command bar(n,A) graphs the bar plot of the values held in A against the one in n – for our problem that is the amplitude spectrum as presented in the figure 9.3(g). Similarly, bar(n,Ph) graphs the phase spectrum as shown in the figure 9.3(h). The command figure generates another figure window once the amplitude spectrum is plotted.

9.4 Fourier transform of a discrete function

Recall that the continuous forward Fourier transform of the section 9.2 contains the continuous angular frequency ω whose domain is from $-\infty$ to $+\infty$. In chapter 6, the formula of the discrete Fourier transform is defined in terms of the sample number n, which is discrete. But the Fourier transform of a discrete signal $x[n]$ is defined as

$X(\Omega) = \sum_{n=-\infty}^{n=+\infty} x[n]e^{-jn\Omega}$ in which Ω is continuous and $X(\Omega)$ is 2π periodic (on the contrary $F(\omega)$ is nonperiodic).

The Ω is called the digital frequency just to distinguish from the continuous ω.

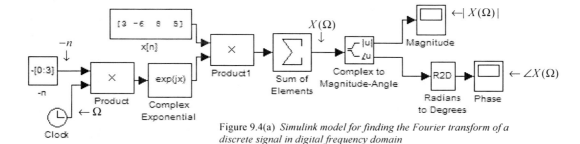

Figure 9.4(a) *Simulink model for finding the Fourier transform of a discrete signal in digital frequency domain*

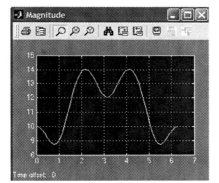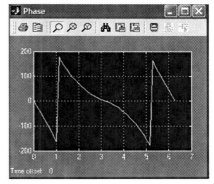

Figure 9.4(b) *Scope output from the model of the figure 9.4(a) for* $|X(\Omega)|$

Figure 9.4(c) *Scope output from the model of the figure 9.4(a) for* $\angle X(\Omega)$

Our objective here is to simulate the Fourier transform $X(\Omega)$ of a discrete signal (means computation of the magnitude and angle parts of the transform) in the digital frequency Ω domain ($0 \le \Omega \le 2\pi$). The computation is a complex one as far as the expression of $X(\Omega)$ is concern.

Let us assume that we have the discrete signal $x[n] =$[3 –6 8 5] that has the sample index domain $0 \le n \le 3$ and n is integer (we chose consistent sequence with the simulink's default setting $n \ge 0$). The Fourier transform of the sequence $x[n]$ is calculated as $X(\Omega) = \sum_{n=0}^{n=+3} x[n]e^{-jn\Omega} = 3 - 6 e^{-j\Omega} + 8 e^{-j2\Omega} + 5 e^{-j3\Omega}$. Figure 9.4(a) presents the simulink model to obtain the magnitude $|X(\Omega)|$ and phase $\angle X(\Omega)$ starting with the $x[n]$. Let us describe the model building process (all block links are given in the table at the end of the chapter):

⇒ *bring two Constant, two Product, one Complex Exponential, one Sum of Elements, two Scope, one Clock, one Radians to Degrees, and one Complex to Magnitude-Angle blocks in a new simulink model file*

⇒ *rename one Constant block as –n, doubleclick the block to enter the Constant value as –[0:3] (it generates 0 –1, –2, and –3 to be used for the –n in the expression $e^{-jn\Omega}$), connect the block and the Clock as the inputs to the block Product assuming that the Clock generates Ω on $0 \le \Omega \le 2\pi$*

193

$\Rightarrow$ *rename the other Constant block as x[n], doubleclick the block to enter the Constant value as [3 –6 8 5] just to enter the discrete signal x[n], and enlarge the block to display its contents*

$\Rightarrow$ *rename the Scope blocks as Magnitude and Phase respectively, place all blocks relatively, and connect them according to the figure 9.4(a)*

The Complex Exponential intakes $-n\Omega$ and returns $e^{-jn\Omega}$ for every Ω and the output of the block is $[1 \quad e^{-j\Omega} \quad e^{-j2\Omega} \quad e^{-j3\Omega}]$. The block Product1 performs element by element product of $x[n]$ and $e^{-jn\Omega}$ hence the output of the block is $[3 \quad -6\,e^{-j\Omega} \quad 8\,e^{-j2\Omega} \quad 5\,e^{-j3\Omega}]$. The Sum of Elements then adds all these to form the required $X(\Omega)$. Again these all happens just for a single Ω generated by the simulink solver. Now enter the solver stop time as 2*pi (for the insertion of $\Omega = 2\pi$) from the simulink model menu bar, doubleclick the block Sum of Elements, and change the output data type mode in the parameter window of the block to double keeping the others as default. Let us run the model and doubleclick the Magnitude and Phase blocks. Both Scope outputs with the autoscale setting are shown in the figures 9.4(b) and 9.4(c) respectively. In these figures the horizontal axes correspond to the digital frequency Ω ($0 \le \Omega \le 2\pi$). To view the phase angle of the transform in degrees (the vertical axis of the figure 9.4(c)), we employed the block Radians to Degrees.

We illustrated the idea of implementation by taking four-element sequence. In practice, we may have hundreds of data existing in MATLAB workspace. Let us say our discrete sequence is a ten element row matrix existing in MATLAB workspace by the variable name test. The modification we need is enter test and $-[0:9]$ in the blocks x[n] and $-n$ of the figure 9.4(a) respectively.

The important point is whatever be the number of elements in a sequence, we map that in the digital frequency Ω domain from 0 to 2π. Also there is no need to analyze the transform magnitude for the whole period 2π only up to π is enough because of the evenness of the function about π.

♦ ♦ Sampling and digital frequency of a sinusoidal signal

The model of the figure 9.4(a) can help us visualize the effect of the sampling frequency on the Fourier transform of a sine signal. Let us say we have a pure sinusoidal signal given by $y = A\sin 2\pi ft$. For sure the sine wave defined by the function is continuous because t can assume any fractional or integer value. Whenever we take the value or sample of the function at some time, we take the discrete values of the wave. The question is what time interval we choose for the wave for the discrete form. Let us say the sine wave has the frequency $f = 10 Hz$ indicating wave period $T = \dfrac{1}{f} = 0.1$ sec and the time interval chosen is 0.01 sec. It means we take the sample values of the sine wave at 0 sec, 0.01 sec, 0.02 sec, 0.03 sec, … and so on. This indicates that our sampling time period is $T_S = 0.01$ sec or the sampling frequency is $f_S = \dfrac{1}{T_S} = 100\,Hz$. One can say that choosing different step or interval indirectly gives the option for selecting different sampling

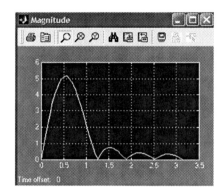

Figure 9.4(d) $|X(\Omega)|$ *output from the model of the figure 9.4(a) for the sampled sine wave*

$\Omega_1\ \Omega_2$

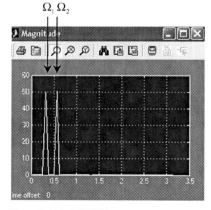

Figure 9.4(e) $|X(\Omega)|$ *output containing two frequency components*

$\Omega_1\ \Omega_2\ \Omega_3$

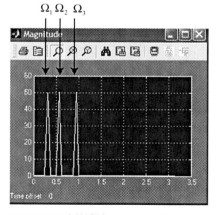

Figure 9.4(f) $|X(\Omega)|$ *output containing three frequency components*

frequency. The relationship among the digital frequency Ω, sine frequency f, and sampling frequency f_s is $\Omega = 2\pi \dfrac{f}{f_s}$. With the aforementioned frequencies our Ω should be $2\pi \dfrac{10}{100} = 0.6283$. We should obtain the Fourier magnitude spectrum $|X(\Omega)|$ as the maximum at $\Omega = 0.6283$ radians.

So to feed the wave in simulink, we code the sine wave as sin(2*pi*10*[0:.01:.1]) but this necessitates to use specific number of samples. The time 0 to 0.1 sec with 0.01 sec step (11 samples are there) needs to write 0, 1, 2, ... up to 10 or in simulink code 0:10. Referring to the model 9.4(a), we enter sin(2*pi*10*[0:.01:.1]) and –[0:10] in the parameter window of the blocks x[n] and –n on doubleclicking each respectively. Just to see the response on $0 \leq \Omega \leq \pi$, let us change the solver stop time to pi and run the model. Figure 9.4(d) shows the Magnitude output from the model with the autoscale setting. As we mentioned, the maximum of the $|X(\Omega)|$ occurs exactly at $\Omega = 0.6283$ radians. If you change the step size or in other words sampling frequency, the maximum will be shifted in accordance with $\Omega = 2\pi \dfrac{f}{f_s}$. If we make the sampling frequency more, the main lobe in the figure 9.4(d) will be narrower and shifted toward the $\Omega = 0$ and the side lobes will go down. The reader can verify that by taking $f_s = 1000\ Hz$ that is passing sin(2*pi*10*[0:.001:.1]) and –[0:100] in the parameter window of the blocks x[n] and –n respectively in the model of the figure 9.4(a).

❖ ❖ *Multiple frequency issue*

Assume that we have the equal amplitude two frequency wave $y = \sin 2\pi 50t + \sin 2\pi 90t$. When the t is in second, the wave has two frequency components $50\ Hz$ and $90\ Hz$. If we choose the sampling frequency as f_s $= 1000\ Hz$, the digital frequency becomes $\Omega_1 = 2\pi \dfrac{50}{1000} = 0.3142$ radians and $\Omega_2 = 2\pi \dfrac{90}{1000} = 0.5655$ radians for $50\ Hz$ and $90\ Hz$ respectively. What we want to focus is there should be two peaks located at Ω_1 and Ω_2 in the $|X(\Omega)|$ versus Ω plot. Let us verify that by the model in the figure 9.4(a). The coding of the double frequency function suitable for the block x[n] in the figure 9.4(a) is sin(2*pi*50*[0:0.001:0.1])+sin(2*pi*90*[0:0.001:0.1]). Also the sample index coding needs to be –[0:100] in the –n block in the same model. The Magnitude scope output with the autoscale setting is shown in the figure 9.4(e) in which you see exactly two peaks at the specified points of Ω_1 and Ω_2. The reader might ask how we should know that the frequencies corresponding to the peaks are exactly 0.3142 radians or 0.5655 radians. The answer is zoom the peak point area with your mouse in the Magnitude Scope. If you zoom once, you see the horizontal or vertical axis marks increased by one decimal. Thus four or five times zooming can help us read off the peak in the vertical axis or the Ω in the horizontal axis with three or four digit accuracy.

Figure 9.4(g) *Model for finding the Fourier transform of a triple frequency signal in digital frequency domain*

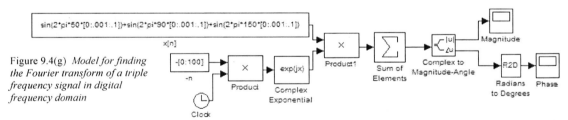

What if we have a three-frequency wave, for example, $y = \sin 2\pi 50t + \sin 2\pi 90t + \sin 2\pi 150t$. The frequencies related with the wave are $50\ Hz$, $90\ Hz$, and $150\ Hz$ and corresponding digital frequencies considering $1000\ Hz$ sampling frequency are $\Omega_1 = 2\pi \dfrac{50}{1000} = 0.3142$ radians, $\Omega_2 = 2\pi \dfrac{90}{1000} = 0.5655$ radians, and $\Omega_3 = 2\pi \dfrac{150}{1000} = 0.9425$ radians respectively. With that the coding of the x[n] block in the model in figure 9.4(a) should be sin(2*pi*50*[0:.001:.1])+sin(2*pi*90*[0:.001:.1])+sin(2*pi*150*[0:.001:.1]). Along with that the –n block in the model should contain –[0:100]. The modified model following the enlargement of the x[n] and –n blocks are presented in the figure 9.4(g) for the reader's convenience. On running the model, one would obtain the simulation result as presented in the figure 9.4(f) with the autoscale setting.

We took the samples up to $t = 0.1$ sec for this reason the argument in the sine functions at the end of the interval contains 0.1 in the preceding simulations.

✦ ✦ *Inference from the study*

This simulation study helps us arrive at some important conclusion of discrete signals. The model does not know any information about the input sequence $x[n]$. Any arbitrary discrete signal $x[n]$ can be passed to the model and its $|X(\Omega)|$ versus Ω plot can be observed. If we find any peak in the plot, one can easily detect the peak and corresponding Ω. Then the back calculation can help us determine the sinusoidal frequency components present in the discrete sequence employing the formula $\Omega = 2\pi \dfrac{f}{f_S}$. This kind of analysis is very useful for the noise removal from a signal.

9.5 Discrete Fourier transform (DFT) of a discrete function

In chapter 6 we introduced the formula for the discrete Fourier transform of a discrete signal. We mention here few more relevant issues on the subject. Recall that the discrete Fourier transform $X[k]$ of a sequence $x[n]$ is

defined as $X[k] = \sum\limits_{n=1}^{N} x[n] e^{-j2\pi(k-1)\frac{(n-1)}{N}}$,

where k can vary from 1 to N and N is the length of the sequence $x[n]$. Both the $x[n]$ and $X[k]$ are discrete in nature and having the same length and the transform is in general complex. The discrete transform possesses a periodicity of N and basically sample number based. The $X[k]$ also has the

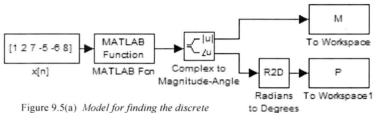

Figure 9.5(a) *Model for finding the discrete Fourier transform of a sequence*

magnitude and phase, both are discrete and of length N. Their plot can best be implemented in MATLAB in terms of the stem plot but for the simulation we can use simulink.

Let us consider the sequence $x[n] = [1 \quad 2 \quad 7 \quad -5 \quad -6 \quad 8]$. Our objective is to obtain the discrete Fourier transform magnitude $|X[k]|$ and phase $\angle X[k]$ and see their discrete plots.

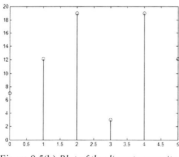

Figure 9.5(b) *Plot of the discrete magnitude spectrum* $|X[k]|$

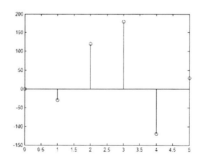

Figure 9.5(c) *Plot of the discrete phase spectrum* $\angle X[k]$ *in degrees*

Figure 9.5(a) depicts the model for the problem at the hand. Bring one Constant, one MATLAB Fcn, one Complex to Magnitude-Angle, one Radians to Degrees, and two To Workspace blocks in a new simulink model file. Doubleclick the Constant block, enter the sequence in the slot of constant value as [1 2 7 −5 −6 8], rename the block as x[n], enlarge the block to show its contents, doubleclick the block MATLAB Fcn, enter the function as fft (as we did in chapter 6), set the output signal type as complex, uncheck the button Collapse 2-D results to 1-D, doubleclick the block To Workspace, change the variable name to M, doubleclick the block To Workspace1, change the variable name to P (just to make consistent with the magnitude and phase), change also the save format to Array for each, and connect the blocks as shown in the figure 9.5(a). Change the solver stop time to 0 from the model menu bar because we do not want simulink time variable to be involved in the simulation on account of the constant generation of the sequence $x[n]$. Now go to the MATLAB command prompt and type M or P to see the discrete magnitude spectrum $|X[k]| = [7 \quad 12.1244 \quad 19 \quad 3 \quad 19 \quad 12.1244]$ and discrete phase spectrum $\angle X[k] = [0^{0} \quad -30^{0} \quad 120^{0} \quad 180^{0} \quad -120^{0} \quad 30^{0}]$. Of coarse, the number of elements in $|X[k]|$ or $\angle X[k]$ must be the same as that in $x[n]$ which is 6 for the given example. Now the reader can choose these six values as sample-based or frequency-based. Let us assume that they are sample based and k varies from 0 to 5. Let us perform the following in the command prompt:

MATLAB Command

```
>>k=0:5; ↵                          >>figure ↵
>>stem(k,M) ↵                       >>stem(k,P) ↵
```

Above commands result the discrete plots of $|X[k]|$ or $\angle X[k]$ as presented in the figures 9.5(b) and 9.5(c) respectively in which the horizontal axis refers to k. That is what we expected.

Let us consider the single frequency sine wave of previous section whose frequency, sampling frequency, and observation time are $f=10\,Hz$, $f_s=100\,Hz$, and $0\le t\le 0.2\sec$ respectively and the coding is

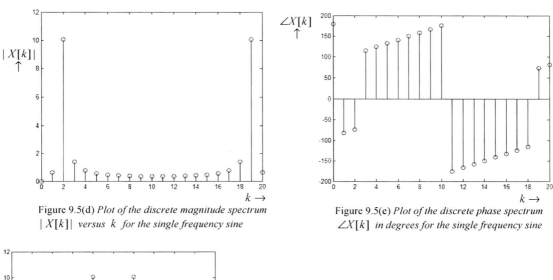

Figure 9.5(d) *Plot of the discrete magnitude spectrum* $|X[k]|$ *versus* k *for the single frequency sine*

Figure 9.5(e) *Plot of the discrete phase spectrum* $\angle X[k]$ *in degrees for the single frequency sine*

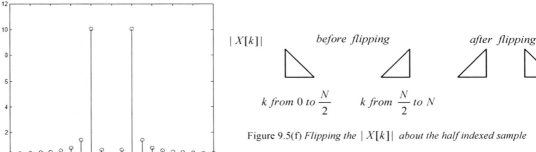

Figure 9.5(f) *Flipping the* $|X[k]|$ *about the half indexed sample*

Figure 9.5(h) *Flipped* $|X[k]|$ *about the half indexed sample*

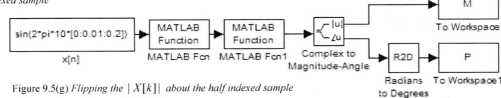

Figure 9.5(g) *Flipping the* $|X[k]|$ *about the half indexed sample*

sin(2*pi*10*[0:0.01:0.2]). Doubleclick the block x[n] of the model in figure 9.5(a), enter the last code in the parameter window, and run the model. Since the sampling frequency is $100\,Hz$, the step size in $0\le t\le 0.2\sec$ becomes $\frac{1}{100}=0.01$ sec and the number of samples is 21. For plotting purpose, one can use $k=0$ to $k=20$. Performing the following commands in the command window, one would obtain the discrete $|X[k]|$ and $\angle X[k]$ spectra as shown in the figures 9.5(d) and 9.5(e) respectively.

```
>>k=0:20; ↵                         >>figure ↵
>>stem(k,M) ↵                       >>stem(k,P) ↵
```

Observing the discrete spectra for the waves containing two or three frequency sines described in the last section is left as an exercise for the reader.

The magnitude spectrum $|X[k]|$ is even about the half-number indexed sample. Sometimes it might be desired to view the magnitude spectrum flipped about the half index sample depending on the significant frequency present in the sequence $x[n]$. The concern flipping is illustrated in the figure 9.5(f). The command fftshift can help us implement that. The modification we need is insert another MATLAB Fcn block between the MATLAB Fcn and Complex to Magnitude-Angle blocks in the model of the figure 9.5(a). We suggest you to copy the existing Fcn to the clipboard and then paste it in the model. Figure 9.5(g) presents the modifications associated with. Doubleclick the MATLAB Fcn1, enter the function fftshift, and run the model. By making the use of the stem plot in command prompt as we did before should return us the plot like the figure 9.5(h).

Knowing the sampling frequency, it is also convenient to display the plot in terms of the discrete frequency rather than the sample index. The relationship is given by $f = \frac{f_S}{N} k$. For the aforementioned example with $\begin{cases} f_S = 100 Hz \\ N = 20 \\ k = 0 \ to \ 20 \end{cases}$, the discrete frequency range is from $0\ Hz$ to

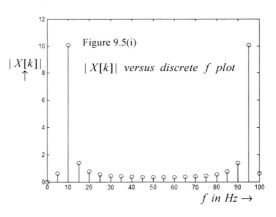

100 Hz with a step of 5 Hz. So when the command stem is used, we conduct the following commands:

```
>>k=0:20; ↵
>>stem(k*100/20,M) ↵
```

The result is the plot of the figure 9.5(i). Our selected sequence $x[n]$ was sampled from a sine wave of frequency 10 Hz and exactly at 10 Hz of the figure 9.5(i) we locate the discrete maximum. If we had had two or three sinusoids present in $x[n]$, we would have two or three discrete maxima like the one in the figure 9.5(i) located at those specified frequency. It is sufficient to analyze the sequence only up to the $\frac{N}{2}$ indexed sample because of the evenness.

❖ ❖ *Inference from the study*

Suppose we do not have any information about the sinusoidal frequency present in our given signal. What we do is we acquire the sample of the given signal at some specific frequency with known f_S, analyze the sampled signal employing the model of the figure 9.5(a) to obtain the plot like the figure 9.5(i), and infer from the figure what are the frequency components in the signal just looking at the discrete maximum.

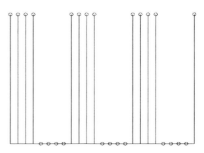

Figure 9.6(a) *A periodic sequence of periodicity 8*

9.6 Discrete Fourier series of a discrete function

A discrete signal or sequence can be periodic as well. Looking into the figure 9.6(a), the sequence is repeating at every after 8 samples hence the sequence is periodic of $N = 8$. The discrete Fourier series is the appropriate tool for the analysis of such signal. The expression in chapter 6 we mentioned for the discrete Fourier transform and the discrete Fourier series are basically identical. For readers convenience we recall them as follows:

Discrete Fourier transform:

$x[n]$ *forward transform* $X[k] = \sum_{n=1}^{N} x[n] e^{-j 2\pi(k-1)\frac{(n-1)}{N}}$

$X[k]$ *inverse transform* $x[n] = \frac{1}{N} \sum_{k=1}^{N} X[k] e^{+j 2\pi(k-1)\frac{(n-1)}{N}}$

Discrete Fourier series (DFS) of period N :

$DFS\ coefficients : c_k = \sum_{n=0}^{N-1} x[n] e^{-\frac{j 2\pi k n}{N}}$

$DFS\ expansion : x[n] = \frac{1}{N} \sum_{k=0}^{N-1} c_k\, e^{+j\frac{2\pi k n}{N}}$

The $X[k]$ in the transform and the c_k in the series are functionally same. In the summation sign of the transform, the lower limit is 1 because we choose the k or n to be from 1 to N. If we chose the variation from 0 to $N-1$, we would have the exponent expression as $e^{-\frac{j 2\pi k n}{N}}$ or $e^{+\frac{j 2\pi k n}{N}}$ proving indistinguishabilty of the two. However, simulink has a block by the name Discrete Fourier (link in the table 9.A) that implements the discrete Fourier series.

When employing the block for the discrete Fourier series analysis, specified data feeding must be taken place to make the block operational because the block is simulink Clock or independent variable dependent. The periodic sequence $x[n]$ can appear in simulink in two styles either expression oriented or data oriented. Let us say that our periodic sequence follows the sine wave variation $x[n] = \sin 2\pi f (nT_S)$ where $t = nT_S$, n is the sample index, T_S is the sampling period or the step size in continuous t. Choosing $f = 2\,Hz$ (fundamental frequency),

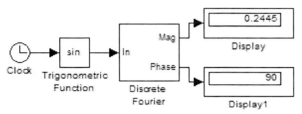

Figure 9.6(b) *Model for finding the discrete Fourier series coefficients for a periodic sequence that follows sine variation*

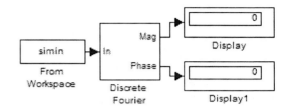

Figure 9.6(c) *Model for the DFS operated in external data*

$f_S = 8\,Hz$ (remembering $f_S \geq 2f$), and observation time $0 \leq t \leq 0.5\,sec$ (exactly one period of the sine wave and it has to be as far the periodicity is concern) tells us $T_S = \dfrac{1}{f_S} = 0.125$ sec and $n = 0, 1, 2, 3,$ and 4 (this is the DFS related n).

We can not apply here Constant block because the Discrete Fourier is designed to include the Clock in its operation. Anyhow figure 9.6(b) illustrates the DFS implementation. Following the construction of the model, doubleclick the Discrete Fourier, change the Fundamental frequency to 2, change the Sample time to 0.125, and set the Harmonic as 0 in the parameter window of the block. Also clicking the model menu bar, change the solver stop time to 0.5, solver option types to fixed step, change the step size to 0.125 instead of auto. Run the model and the Display blocks indicate the DFS related c_k magnitude and phase. The model is displaying the $|c_0| = 0.2445$ and $\angle c_0 = 90^0$ of the DFS expansion in Display and Display1 respectively. You can doubleclick the Discrete Fourier and set other harmonic (say $k = 3$, the k in DFS expansion is equivalent to the harmonic n in the parameter window of the Discrete Fourier). On running the model with the last k should provide us the $|c_3| = 0.1209$ and $\angle c_3 = -1.823^0$ of the DFS expansion as the Display outputs in the model.

In practical situations we may not have deterministic function like the sine instead our data might be taken on a particular sampling interval and picked up from one period of the sequence $x[n]$ in our observation time. Let

$$\begin{Bmatrix} t & x[n] \\ 0 & 0 \\ 0.125 & 1 \\ 0.25 & 0 \\ 0.375 & -1 \\ 0.5 & 0 \end{Bmatrix}$$

say our data taken in some observation appears to be in one period. Enter the data in MATLAB command window as follows and then employ the model of the figure 9.6(c) with appropriate block and solver settings.

```
>>t=[0:0.125:0.5]';       ← Entering the time information data as a column matrix to t
>>x=[0 1 0 -1 0]';        ← Entering the sequence data as a column matrix to x
>>simin=[t x];            ← Forming a matrix simin taking the 1st and 2nd columns as the t and x respectively
```
However, we bring an end to the section with the DFS implementation.

9.7 Block links used in this chapter

The subject matter in this chapter is to describe the procedural steps for Fourier analyses of the continuous and discrete functions both to the context of periodic and nonperiodic in simulink. Blocks placed in the table 9.A help us facilitate the analyses.

Table 9.A   Necessary blocks for modeling the Fourier analyses problems found in simulink library (not arranged in the alphabetical order)

Block name	Representative Symbol/Function	Icon Outlook	Block name	Representative Symbol/Function	Icon Outlook
Clock	It generates solver times or independent variable of the functions		Scope	It displays the function of the connected line	
Link: *Simulink → Sources → Clock*			Link: *Simulink → Sinks → Scope*		

Continuation of previous table:

Block name	Representative Symbol/Function	Icon Outlook	Block name	Representative Symbol/Function	Icon Outlook
Complex to Real-Imag	It separates the complex functional values into real and imaginary parts	Re{u} Im{u} Complex to Real-Imag	Complex to Magnitude-Angle	It separates the complex functional values into the magnitude and phase angle parts	|u| ∠u Complex to Magnitude-Angle
Link: *Simulink → Math Operations → Complex to Real-Imag*			Link: *Simulink → Math Operations → Complex to Magnitude-Angle*		
dB Conversion	It converts the value at its input port (let us say V) to $20 \log_{10} V$	dB (1 ohm) dB Conversion	MATLAB Fcn	It invokes any built-in MATLAB or executes the user defined MATLAB coded function	MATLAB Function MATLAB Fcn
Link: *DSP Blockset → Math Functions → Math Operations → dB Conversion*			Link: *Simulink → User-Defined Functions → MATLAB Fcn*		
Radians to Degrees	It converts angle in radians to angle in degrees	R2D Radians to Degrees	Pulse Generator	It generates pulses of swing from 0 to some amplitude of different frequencies and phases	Pulse Generator
Link: *Simulink Extras → Transformations → Radians to Degrees*			Link: *Simulink → Sources → Pulse Generator*		
Fourier	It computes the Fourier series magnitude and phase coefficients for one harmonic taking a periodic wave as input	magnitude signal angle Fourier	Display	It displays the instantaneous functional value of the concern functional line	0 Display
Link: : *SimPowerSystems → Extra Library → Measurements → Fourier*			Link: *Simulink → Sinks → Display*		
Sine Wave	It generates sine waves of various frequencies, amplitudes, and phases	Sine Wave	Product	It multiplies two functions on the common time or horizontal scale	× Product
Link: *Simulink → Sources → Sine Wave*			Link: *Simulink → Math Operations → Product*		
Constant	It generates constant value (s) or constant built-in functions	1 Constant	Integrator	It integrates the function entering to its input port	$\frac{1}{s}$ Integrator
Link: *Simulink → Sources → Constant*			Link: *Simulink → Continuous → Integrator*		
Sum	It sums two or more functional values or constants or combinations of them	+ ×	To Workspace	It exports simulink generated data to MATLAB Workspace by the name simout	simout To Workspace
Link: *Simulink → Math Operations → Sum*			Link: *Simulink → Sinks → To Workspace*		
Complex Exponential	It takes θ as the input and returns $\cos\theta + j \sin\theta$	exp(jx) Complex Exponential	Sum of Elements	It sums all constant generated values flowing in a functional line	Σ Sum of Elements
Link: *DSP Blockset → Math Functions → Math Operations → Complex Exponential*			Link: *Fixed-Point Blockset → Math → Sum of Elements*		
Trigonometric Functions	It returns trigonometric output taking input	sin Trigonometric Function	From Workspace	It imports external data into simulink by the name simin	simin From Workspace
Link: *Simulink → Math Functions → Trigonometric Function*			Link: *Simulink → Sources → From Workspace*		
Discrete Fourier	It computes discrete Fourier series for one harmonic at a time	In Mag Phase Discrete Fourier			
Link: *SimPowerSystems → Extra Library → Discrete Measurements → Discrete Fourier*					

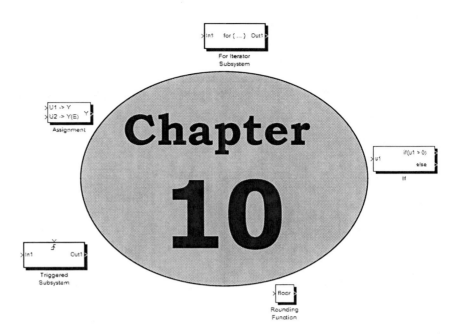

For Iterator
Subsystem

Assignment

If

Triggered
Subsystem

Rounding
Function

Miscellaneous Modeling and Some Programming Issues

As the name implies, there are two main subjects in this chapter. As miscellaneous modeling, we outline algebraic equation solving, single definite integration, dot and cross products, and rounding functions in the earlier part of the chapter. In the next addressing, our focus is on some programming issues. The programming in SIMULINK obligates to cover a lot of titles ranging from the base programming to the graphical user interface. Again application-based blocksets extended the scope of SIMULINK to a further extent. We confined our approach throughout the whole text as example-modeling-implementation. In this perspective, we do not wish to include all aspects of SIMULINK. Nevertheless we highlight subsystem, masking, for looping, and some other relevant programming issues in the latter part of the chapter.

10.1 Solving algebraic equations

We present how simulink can be used to solve various algebraic equations in this section. The approach of simulink is completely numerical and the whole equation is modeled through various blocks and functional lines related with the equation. In subsequent discussions, the block links related with the modeling can be found in the table 10.A.

♦ ♦ Equation of the type $f(x) = 0$

To solve the equation of the type $f(x) = 0$, we first need to rearrange the equation so that x is on the left side of the equation. As an elementary example, let us say we want to solve the equation $5x = 13$ whose solution is 2.6. Keeping x on the left side, one can write the equation as $x = 13 - 4x$. We maintain the equality of the equation in the model. We see that the gain of the x is -4 and $-4x$ is added with the constant 13. Let us construct the model of the figure 10.1(a) by bringing one Constant, one Gain, one Sum, and one Display blocks in a new simulink model file. Set the constant value and gain as 13 and -4 on doubleclicking the blocks Constant and Gain respectively. Run the model, and the output is shown in the Display block. The functional flow is also presented in the same figure for clarity. In the following we attached more examples so that the reader feels comfortable with the modeling.

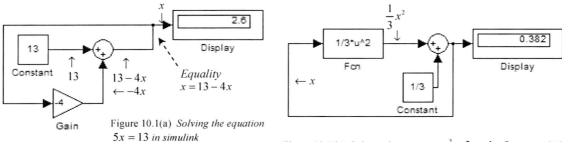

Figure 10.1(a) *Solving the equation* $5x = 13$ *in simulink*

Figure 10.1(b) *Solving the equation* $x^2 - 3x + 1 = 0$ *in simulink*

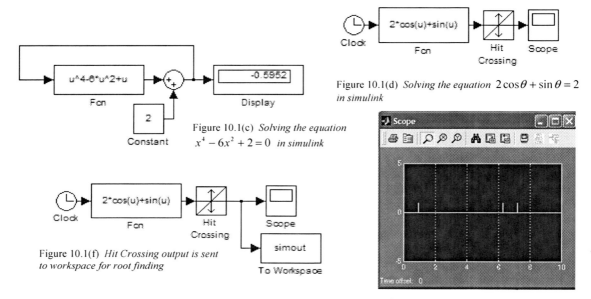

Figure 10.1(c) *Solving the equation* $x^4 - 6x^2 + 2 = 0$ *in simulink*

Figure 10.1(d) *Solving the equation* $2\cos\theta + \sin\theta = 2$ *in simulink*

Figure 10.1(f) *Hit Crossing output is sent to workspace for root finding*

Figure 10.1(e) *Scope output from the model of the figure 10.1(d)*

🗗 Example 1

Solve the quadratic equation $x^2 - 3x + 1 = 0$ in simulink whose solution is given by $\begin{Bmatrix} x = 2.6180 \\ x = 0.3820 \end{Bmatrix}$.

Solution:

Rearranging the equation, we have $x = \frac{1}{3}x^2 + \frac{1}{3}$. Even though there should be two roots but we get only the first one due to the numerical approach of simulink. The model is presented in the figure 10.1(b) in which you find the solution in the Display block. The Fcn in the model contains the code of the variable part $\frac{1}{3}x^2$ in the rearranged form considering the independent variable as u.

🗗 Example 2

Let us find the solution of the fourth degree equation $x^4 - 6x^2 + 2 = 0$ in simulink.

Solution:

There is no term related to the x in the given equation. Under this kind of circumstance, we apply some mathematical artifice as $x = x + x^4 - 6x^2 + 2$. Now the right hand side of the equation is separated as the variable part $x + x^4 - 6x^2$ and the constant part 2. Figure 10.1(c) shows the model and it output which is $x = -0.5952$.

🖰 **Example 3**

Find the solution of the trigonometric equation $2\cos\theta + \sin\theta = 2$.

Solution:

In this example we introduce another approach. In the left side of the equation there is no constant term. We provide θ and compute the left hand side. As soon as the left hand side is 2, we have the θ solution. We code the left hand expression as 2*cos(u)+sin(u) and pass it to the Fcn block. A block called Hit Crossing can detect the crossing of the function when it is 2 (either from more than 2 or from less than 2). Figure 10.1(d) shows the complete modeling, where the Clock generates θ, the block Fcn contains the code of the expression $2\cos\theta + \sin\theta$ (doubleclick the block and enter the code in the parameter window), and Hit Crossing should have the offset 2 (doubleclick the block and enter the Hit Crossing Offset as 2 in the parameter window). The reader should see the Scope output as shown in the figure 10.1(e) on running the model. The spikes in the figure correspond to the roots of the equation $2\cos\theta + \sin\theta = 2$. It seems that multiple roots are there in the equation within the default setting from 0 to 10 of the horizontal axis, which is θ. The next legitimate query is how we read off the value of the root. The answer is zoom the area around each spike towards the horizontal axis multiple times (at least four or five times) and read off the value from the horizontal axis. We advise you to zoom the spike zone in maximized window position otherwise the adjacent values will overlap. Once you are done with finding one root, click the autoscale icon of the Scope and zoom the zone for the next root. We found the θ roots to be 0, 0.9273, 6.2832, and 7.2105 radians for $0 \leq \theta \leq 2\pi$. The root at $\theta = 0$ is difficult to find by zooming.

As another convenience, to acquire the roots without zooming needs little programming. But first we modify the model of the figure 10.1(d) by connecting one To Workspace block as shown in the figure 10.1(f). Doubleclick the block to change its Save format to Array. Run the model, go to MATLAB command prompt, and perform the following:

MATLAB Command

```
>>p=find(simout==1); ↵
>>tout(p) ↵

ans =
          0
     0.9273
     6.2832
     7.2105
```

Having run the model, the workspace acquires two variables tout and simout. The simout and tout possess the dependent and independent variable values respectively. The Hit Crossing output is either 0 or 1. The 1 corresponds to a root. The first line of the command finds at what index the 1 is appearing in the array simout, those indexes are stored to p, and then selecting tout values at those indexes gives us the expected roots.

🖰 **Example 4**

Solve the equation $x\tan x + 4 = 0$ for $0 \leq x \leq 9$.

Solution:

The model we employed in the example 3 can in fact applies to any equation of the form $f(x) = 0$. All we need is pass the code of the equation as u*tan(u)+4 to the Fcn block of the figure 10.1(f) on doubleclicking, set the Hit Crossing offset as 0 on doubleclicking, change the solver stop timer to 9 because now the interval is different, run the model, and execute the following in the command window:

```
>>p=find(simout==1); ↵
>>tout(p) ↵

ans =
     1.5708
     2.0430
     4.7124      } roots of the equation x tan x + 4 = 0
     5.6687
     7.8540
```

♦ ♦ **Equation of the type** $f(x, y) = 0$

The equation of the type $f(x, y) = 0$ has two independent variables, x and y. The solution procedure is very similar to the single variable. There must be two equations and each independent variable is expressed by the other or the function of both. Let us see the following examples on the two variable case.

🖰 **Example 1**

Solve the equations $2x + 3y = -9$ and $5x - y = 8$ in simulink whose solution is given by $\begin{cases} x = 0.8824 \\ y = -3.5882 \end{cases}$.

Solution:

First we need to rearrange the equations for x and y on the left side which are $x = -\dfrac{3}{2}y - \dfrac{9}{2}$ and $y = 5x - 8$. As if we are solving two equations, they are just coupled with the output of the other. Figure 10.2(a) shows the modeling in which the Constant, upper Sum, and Gain1 form the first equation. The Constant1, lower Sum, and Gain form the second equation. The Display and Display1 are connected to the functional lines of x and y respectively. Some functional flows are also presented in the model.

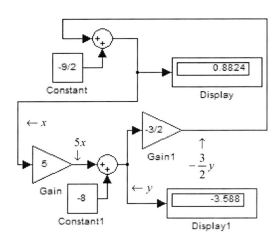

Figure 10.2(a) *Simulink model for solving the equations* $2x + 3y = -9$ *and* $5x - y = 8$

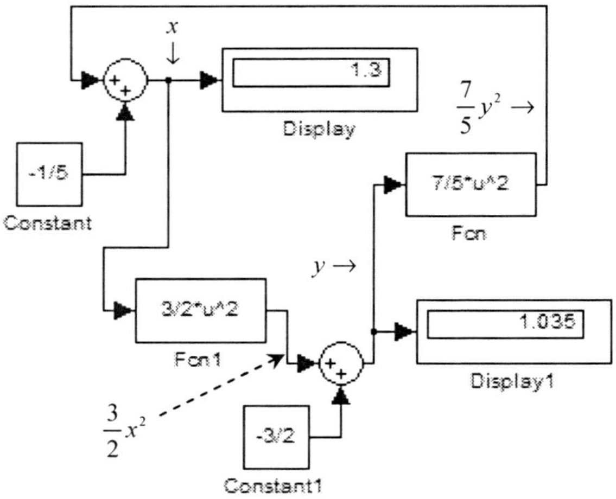

Figure 10.2(b) *Simulink model for solving the equations* $7y^2 - 5x = 1$ *and* $3x^2 - 2y = 3$

⊟ **Example 2**

Solve the equations $7y^2 - 5x = 1$ and $3x^2 - 2y = 3$ in simulink whose solution is given by $\begin{Bmatrix} x = 1.3 \\ y = 1.0351 \end{Bmatrix}$.

Solution:

The rearranged equations are $x = -\dfrac{1}{5} + \dfrac{7}{5}y^2$ and $y = \dfrac{3}{2}x^2 - \dfrac{3}{2}$. Figure 10.2(b) is presented for the modeling in which $\begin{Bmatrix} \text{Display} \\ \text{Display1} \end{Bmatrix}$ corresponds to the solution for $\begin{Bmatrix} x \\ y \end{Bmatrix}$. The blocks Fcn and Fcn1 contain the codes for $\dfrac{7}{5}y^2$ and $\dfrac{3}{2}x^2$ as 7/5*u^2 and 3/2*u^2 respectively.

After running the model in previous examples does the reader find some warning message in the MATLAB command window. The default setting of the solver gives warning if any algebraic equation is formed in the simulink model. To prevent from seeing the warning message, click *'Simulation → Simulation parameters → Diagnostics'* from the model file menu bar. You find solver performance in the diagnostics window. Select the algebraic loop action as none in that window. However, the reader can extend the procedure we adopted in the preceding examples to any other algebraic equations.

10.2 Single definite integration of function

Simulink can perform the single definite integration numerically with the aid of the operator Integrator (link: *'Simulink → Continuous → Integrator'*). Let us start with the simplest example of $\int_{x=0}^{x=6} x\,dx = 18$. The integrand x can be simulated via the block Clock.

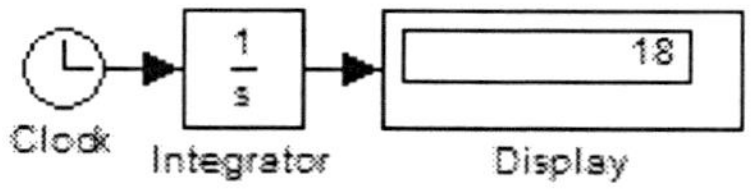

Let us construct the model of the figure 10.3(a). The lower and upper limits of the integration are entered through the solver start

Figure 10.3(a) *Implementation of the single integration* $\int_{x=0}^{x=6} x\,dx = 18$

and stop time respectively. Hence click the *'Simulation → Simulation parameters → Solver'* from the model menu bar and enter the stop time as 6 for the upper limit. On running the model you should see the result as shown in the Display block.

Compute the following single definite integrations (attached in the examples are the results of the symbolic computation, we omitted the computations for space reason) in simulink.

A. $\int_{\theta=-\frac{\pi}{4}}^{\theta=\frac{\pi}{3}} (\tan\theta + \cos\theta)^2 \, d\theta = \frac{9\sqrt{3}}{8} - \frac{7\pi}{24} + \frac{1}{4} + \sqrt{2} = 2.6965$

B. $\int_{x=3}^{x=9} \frac{1}{x(x^2+1)^2} \, dx = \ln 3 - \frac{1}{2}\ln\frac{41}{5} - \frac{9}{205} = 0.0026$

C. $\int_{x=1}^{x=2} (\sinh x + \ln x)dx = \cosh 2 - \cosh 1 + \ln 4 - 1 = 2.6054$

⊟ Example A

Figure 10.3(b) presents the modeling. The integrand $(\tan\theta + \cos\theta)^2$ is the function of the independent variable θ. For this reason we bring the Fcn block which contains the code $(\tan(u)+\cos(u))^\wedge 2$ of the integrand assuming the independent variable u. One can say that now the Clock generates the independent variable θ. The lower and upper limits should be set as –pi/4 and pi/3 as the solver start and stop times respectively.

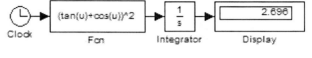

Figure 10.3(b) *Model for the integration A*

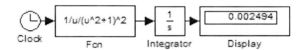

Figure 10.3(c) *Model for the integration B*

⊟ Example B

The code for the function $\frac{1}{x(x^2+1)^2}$ is $1/u/(u^\wedge 2+1)^\wedge 2$ considering the independent variable as u. We passed this code to the Fcn block of the figure 10.3(c) for which the solver start and stop times should be 3 and 9 respectively.

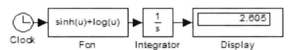

Figure 10.3(d) *Model for the integration C*

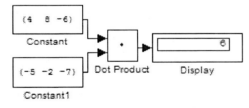

Figure 10.3(e) *Model for the dot product of two vectors*

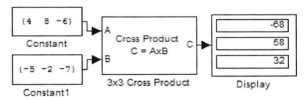

Figure 10.3(f) *Model for the cross product of two vectors*

⊟ Example C

For the integration, the necessary integrand code is sinh(u)+log(u) along with the solver start and stop times as 1 and 2 respectively. The Display block of the model in figure 10.3(d) shows the result.

The reader is referred to chapter 2 for the functional code writing.

10.3 Dot and cross products of vectors

Suppose we have two vectors $\bar{A} = 4i + 8j - 6k$ and $\bar{B} = -5i - 2j - 7k$ and their dot and cross products are given by $\bar{A} \bullet \bar{B} = 6$ and $\bar{A} \times \bar{B} = -68i + 58j + 32k$ respectively. We wish to have them implemented in simulink. Figures 10.3(e) and 10.3(f) present their implementations. The blocks Constant and Constant1 contain the vector coefficients as a three-element row matrix for $\bar{A}$ and $\bar{B}$ respectively. The blocks Dot Product and 3×3 Cross Product can be reached via *'Simulink → Math Operations → Dot Product'* and *'Aerospace Blockset → Transformations → Axes → 3×3 Cross Product'* respectively. The Constant and Display blocks need to be enlarged.

10.4 Rounding functions

The numbers flowing in a functional line may need to be rounded or truncated. Let us bring one Rounding Function (link: *'Simulink → Math Operations → Rounding Function'*) block in a new simulink model file. On

205

doubleclicking, the reader finds the functions $\begin{Bmatrix} \text{Floor} \\ \text{Ceil} \\ \text{Round} \\ \text{Fix} \end{Bmatrix}$ in the parameter window of the block. The function Fix

discards the fractional part of a decimal number regardless of the magnitude, for example, it turns 12.001 or 1.9999 to 12 or 1 respectively. If the fractional part in the decimal number is more than 0.5 and less than 0.5, that is taken as 1 and 0 respectively by the function Round, for example, it turns 12.4999 or –71.50001 to 12 or –72 respectively. The function Ceil rounds a decimal number towards the positive infinity whereas Floor does toward the negative infinity. Given figures 10.4(a), 10.4(b), 10.4(c), and 10.4(d) present the implementation of the four functions on the numbers –12.001, –1.9999, 0.49999, and 0.5001 placed in a row matrix inside the Constant block. Both the Constant and Display blocks need to be enlarged horizontally. Also you need to uncheck the button interpret vector parameters as 1-D in the parameter window of Constant to show the output as a row matrix.

Figure 10.4(a) *Fix applied on the numbers -12.001, -1.9999, 0.49999, and 0.5001*

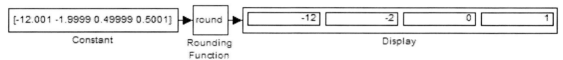

Figure 10.4(b) *Round applied on the numbers -12.001, -1.9999, 0.49999, and 0.5001*

Figure 10.4(c) *Floor applied on the numbers -12.001, -1.9999, 0.49999, and 0.5001*

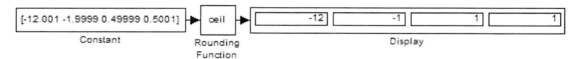

Figure 10.4(d) *Ceil applied on the numbers -12.001, -1.9999, 0.49999, and 0.5001*

10.5 Subsystem formation

When we have dozens of blocks present in a single simulink model file, the model becomes clumsy and it is not easy to check the blocks and functional line connections in a complicated model. In that case we can model our problems module by module considering several blocks as one module. We encountered the problem once in chapter 5. Now we address the subsystem formation separately and let us see the following examples.

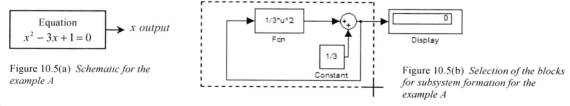

Figure 10.5(a) *Schematic for the example A*

Figure 10.5(b) *Selection of the blocks for subsystem formation for the example A*

⌸ Example A

Let us say we want to have only the solution output for x from the equation $x^2 - 3x + 1 = 0$. It means we should have a block like the figure 10.5(a) that provides only one scalar output. We modeled the equation in the figure 10.1(b). Referring to the model, the blocks Fcn, Sum, and Constant are used for the modeling and the solution is being fed to the Display block. Now bring the mouse pointer to the upper left corner of the model and move on to the lower right corner in the model keeping the left button of the mouse pressed so that all blocks in the model except

Display are selected as shown in the figure 10.5(b). Release the mouse button and you see the selection of the required blocks. From the model menu bar, click Edit, find the Create Subsystem in the pull down menu, and click

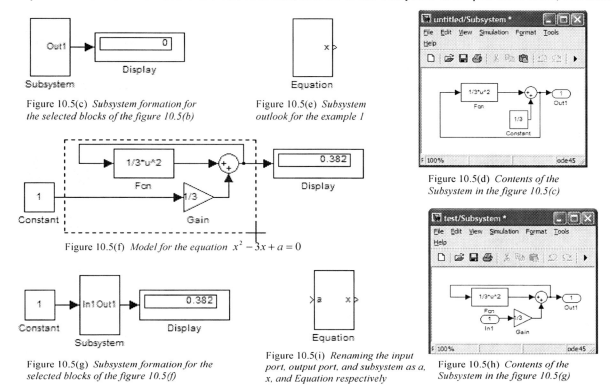

Figure 10.5(c) *Subsystem formation for the selected blocks of the figure 10.5(b)*

Figure 10.5(e) *Subsystem outlook for the example 1*

Figure 10.5(d) *Contents of the Subsystem in the figure 10.5(c)*

Figure 10.5(f) *Model for the equation* $x^2 - 3x + a = 0$

Figure 10.5(g) *Subsystem formation for the selected blocks of the figure 10.5(f)*

Figure 10.5(i) *Renaming the input port, output port, and subsystem as a, x, and Equation respectively*

Figure 10.5(h) *Contents of the Subsystem in the figure 10.5(g)*

that. You see the subsystem formed like the figure 10.5(c). You may need to move the Subsystem block to see the seemly placement of the blocks. On doubleclicking the Subsystem, you find the model like the figure 10.5(d). You can run the model in the figure 10.5(c) and the Display block shows the same output what we found from the model of the figure 10.1(b). The Display and the connecting functional line can be deleted, the Subsystem can be renamed as Equation, and the block Out1 of the figure 10.5(d) can be renamed as x so that we have the block outlook as shown in the figure 10.5(e) that is what our objective is. The block has only one output port.

⏎ Example B

Let us say that the quadratic equation in the example A is now modified as $x^2 - 3x + a = 0$. Our objective is to form a subsystem that takes a as the input and returns the solution of the equation.

The rearranged form of the equation for simulink is $x = \dfrac{x^2}{3} + \dfrac{a}{3}$ but a is user defined. We wish to feed the a to the model outside the Subsystem block. The model in the figure 10.1(b) is modified as shown in the figure 10.5(f). Construct the model, select the targeted area of the figure 10.5(f), and click Create subsystem under the pull down menu of the Edit from the model menu bar. With displacement of the blocks, you find the model of the figure 10.5(g). The targeted area indicates that there should be one input and one output port. Doubleclick the block Subsystem in the figure 10.5(g), you see the hidden model under the Subsystem as shown in the figure 10.5(h). The blocks In1 and Out1 are there in the model that correspond to the input port and the output port respectively. We deleted the functional lines and the Constant and Display blocks of the model in the figure 10.5(g). We renamed the In1 and Out1 blocks of the model in the figure 10.5(h) as a and x respectively and the Subsystem in the figure 10.5(g) as Equation thereby giving the block outlook consistent with the problem as shown in the figure 10.5(i). The block Equation has one input port to which we can input any constant a and one output port from which we can see the solution of the equation by connecting a Display – that is what we expected.

⏎ Example C

Let us consider the model in the figure 10.2(a) which provides the solution of x and y from the equations $2x + 3y = -9$ and $5x - y = 8$. Let us say all we want is the solution from the block like the figure 10.6(a).

207

When we select the targeted blocks for a subsystem with the mouse pointer, the placement of the blocks should be such that the input and output ports are explicit in the way we want. Referring to the figure 10.2(a), our required *x* and *y* outputs are exhibited by the Display and Display1 respectively. That is why we shifted the two

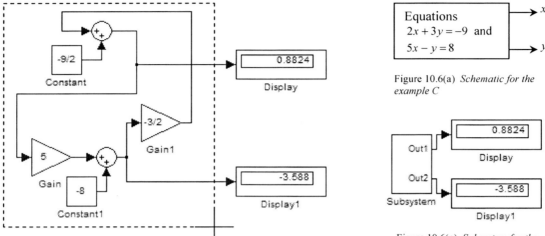

Figure 10.6(a) *Schematic for the example C*

Figure 10.6(b) *Subsystem selection for the example C*

Figure 10.6(c) *Subsystem for the example C*

Display blocks to the right and moved the vertical functional line connecting the Gain1 and the upper Sum block to the left. Now the targeted area makes sense with our requirement as shown in the figure 10.6(b). Also it is worthy to mention that the overlapping lines do not indicate the connection. A bold dot or node only establishes a connection. After selecting the blocks like the figure 10.6(b), click the Cerate subsystem via the Edit menu. Another option is rightclick the mouse following the selection and click the Create subsystem in the prompt window. But make sure your mouse pointer is in the target area. We see the Subsystem block as shown in the figure 10.6(c) due to previous action with the displacement of the blocks. Referring to the figure 10.6(c), we delete the functional lines and the Display blocks, doubleclick the

Figure 10.6(d) *Outlook of the Subsystem following deletion and renaming*

Subsystem, and rename the blocks Out1 and Out2 as x and y in the Sub model window. Hence our final block appears as in the figure 10.6(d) which is what we want.

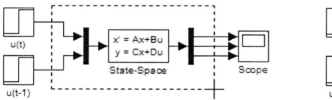

Figure 10.6(e) *Subsystem selection for the multiple inputs and multiple outputs*

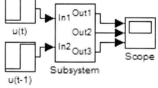

Figure 10.6(f) *Subsystem with multiple inputs and multiple outputs*

⧉ Example D

Referring to the model of the figure 5.7(b), we have two inputs and three outputs. Selecting the blocks with the target area as shown in the figure 10.6(e) provides the subsystem in the figure 10.6(f). You can rename the Subsystem or its input-output ports with your convenient name as we did before.

10.6 Execution order in a model

Once you construct a model, it might be necessary to view which functional lines are being executed first. Considering the model of the figure 10.2(a), click the *Format → Execution order* from the model menu bar. Simulink responds with the model of the figure 10.7(a). Looking into the numbers, one can easily identify that the execution order of the blocks is Constant1 → Constant → Fcn1 .. and so on. If you perform the same action, the execution order numbers will be removed.

10.7 Enabled subsystem

Let us construct the model of the figure 10.7(b) by bringing one Constant, one Sine Wave, one Enabled Subsystem, and one Scope blocks (block links are in the table 10.A) in a new simulink model file. Doubleclick the Constant block, enter the constant value as –1 in the parameter window of the block, run the model, and doubleclick the Scope. You should see the Scope output as 0. Again doubleclick the Constant block, enter the constant value as 0, run the model, and doubleclick the Scope. Still you find the Scope output as 0. Now let us enter 0.00001 in the Constant block and run the model. You should see the sine wave now in the Scope. Referring to the figure 10.7(b), the port to which the Constant block is connected is called the Enable input port. If the value entering to the Enable input port is more than zero or positive, the function or wave available to the input port In1 is passed to the output port Out1.

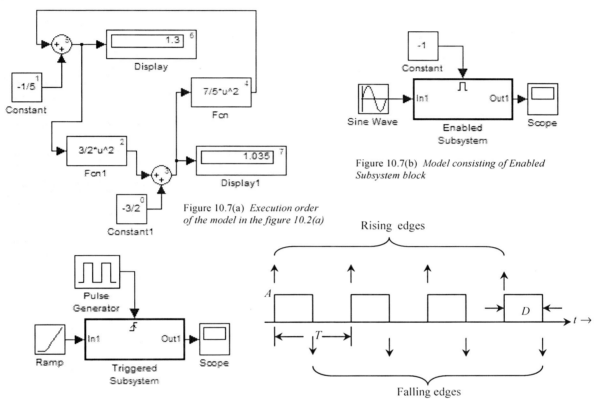

Figure 10.7(a) *Execution order of the model in the figure 10.2(a)*

Figure 10.7(b) *Model consisting of Enabled Subsystem block*

Figure 10.7(d) *A Triggered Subsystem block operates on a ramp function*

Figure 10.7(c) *A rectangular pulse indicating its rising and falling edges*

10.8 Triggered subsystem

Let us concentrate on the rectangular pulse of the figure 10.7(c) in which the wave takes a transition from 0 to A and from A to 0. The former and latter are termed as the rising and falling edges respectively. The up and down arrows of the figure correspond to the

Rising edge   Falling edge   Both edges

Figure 10.7(e) *Various edge indications in a triggered subsystem*

rising and falling edges respectively. These rising and falling edges can act as a switch. Let us construct the model of the figure 10.7(d) (block links are provided at the end of the chapter). The ramp or straight-line function at the input port In1 of the Triggered Subsystem block becomes available at the output port Out1 only when the rising and/or falling edge happen(s). The block Pulse Generator simulates the figure 10.7(c). Both the Ramp and the Pulse Generator share the common time axis. It is also assumed that between the edges the functional values are kept constant. On running the model of the figure 10.7(d), one would get the stair case wave in the Scope. If you doubleclick the block Trigger Subsystem, you find Trigger block in the prompt window. Again doubleclick the block Trigger and you find the option for the edge selection in a popup menu whether it is rising, falling, or either in the parameter window. Each time you change the selection, the icon outlook of the block becomes consistent making sense with the modeling. Figure 10.7(e) shows the possible edge indication in a triggered subsystem. For instance, the edge associated with the figure 10.7(d) is the rising one.

10.9 Assigning values or functions to a functional line

Let us say we have the row matrix $R = [7 \quad -5 \quad 4 \quad 2 \quad -7]$. We can assign any elements to R with the help of the block Assignment (link: *'Simulink → Math Operations → Assignment'*). Bring two Constant, one Assignment, and one Display blocks in a new simulink model file, rename one Constant block as R, enter the code of the row matrix as $[7 \; -5 \; 4 \; 2 \; -7]$ in R, and connect the blocks as shown in the figure 10.7(f). On doubleclicking the block Assignment, we find the parameters option as $\left\{\begin{array}{l}\text{Input type}\\\text{Source of element indices}\\\text{Elements}\end{array}\right\}$. If we have row or column

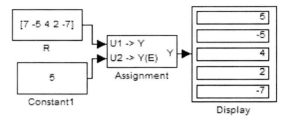

Figure 10.7(f) *Assigning the elements of the functional line U2 to the functional line U1 considering U1 as a vector*

matrix, we choose the Input type as vector and for the matrix input we choose Matrix as the Input type. The Source of element indices has two options internal and external. Let us say we want to change the first element of R by 5 hence the index is 1. If we type it in the parameter window, that is internal but if we seek for the index or element through another input port, that is external. Let us see some assignment on the model. With the Input type:Vector and Source of element indices:Internal and

for Elements:1, Display shows $[5 \quad -5 \quad 4 \quad 2 \quad -7]$ i.e. replacement of the first element of R by 5

for Elements:−1, Display shows $[5 \quad 5 \quad 5 \quad 5 \quad 5]$ i.e. replacement of all elements of R by 5

for Elements:[2 3], Display shows $[7 \quad 5 \quad 5 \quad 2 \quad -7]$ i.e. replacement of the 2nd and 3rd elements of R both by 5

for Elements:[2 3], Display shows $[7 \quad 6 \quad 0 \quad 2 \quad -7]$ i.e. replacement of the 2nd and 3rd elements of R by 6 and 0 respectively but the block Constant1 in the model of the figure 10.7(f) should hold [6 0]

Let us say that now R represents a matrix $\begin{bmatrix} 6 & 5 & 8 \\ 4 & 9 & 0 \end{bmatrix}$. Doubleclick the block R of the model in the figure 10.7(f), enter the code of the matrix as [6 5 8;4 9 0], doubleclick the block Assignment, and select the Input type as matrix. In doing so one obtains the model of the figure 10.7(g). Now the Assignment block contains the

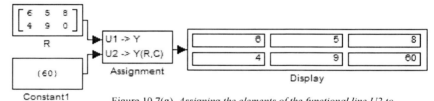

Figure 10.7(g) *Assigning the elements of the functional line U2 to the functional line U1 considering U1 as a matrix*

parameters $\left\{\begin{array}{l}\text{Input type}\\\text{Source of row indices}\\\text{Rows}\\\text{Source of column indices}\\\text{Columns}\end{array}\right\}$. In the last matrix the element (2,3) is 0 and we wish to replace the element

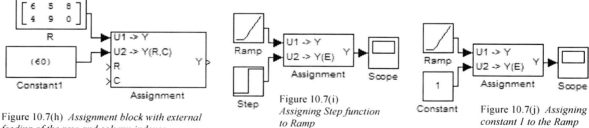

Figure 10.7(h) *Assignment block with external feeding of the row and column indexes*

Figure 10.7(i) *Assigning Step function to Ramp*

Figure 10.7(j) *Assigning constant 1 to the Ramp*

by 60. So we enter $\left\{\begin{array}{l}\text{Rows}:2\\\text{Columns}:3\end{array}\right\}$ in the parameter window of the Assignment and the Display output of the figure 10.7(g) is the result. Let us see more examples on this:

When we select $\left\{\begin{array}{l}\text{Rows}:2\\\text{Columns}:-1\end{array}\right\}$ and Constant1 contents as 60, Display shows $\begin{bmatrix} 6 & 5 & 8 \\ 60 & 60 & 60 \end{bmatrix}$ i.e. all elements of the second row of R is replaced by 60

When we select $\left\{ \begin{array}{l} \text{Rows}: -1 \\ \text{Columns}: 2 \end{array} \right\}$ and Constant1 contents as 60, Display shows $\begin{bmatrix} 6 & 60 & 8 \\ 4 & 60 & 0 \end{bmatrix}$ i.e. all elements

of the second column of R is replaced by 60.

In the parameter window of the Assignment if we select External for the Source of row indices and Source of column indices, the block outlook of the Assignment appears as shown in the figure 10.7(h) indicating R for row index and C for column index. We can feed -1 and 2 via two Constant blocks to R and C input ports of the Assignment respectively for the last example. This sort of external feeding is also there for the vector input presented in the figure 10.7(f).

Figure 10.8(a) *Icon outlook of the Selector*

So far we mentioned only the vector or matrix assignment. Referring to the models of the figures 10.7(i) and 10.7(j) in which we assigned a Step function to a Ramp and constant 1 to the Ramp function respectively.

10.10 Selector, Terminator, Ground, and Width blocks

The block Selector (link: *'Simulink → Signal Routing → Selector'*) selects or picks up elements from the row, column, or rectangular matrix in accordance with the user requirement. The parameter window of the block intakes the parameters in a similar fashion as that of the Assignment one does. The reader is referred to previous section for the window parameters' description. Figure 10.8(a) shows the default icon outlook of the Selector block. You find three and two square dots in the input and output port sides of the block respectively. That means the input must be a three-element vector and we are selecting the first and the third elements from the vector. In the figure 10.8(b) the input must be a 5-element vector and we are selecting the second and the fourth elements from the vector. For more than 5 elements this sort of indication is not exhibited by the block instead it exhibits the view as shown in the figure 10.8(c). In the parameter window of the block, you find the Input port width that is actually the number of elements in the vector.

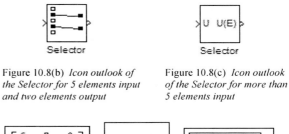

Figure 10.8(b) *Icon outlook of the Selector for 5 elements input and two elements output*

Figure 10.8(c) *Icon outlook of the Selector for more than 5 elements input*

Figure 10.8(d) *Element selection by the Selector from a rectangular matrix*

Having selected the Input type as matrix in the parameter window of the block, we provide a matrix to the input port of the Selector and specific user required row and column numbers in the parameter window (as we did for the Assignment block) to pick up the elements from the matrix. For instance, we picked up the element 65 from the matrix $\begin{bmatrix} 6 & 7 & 0 \\ 8 & 9 & 65 \end{bmatrix}$ whose row and column indexes are (2,3). Model in the figure 10.8(d) depicts the selection via the Selector block for which reasons the settings in the parameter we entered are $\left\{ \begin{array}{l} \text{Input type}: \text{Matrix} \\ \text{Source of row indices}: \text{Internal} \\ \text{Rows}: 2 \\ \text{Source of column indices}: \text{Internal} \\ \text{Columns}: 3 \end{array} \right\}$. During the

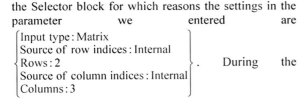

Figure 10.8(e) *The Constant block is connected with a Terminator*

Figure 10.8(f) *The Scope block is connected with a Ground*

modeling you need to enlarge the Selector block.

Let us bring one Constant block in a new simulink model file, run the model, and go to MATLAB command window. Since the output port of the block Constant is not connected to any functional line, simulink leaves some warning in the Command window. What if we connect the Constant block with a Terminator one (link: *'Simulink → Sinks → Terminator'*) as shown in the figure 10.8(e) and run the

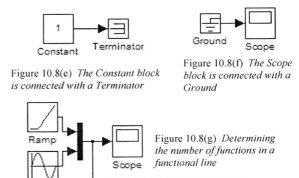

Figure 10.8(g) *Determining the number of functions in a functional line*

model, there should not be any warning message in the command window. Any unconnected output port can be connected to the Terminator to prevent from seeing the warning message.

Again let us bring one Scope block in a new simulink model and run the model. You find warning in the command window but if you connect the Ground (link: *'Simulink → Sources → Ground'*) block with the Scope as shown in the model of the figure 10.8(f) and run the model, you do not see warning in the command window. So any unconnected input port can be connected to the Ground to prevent from displaying the warning message.

By now we know that the Ramp and Sine wave blocks generate the functions *t* versus *t* and sin*t* versus *t* respectively. Let us say we multiplex the two functions according to the model of the figure 10.8(g). When we run the model and doubleclick the Scope block, we see the two functions on common *t* . But looking into the functional line connecting the Mux and the Scope, there is no way of knowing that the line contains two dissimilar functions. The block Width (link: *'Simulink → Signal Attributes → Width'*) as connected in the model can show how many functions are flowing in a functional line. In this example we have two functions that is why the Display output is 2.

10.11 Modeling a mask block

The masking facility gives the provision for designing user-defined icon, block, and help but the precondition is we have to have a subsystem formed in the simulink model. When we click a subsystem block, we see the sub model inside the subsystem. On the contrary when we click the masked block, we only see the parameter window the way we design for and the sub model remains hidden. Let us see the following examples on mask block formation.

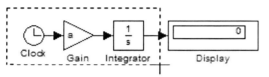

Figure 10.9(a) *Submodel selection for the masking example A*

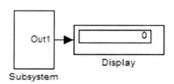

Figure 10.9(b) *Subsystem for the example A*

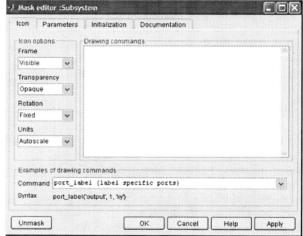

Figure 10.9(c) *Mask Editor Window for the Subsystem*

⎋ Example A

Let us say we want to integrate the function $\int_{t=0}^{t=10} at \, dt$ where *a* is a variable. Our objective is we will provide the value *a* without looking into the model through a parameter window. So when *a* is 2 or 4, the integration result should be 100 or 200.

We discussed simulink implementation for the single definite integration in the section 10.2. However, the procedure for the problem at hand is as follows:

⇒ *bring one Clock, one Gain, one Integrator, and one Display blocks in a new simulink model file and connect them as shown in the figure 10.9(a) for the computation of the integration*

⇒ *doubleclick the Gain block and type a in the slot of Gain in the parameter window of the block*

⇒ *select the targeted area of the figure 10.9(a) with your mouse and click the Create subsystem from the Edit menu of the model to form the Subsystem of the figure 10.9(b)*

⇒ *select the Subsystem on clicking the left button of the mouse, click the Edit menu, click the Mask subsystem from the pulldown menu of Edit to view the Mask Editor Window of the figure 10.9(c)*

As you see in the Mask Editor window, there are four subwindow options namely $\begin{cases} \text{Icon} \\ \text{Parameters} \\ \text{Initialization} \\ \text{Documentation} \end{cases}$. We are focused

on the theoretical part for the modeling hence we click the Parameters under the Mask Editor window to see the subwindow as shown in the figure 10.9(d). Again under the Dialog parameters of the figure 10.9(d), we find the sub

$$\text{parameters} \begin{Bmatrix} \text{Prompt} \\ \text{Variable} \\ \text{Type} \\ \text{Evaluate} \\ \text{Tunable} \end{Bmatrix}$$. The parameter a of the integration is related with the Gain block. Now click the Add button

of the figure 10.9(d) to see the slot for different sub parameters. In the response slot, enter Gain and a under Prompt and Variable respectively as shown in the figure 10.9(e) leaving the others as default, and click OK to return to the model like the figure 10.9(b). You finished forming the Mask block. The sub parameters Prompt and Variable correspond to the exact associated block name and variable respectively, for instance, here they are Gain and a respectively for the integration problem. Now doubleclick the block Subsystem of the model in the figure 10.9(b) and you do not see the sub model – that is what we mentioned before. Instead you find the block parameter window like the figure 10.9(f) the one you designed. If you type 2 in your parameter window, that means $a = 2$. As we stated before, we should see the Display output as 100 on running the model. Or, when we set 4, we should get Display output as 200. These simulations confirm that our design is perfectly okay.

The name of the associated block does not have to be Gain, we can rename the

Add Button

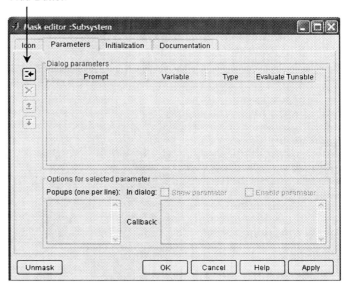

Figure 10.9(d) *Subwindow Parameters under the Mask Editor*

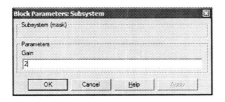

Figure 10.9(f) *Parameter window of your masked subsystem*

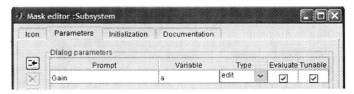

Figure 10.9(e) *Upper portion of the window in the figure 10.9(d)*

block according to our choice, for example, test. In that case the Prompt of the figure 10.9(e) would contain test. You can also rename your subsystem or output-input ports as we did in the section 10.5. To undo the masking, the action we need is select the Subsystem, click the Edit menu, click Edit mask in the pull down menu (figure 10.9(d) appears), and click unmask. So now if you doubleclick the subsystem, you can have your sub model of the figure 10.9(a).

⌗ Example B

Let us integrate the quadratic function $\int_{t=0}^{t=10}(at^2+b)dt$ where a and b are user defined

Figure 10.9(g) *Subsystem selection for the mask example B*

constants and which should be fed through a mask block. The mask block should have a name Integration and the output port of which should be named as Value. When $a = 3$ and $b = -30$, $\int_{t=0}^{t=10}(at^2+b)dt$ becomes 700. On forming the mask block and running the masked subsystem, we should have the result 700 in a Display block. Figure 10.9(g) presents the model to compute the integration. However, the procedure we adopted to form the mask block is as follows:

⇒ *bring one Clock, one Fcn, one Sum, one Constant, one Integrator, and one Display blocks in a new simulink model file*

⇒ *rename the Fcn block as a Value, doubleclick the block, set the expression as a*u^2 for the coding of the function at^2 pertaining to the integration*

⇒ *doubleclick the Constant block, enter the constant value as b, rename the block as b Value, connect the blocks as shown in the figure 10.9(g)*

⇒ *select the blocks in the targeted area of the figure 10.9(g) with your mouse, click Edit from the model menu, and click Create subsystem in the pull down menu of Edit to form the subsystem like the figure 10.9(b)*

⇒ *doubleclick the Subsystem to see the output port Out1 in the submodel, click the mouse pointer on Out1, delete Out1, type Value (user required name), click left button of the mouse outside the output port in the submodel, close the Subsystem window, and rename the Subsystem as Integration (user required name) to have the outlook of the mask block like the figure 10.9(h)*

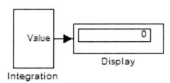

Figure 10.9(h) *Outlook of the masked block that user requires*

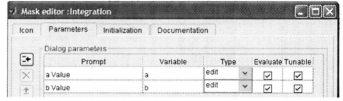

Figure 10.9(i) *Upper portion of the window in the figure 10.9(d) for the mask example B*

⇒ *select just designed block Integration, click the Edit of the menu bar, click Edit mask in the Pull down menu of Edit to see the design window of the figure 10.9(c) but now the design window appears by the name Integration because of our choice, click Parameters in the last window to see the window of the figure 10.9(d), click Add icon to type a Value (associated with the a Value block of the figure 10.9(g)) and a (since we set a in the expression a*u^2) in the Prompt and Variable slots respectively, click the Add icon again to enter b Value (associated with the b Value block of the figure 10.9(g)) and b (since we set b in constant value slot) in the Prompt and Variable slots respectively like the figure 10.9(i), and click OK to finish the design of the mask block*

⇒ *doubleclick just designed mask Integration block to see the parameter window as presented in the figure 10.9(j)*

As you see in the figure 10.9(j), the block parameter is named as Integration. The parameters have two slots: a Value and b Value. That is what we expected. Just to verify our design, let us enter 3 and −30 in the slots for the a Value and b Value in our designed parameter window of the Integration respectively and run the model. For sure you find 700 in the Display block thereby confirming the accurate design.

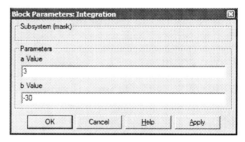

Figure 10.9(j) *Parameter window of the mask block for the example B*

Figure 10.10(a) *Contents of the For Iterator Subsystem submodel window*

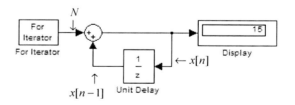

Figure 10.10(b) *Contents of the submodel for successive summation*

However, we close the section of the mask block design with this example. For further study on the mask block, the reader is referred to [14].

10.12 For looping in a model

Simulink can also implement the For looping similar to the conventional programming through block and functional lines. We know that a For loop has an integer counter index. Let us say that the counter index is 5 and we want to add the integers from 1 to 5 that should be 15 – that is our beginning problem statement. The block For Iterator Subsystem (link: *'Simulink → Ports & Subsystems → For Iterator Subsystem'*) can help us implement the For looping. Let us bring one For Iterator Subsystem block in a new simulink model file whose icon outlook is presented in the table 10.A. We do not need any input or output port because we are using just the index of the For looping. So doubleclick the block and find the submodel window contents like the figure 10.10(a). We delete the blocks In1 and Out1 and the functional line connecting them because of their irrelevancy. So we should have only the For Iterator left in the submodel. Bring one Sum, one Unit Delay, and one Display blocks (links in the table 10.A) in the last submodel, flip the Unit Delay block, and connect them as shown in the figure 10.10(b). We do not want to be overwhelmed with the mathematics but delicate mathematics is always associated with the modeling. Referring to the last model, the difference equation is given by

$$x[n] = N + x[n-1]$$

For Iterator generates integer from 1 to N (for this example 5). Let us plug $N = 1, 2, 3, \ldots 5$ in the difference equation:

$$x[1] = 1 + x[0] \qquad \text{when } N = 1$$
$$x[2] = 2 + x[1] \qquad \text{when } N = 2$$
$$x[3] = 3 + x[2] \qquad \text{when } N = 3, \text{ etc.}$$

If we set $x[0] = 0$, continuing above induction method provides ultimately $x[5] = 5+4+3+2+1 = 15$ that is what we want. However, let us get

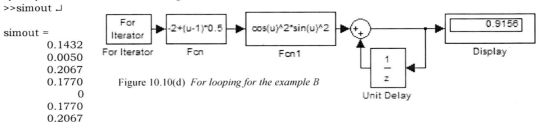

Figure 10.10(c) *For looping for the example A*

down to the simulink submodel, doubleclick the Unit Delay block, enter the Sample time as −1 in the parameter window of the block (since we have the successive addition), click *'Simulation → Simulation parameters → Solver'* from the submodel menu bar, change the solver option type from the Variable to Fixed type because of the integer situation, change the solver stop time as 0 otherwise the operation happens 10 times, and run the submodel. You should see 15 in the Display block validating our modeling. On doubleclicking the For Iterator, you find the default index as 5 (indicated by the number of iterations) that is why we did not mention about its entering. Let us see two more examples on For looping.

Example A

Let us compute the expression $\cos^2 x \sin^2 x$ for $x = -2$ to 1 with a step size of 0.5. Following the computation one should obtain [0.1432   0.0050   0.2067   0.1770   0   0.1770   0.2067].

The important point is looping in the For Iterator happens through integer numbers but this problem involves some decimal number. If we assume that x is changing from a to b (where $b > a$) with a step h, there should be $N = 1 + \dfrac{b-a}{h}$ numbers in the x domain. Any x is given by $a + (n-1)h$ where n varies from 1 to N and that is what the For Iterator does. For the example at hand, the variable x should be $-2 + (n-1)0.5$ whose simulink's u dependent code is $-2 + (u-1)*0.5$. The given function also has the u dependent code as $\cos(u)^2 * \sin(u)^2$. You can reach to the For Iterator as we did in the model 10.10(b) and then bring two Fcn and one To Workspace blocks in the submodel. Doubleclick the For Iterator block and you find there the number of iteration in the parameter window. We want simulink to find the For index hence we type length(−2:0.5:1) in the slot of the number of iterations and set the output data type as double in the parameter window of the For Iterator. Doubleclick the blocks Fcn and Fcn1 and enter the codes $-2 + (u-1)*0.5$ and $\cos(u)^2 * \sin(u)^2$ respectively in their parameter windows. Doubleclick the block To Workspace, set its Save format as array, change the Solver stop time to 0, run the model, go to MATLAB command prompt, and execute the following:

```
>>simout ↵

simout =
    0.1432
    0.0050
    0.2067
    0.1770
         0
    0.1770
    0.2067
```

Figure 10.10(d) *For looping for the example B*

That is what we expected from the modeling. The reason we did not employ the Display block is it gives the instantaneous value of the concern functional line. We would get the last value if it were used.

Example B

In the example A, adding all computed values for the function yields 0.9156 – we wish to have it implemented. The beginning example model in the figure 10.10(b) presents how one can add successively the For

Iterator outputs. We inject the output of the model in the figure 10.10(c) to the one in the figure 10.10(b). Figure 10.10(d) shows the model for the example B and its simulated output in the Display block.

10.13 If-then-else modeling

Employing the block and the functional line, the if-then-else logical structure is simulated in simulink. Let us consider the figures 2.3(c) and 10.7(c) in which we have a sine wave and a rectangular pulse wave respectively. Assume that the sine wave has the amplitude swing from –1 to 1 and frequency 10 *Hz* . The rectangular pulse also has the same frequency but the amplitude swing is from 0 to 1 and with a duty cycle of 50% (the reader is referred to chapter 2 for the wave modeling). If we say that when the rectangular pulse is 1, we let the sine wave pass to one functional line. Otherwise, we pass the sine wave to another functional line. Because both waves share common period and time, we separate the positive and negative halves of the sine wave that is what we expect from simulink.

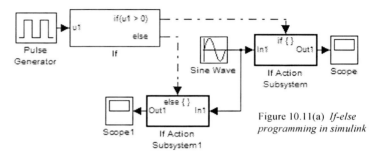

Figure 10.11(a) *If-else programming in simulink*

To mention about the modeling procedure, let us bring one If (link: *'Simulink → Ports & Subsystems → If'*) block in a new simulink model file (icon outlook can be seen in the table 10.A). To perform any operation following the logical decision taken by the If block, you need If Action Subsystem (link: *'Simulink → Ports & Subsystems → If Action Subsystem'*) block (icon outlook in the table 10.A). Bring one If, two If Action Subsystem, one Sine Wave, one Pulse Generator, and two Scope blocks in a new simulink model file. Flip the blocks Scope1 and If Action Subsystem1 and place the blocks relatively and connect them as shown in the figure 10.11(a). Doubleclick the Sine Wave block and enter its frequency as 2*pi*10 to feed the frequency information. Doubleclick the Pulse Generator block and enter its period as 0.1 to feed the given information. Change the solver stop time to 0.2 from the model menu bar (*'Simulation → Simulation parameters → Solver'*) to see two

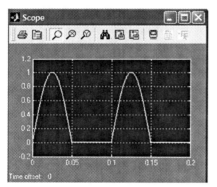

Figure 10.11(b) *Positive half of the sine wave displayed by the Scope*

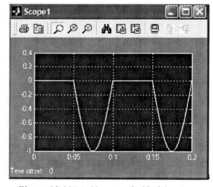

Figure 10.11(c) *Negative half of the sine wave displayed by the Scope1*

cycles of the wave and run the model. Doubleclick the Scope and Scope1 to see the positive and negative halves of the sine wave respectively. Looking into the Scope outputs, you should not be satisfied because the sine waves so displayed will not have the perfect sine shape. The reason is simulink automatically takes some samples on the output and shows you the sampled output to reduce the simulation time. Theoretically we need infinite points to describe a continuous system. However, let us change the solver option type from the default Variable to the Fixed setting and the step size from the auto to 0.0001 in the solver window. Now run the model and doubleclick each Scope. You must see the figures 10.11(b) and 10.11(c) as Scope and Scope1 outputs with the autoscale setting, the first and second of which contain the positive and the negative halves of the sine wave respectively. That is what the beginning statement was.

As we simulated, the If block has one input and two output ports. Any function can appear at the input port labeled by u1. The logical statement If has two states $\begin{Bmatrix} Then \\ Else \end{Bmatrix}$ in the default block. Out of the two output ports of the

If block, the upper one is the Then and the lower one is the Else. Each of them needs one If Action Subsystem block to carry out some simulation. On doubleclicking the If block, the parameter window of the figure 10.11(d) appears which contains the slot for $\left\{\begin{array}{l}\text{Number of inputs}\\ \text{If expression}\\ \text{Elseif expressions}\end{array}\right\}$. The default number of

input of the If block is 1 but, for instance, if we enter 3 in the parameter window, the icon outlook appears as shown in the figure 10.11(e) enabling the block to accept three input functions labeled by u1, u2, and

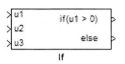

Figure 10.11(e) *If block with three input ports*

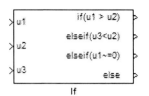

Figure 10.11(f) *If block adapted with the Elseif situation*

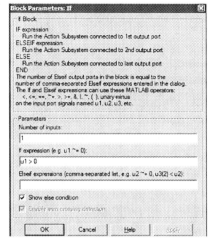

Figure 10.11(d) *Parameter window for the If block*

u3. In some situations we may need only one checking – that is no else is necessary. You can have that by unchecking the *Show else condition* in the parameter window of the figure 10.11(d). The default logical expression in the If block is u1>0 (means any function entering to u1 is greater than zero). But other logical expression can also appear. Simulink logical notations are as follows:

greater than	>
greater than or equal to	>=
less than	<
less than or equal to	<=
equal to	==
not equal	~=

In the slot of If expression of the figure 10.11(d), you can enter above mentioned operator in conjunction with the input labeling depending on the modeling problem you handle. Each time you change the logical expression, the If block icon becomes consistent with your change. Let us say that we entered u1>u2 indicating functional values of u1 is greater than those of u2 at every instant of time considering both share the same time axis. If we had u1==u3, it would mean that functions entered to u1 and u3 are identical at all instant of time. Again if you enter u1>u2 requiring two functions, there should be at least 2 in the slot of number of inputs otherwise simulink would give you error message.

That is not everything the block can offer. Sometimes we need multiple checking that can occur through elseif or nested if statement. This provision is also there in the parameter window of the figure 10.11(d). The general

form of the nested if can be $\left\{\begin{array}{l}\text{If (logical expression}\\ \text{elseif (logical expression)}\\ \text{elseif (logical expression)}\\ \quad\vdots\\ \text{else}\\ \text{end}\end{array}\right\}$. Considering three input functions – u1, u2, and u3, we

would like to check $\left\{\begin{array}{l}\text{functional values of u1 are greater than those of u2}\\ \text{functional values of u3 are less than those of u2}\\ \text{functional values of u1 are not equal to zero}\end{array}\right\}$, where the three functions are jointly

having the same horizontal axis. In the parameter window of the figure 10.11(d), we enter the settings as $\left\{\begin{array}{l}\text{Number of inputs}:3\\ \text{If expression}: u1>u2\\ \text{Elseif expression}: u3<u2, u1\sim=0\end{array}\right\}$. In doing so, the icon outlook appears as shown in the figure 10.11(f) thereby

making consistency with the modeling. Note that the If procures the first logical expression and the subsequent ones are taken care of by the else-if and else. The else-if expressions are separated by a comma. If there are three logical expressions, there is no need to enter two else-if expressions – writing one else-if expression is enough and the rest one is taken care of by the else. For the example at hand, we could have written the else-if expression as u3<u2 and then the else would have dealt with the expression u1~=0. Whatsoever be the number of the output ports of the If block, all of them needs one If Action Subsystem block as we modeled earlier.

In the same link where If block resides, the reader can find other programming blocks such as the switch-case, the while-Iterator, … etc. Their implementation can be conducted by gaining experience from previous programming discussion and noticing the subsystem examples presented in the same link. We keep that an exercise for the reader.

It goes without saying that the modeling resources available in simulink are ample. It is feasible for the multi-faceted real problems in most branches of science and engineering ranging from the elementary to advanced ones. Additionally the application-based blocksets enhanced the versatility of the package to a greater extent. It is not possible to include all aspects of simulink in this short context. Once the reader is well acquainted with the fundamental modeling style, advanced modeling can easily be devised with modification and scheming. We guess the reader at least has had some escorting know-how of simulink. Minute details in the text might seem to be repetitive for the advanced users but those details are very essential for a novice or average reader. Reference [14] can be considered for the advanced topics on simulink. Our effort would be a success if the reader benefited slightly.

However, we bring an end to the text by presenting the block links employed in this chapter.

10.14 Block links used in this chapter

In this chapter we described miscellaneous modeling but some programming issues are also addressed. The blocks organized in the table 10.A comfort the modeling.

Table 10.A   Necessary blocks for modeling the problems of this chapter as found in simulink library (not arranged in the alphabetical order)

Block name	Representative Symbol/Function	Icon Outlook	Block name	Representative Symbol/Function	Icon Outlook
Clock	It generates solver times or independent variable of the functions	Clock	Scope	It displays the function of the connected line	Scope
Link: *Simulink → Sources → Clock*			Link: *Simulink → Sinks → Scope*		
Enabled Subsystem	It acts like a switch. If the enable input value is positive, the function at the In1 becomes available to Out1	Enabled Subsystem	Hit Crossing	It can detect the functional crossing through a particular setting either increasing or decreasing direction	Hit Crossing
Link: *Simulink → Ports & Subsystem → Enabled Subsystem*			Link: *Simulink → Discontinuities → Hit Crossing*		
Fcn	It codes the user defined function in terms of the independent variable u	Fcn	Display	It shows the instantaneous value of the function which it is connected to	Display
Link: *Simulink → User-Defined Functions → Fcn*			Link: *Simulink → Sinks → Display*		
Sum	It sums two or more functions		Constant	It generates constant value	Constant
Link: *Simulink → Math Operations → Sum*			Link: *Simulink → Sources → Constants*		
Sine Wave	It generates sine waves of different frequency, amplitude, and phases	Sine Wave	Integrator	It integrates the function to its input port in continuous sense	Integrator
Link: *Simulink → Sources → Sine Wave*			Link: *Simulink → Continuous → Integrator*		
Dot Product	It finds the dot product of two vectors	Dot Product	3×3 Cross Product	It finds the cross product of two vectors	3x3 Cross Product
Link: *Simulink → Math Operations → Dot Product*			Link: *Aerospace Blockset → Transformations → Axes → 3×3 Cross Product*		
Rounding Function	It rounds or discards the fractional part of a decimal number	Rounding Function	To Workspace	It sends simulink generated data to MATLAB workspace	To Workspace
Link: *Simulink → Math Operations → Rounding Function*			Link: *Simulink → Sinks → To Workspace*		
Gain	It multiplies the function to its input port by a constant value	Gain	Assignment	It performs the assignment of one functional line contents to another	Assignment
Link: *Simulink → Math Operation → Gain*			Link: *Simulink → Math Operation → Assignment*		

Continuation of previous table:

Block name	Representative Symbol/Function	Icon Outlook	Block name	Representative Symbol/Function	Icon Outlook
Selector	It selects elements from vectors or matrices on user requirement	Selector	For Iterator Subsystem	It performs the For looping operation like the conventional programming	In1 for { ... } Out1 / For Iterator Subsystem
Link: *Simulink → Signal Routing → Selector*			Link: *Simulink → Ports & Subsystem → For Iterator Subsystem*		
Terminator	It prevents from displaying warning message if connected to unused output port of any block	Terminator	If	It only performs logical decision taking like if-then-else situation	if(u1 > 0) else / If
Link: *Simulink → Sinks → Terminator*			Link: *Simulink → Ports & Subsystem → If*		
Ground	It prevents from displaying warning message if connected to unused input port of any block	Ground	Unit Delay	It delays any sequence by one sample, for example, $x[n]$ to $x[n-1]$	$\frac{1}{z}$ / Unit Delay
Link: *Simulink → Sources → Ground*			Link: *Simulink → Discrete → Unit Delay*		
Width	It determines how many functions are hidden in a functional line	Width	Pulse Generator	It generates rectangular pulse of swing from 0 to some value and of different period, duty cycle, and phase	Pulse Generator
Link: *Simulink → Signal Attributes → Width*			Link: *Simulink → Sources → Pulse Generator*		
Ramp	It generates straight line function like t versus t of different slopes and intercepts	Ramp	If Action Subsystem	Once logical If-then-else is decided by the If block, it performs the modeling on the if-then-else condition	Action In1 Out1 / If Action Subsystem
Link: *Simulink → Sources → Ramp*			Link: *Simulink → Ports & Subsystem → If Action Subsystem*		
Mux	It multiplexes two or more functions to pass to a single functional line		Step	It generates unit step function of different final values and steps	Step
Link: *Simulink → Signal Routing → Mux*			Link: *Simulink → Sources → Step*		

219

Appendix

Block parameter window reference

The programming convenience in simulink happens through the dialog window or block parameter window. In every model the reader has to enter the block parameter data. A particular problem may not require entering all parameters. The block links for various chapter-title-oriented problems are provided at the end of each chapter. We summarize the parameter window figure reference for various blocks employed in the text in table A so that the reader can easily enter the block parameter data during modeling.

Table A   Block parameter window reference

Block name	Parameter window figure number	Page number (s)
Analog Filter Design	Figure 6.4(a)	119
Band-Limited White Noise	Figure 8.6(n)	177
Constant	Figure 7.1(f)	142
Constant Diagonal Matrix	Figure 7.1(i)	142
Discrete Filter	Figure 6.5(d)	121
Digital Filter Design	Figure 6.9(h)	130
Discrete Transfer Fcn	Figure 6.6(d)	122
Discrete Zero-Pole	Figure 6.7(d)	123
Fcn	Figure 3.2(e)	38
Fourier	Figure 9.3(c)	191
Gain	Figure 1.4(a)	5
Identity Matrix	Figure 7.1(k)	143
If	Figure 10.11(d)	217
Integrator	Figure 3.1(d)	35
Least Squares Polynomial Fit	Figure 8.4(b)	171
Mask Editor	Figure 10.9(c)	212
MATLAB Fcn	Figure 7.11(k)	155
Math Function	Figure 7.4(a)	145
Matrix Multiply	Figure 7.2(d)	143
Overwrite Values	Figure 7.12(b)	155
Ramp	Figure 3.2(b)	37
Random Source	Figure 8.6(g)	175
Reshape	Figure 7.10(c)	152
Scope	Figure 3.15(c)	60
Signal Builder	Figure 2.6(c)	18
Slider Gain	Figure 3.16(d)	64
Solver	Figure 3.1(e)	36
State-Space	Figure 5.4(f)	97
Step	Figure 3.7(e)	46
Submatrix	Figure 7.11(e)	153
To Workspace	Figure 3.16(g)	65
Transfer Fcn	Figure 5.1(e)	93
Transport Delay	Figure 5.8(c)	101
Trigonometric Function	Figure 2.11(a)	27
Unit Delay	Figure 4.2(d)	73
Velocity Conversion	Figure 8.9(b)	181
Zero-Pole	Figure 5.2(e)	94

References

[1]   Peter V. O'Neil, "*Advanced Engineering Mathematics*", Third Edition, 1991, Wadsworth Publishing Company, Belmont, California.

[2]   Mohammad Nuruzzaman, "*Tutorials on Mathematics to MATLAB*", 2003, 1st Books Library, Bloomington, Indiana.

[3]   Alberto Cavallo, Roberto Setola, and Francesco Vasca, "*Using MATLAB, Simulink and Control Systems Toolbox – A Practical Approach*", 1996, Prentice Hall, London.

[4]   Marcus, Marvin, "*Matrices and MATLAB – A Tutorial*", 1993, Prentice Hall, Englewood Cliffs, New Jersey.

[5]   Ogata, Katsuhiko, "*Solving Control Engineering Problems with MATLAB*", 1994, Prentice Hall, Englewood Cliffs, New Jersey.

[6]   Prentice Hall, Inc., "*The Student Edition of MATLAB for MS-DOS Personal Computers*", 1992, Prentice Hall, Englewood Cliffs, New Jersey.

[7]   Saadat, Hadi., "*Computational Aids in Control Systems Using MATLAB*", 1993, McGraw–Hill, New York.

[8]   Gander, Walter. and Hrebicek, Jiri., "*Solving Problems in Scientific Computing Using MAPLE and MATLAB*", Third Edition, 1997, Springer–Verlag, New York.

[9]   D. M. Etter, "*Engineering Problem Solving with MATLAB*", 1993, Prentice Hall, Englewood Cliffs, New Jersey.

[10] Shahian, Bahram. and Hassul, Michael., "*Control System Design Using MATLAB*", 1993, Prentice Hall, Englewood Cliffs, New Jersey.

[11] Ogata, Katshuiko, "*Designing Linear Control Systems with MATLAB*", 1994, Prentice Hall, Englewood Cliffs, New Jersey.

[12] Jackson, Leland B., "*Digital Filters and Signal Processing with MATLAB Exercises*", Third Edition, 1996, Kluwer Academic Publishers, Boston.

[13] Kuo, Benjamin C. and Hanselman, Duanec., "*MATLAB Tools for Control System Analysis and Design*", 1994, Prentice Hall, Englewood Cliffs, New Jersey.

[14] James B. Dabney and Thomas L. Harman, "*Mastering Simulink® 4*", 2001, Prentice Hall, New Jersey.

[15] Richard C. Dorf and Robert H. Bishop, "*Modern Control Systems*", Ninth Edition, 2001, Prentice Hall, New Jersey.

[16] A. C. Bajpai, L. R. Mustoe, and D. Walker, "*Advanced Engineering Mathematics*", 1977, John Wiley & Sons, New York.

[17] Duffy, Dean G., "*Advanced Engineering Mathematics with MATLAB*", Second Edition, 2003, Chapman & Hall, CRC, Boca Raton.

[18] Kreyszig, Erwin, "*Advanced Engineering Mathematics*", Fifth Edition, 1983, Wiley, New York.

[19] Gustafson, G. B. and Wilcox, Calvin H., "*Analytical and Computational Methods of Advanced Engineering Mathematics*", 1998, Springer, New York.

[20] Zill, Dennis G. and Cullen, Michael R., "*Advanced Engineering Mathematics*", Second Edition, 2000, Jones & Bartlett, Boston.

[21] Wilson, Howard B., Turcotte, Louis H., and Halpern, David, "*Advanced Mathematics and Mechanics Applications Using MATLAB*", Third Edition, 2003, CRC, Boca Raton.

[22] Hagin, Frank G. and Cohen, Jack K., "*Calculus with MATLAB*", 1996, Prentice Hall, Upper Saddle River, New Jersey.

[23] Fausett, Laurene V., "*Applied Numerical Analysis Using MATLAB*", 1999, Prentice Hall, London.

[24] Martin Golubitsky and Michael Dellnitz, "*Linear Algebra and Differential Equations Using MATLAB*", 1999, Brookes/Cole Publishing Company, Boston.

[25] Alan V. Oppenheim and Ronald W. Schafer, "*Discrete-Time Signal Processing*", 1989, Prentice Hall, New Jersey.

[26] Joyce Van de Vegte, "*Fundamentals of Digital Signal Processing*", 2002, Prentice Hall, New Jersey.

[27] Robert L. Boylestad, "*Introductory Circuit Analysis*", Ninth Edition, 2000, Prentice Hall, New Jersey.

[28] Ali S. Hadi, "*Matrix Algebra – As A Tool*", 1996, Duxbury Press, California.

[29] Serge Lang, "*Calculus of Several Variables*", Second Edition, 1979, Addison Wesley Publishing Company, Massachusets.

[30] Rogers, Gerald Stanley, "*Matrix Derivatives*", 1980, M. Dekker, New York.

[31] Part-Enander, Eva, "*The MATLAB Handbook*", 1998, Addisson Wesley, Harlow.

[32] Biran, Adrian B and Breiner, Moshe, "*MATLAB for Engineers*", 1997, Addison Wesley, Harlow.

[33] Bishop, Robert H., "*Modern Control Systems Analysis and Design Using MATLAB*", 1993, Addison Wesley, MA.

[34] Moscinski, Jerzy and Ogonowski, Zbigniew, "*Advanced Control with MATLAB and Simulink*", 1995, E. Horwood, Chichester.

[35] Gene Howard Golub and Charles F. Van Loan, "*Matrix Computations*", 1983, Johns Hopkins University Press, Baltimore.

[36] I. Gohberg, P. Lancaster, and L. Rodman, "*Matrix Polynomials*", 1982, Academic Press, New York.

[37] Chipperfield, A. J. and Fleming, P. J., "*MATLAB Toolboxes and Applications for Control*", 1993, Peter Peregrinus, London.

[38] Math Works Inc., "*MATLAB Reference Guide*", 1993, Natick, Massachusets.

[39] Cleve Moler and Peter J. Costa, "*MATLAB Symbolic Math Toolbox*", 1997, User's Guide, Version 2.0, Natick, Massachusets.

[40] R. Braae, "*Matrix Algebra for Electrical Engineers*", 1963, I. Pitman, London.

[41] Hanselman, Duane C. and Littlefield, Bruce R., "*Mastering MATLAB 5: A Comprehensive Tutorial*", 1998, Prentice Hall, Upper Saddle River, New Jersey.

[42] Shampine, Lawrence F. and Reichelt, Mark W., "*The MATLAB ODE Suite*", 1996, The Math-Works, Inc., Natick, MA.

Subject Index